Photoinduced Intramolecular Charge Transfer in Donor-Acceptor Biaryls and Resulting Applicational Aspects Regarding Fluorescent Probes and Solar Energy Conversion

by

Michael Maus

ISBN: 1-58112-030-3

DISSERTATION.COM

1998

Copyright © 1998 Michael Maus
All rights reserved.

ISBN: 1-58112-030-3
Dissertation.com
1998

Photoinduced Intramolecular Charge Transfer in Donor-Acceptor Biaryls and Resulting Applicational Aspects Regarding Fluorescent Probes and Solar Energy Conversion

Dissertation

zur Erlangung des akademischen Grades

doctor rerum naturalium

(Dr. rer. nat.)

im Fach Chemie

eingereicht an der

Mathematisch-Naturwissenschaftlichen Fakultät I

der Humboldt-Universität zu Berlin

von

Diplom-Chemiker Michael Maus

geboren am 29.09.1966 in Köln

Präsident der Humboldt-Universität zu Berlin

Prof. Dr. H. Meyer

Dekan der Mathematisch-Naturwissenschaftlichen Fakultät I

Prof. Dr. J. Rabe

Gutachter: 1. Prof. Dr. W. Rettig
2. Dozent Dr. R. Lapouyade
3. Dozent Dr. J. Bendig

Tag der mündlichen Prüfung: 02. Juli 1998

Danksagungen

In erster Linie bedanke ich mich bei meinem Doktorvater Herrn Prof. W. Rettig, der mir die Arbeit erst ermöglichte und mich durch viele interessante Diskussionen unterstützte. Außerdem danke ich Ihm dafür, daß er mir den Besuch zahlreicher Tagungen erlaubte.

Natürlich gilt mein Dank entspechend Dr. R. Lapouyade, der als langjähriger „Weggefährte" von Prof. Rettig, die Idee das DMABN durch einen Benzolring zu erweitern mitausbrütete und die Donor-Akzeptor Biphenyle schließlich zur Verfügung stellte. Außerdem bedanke ich mich bei ihm, daß er die Funktion als Gutachter übernommen hat.

Herrn Dr. Bendig danke ich sehr dafür, daß er sich trotz des enormen Zeitdrucks bereit erklärt hat, meine Arbeit zu begutachten.

Der gesamten Arbeitsgruppe danke ich für das tolle Arbeitsklima und viele anregende, hilfreiche Diskussionen.

Sehr wichtige und teilweise überaschende Ergebnisse wurden in der Gruppe von Dr. Rullière zusammen mit seinem Laserexperten Gediminias Jonusauskas erzielt, denen ich nicht nur für die wissenschaftliche Zusammenarbeit danken möchte, sondern insbesondere für die äußerst herzliche Betreuung. Die Wochenendausflüge mit Claude und seinen wissenschaftlichen Gästen werde ich immer in lebhafter Erinnerung behalten. Viel Spaß hatte ich auch mit allen Mitarbeitern von Claude, insbesondere mit Emmanuelle dem Weinexperten, der auch im Squash gar nicht so schlecht war.

Bei Prof. F.C. DeSchryver möchte ich mich dafür bedanken, daß ich an „seinen" Phenanthren Derivaten mitarbeiten durfte, wodurch ich viele interessante Erkenntnisse gewinnen konnte.

Bei Knut Rurack bedanke ich mich für wertvolle Kommentare, insbesondere zu seinem Spezialgebiet der Fluoreszenzsonden, nachdem er sich durch einige Seiten meiner Dissertation und „Monstersätze" durchgeschlagen hatte.

Herrn Dr. Moritz danke ich für die Hilfsbereitschaft, mir eine funktionstüchtige pH Glaselektrode zur Verfügung zu stellen.

Thanks to Dr. Murthy S. Gudipati who always had time to discuss and to give valuable and sometimes philosophic advices.

Der Werkstatt des Ivan-N. Stranski Instituts als auch des Walther Nernst Instituts danke ich für die unkomplizierte Umsetzung meiner Anfragen.

Mein besonderer Dank gilt Anja, die mich in der entscheidenden Phase von allen täglichen Pflichten befreit hat und mich auch ansonsten voll unterstützt hat. Obwohl sie es noch nicht wahrnehmen können, danke ich auch Sara und Janina, die mich immer bei guter Laune gehalten haben und im übrigen nie einen Zweifel daran gelassen haben, wer die neuen „Herrscher" im Haus sind.

Wir können einem Menschen verzeihen, daß er etwas Nützliches schafft, solange er es nicht bewundert. Die einzige Entschuldigung dafür, etwas Nutzloses zu schaffen, besteht darin, daß man es über jedes Maß bewundert.
    Oscar Wilde

Wenn man dir liniertes Papier gibt, schreibe quer über die Zeilen.
    Juan Ramón Jiminéz

Zusammenfassung

Photoinduzierter intramolekularer Ladungstransfer in Donor-Akzeptor Biarylen und daraus resultierende Anwendungsaspekte in Hinsicht auf Fluoreszenzsonden und Solarenergieumwandlung

Im Mittelpunkt der vorliegenden Arbeit stehen die Untersuchungen des photoinduzierten intramolekularen Ladungstransfer in drei unterschiedlich verdrillten Donor-Akzeptor (D-A) Biphenylen. Unter Zuhilfenahme eines weiteren Paares unterschiedlich verdrillter D-A Biaryle werden dabei zum einen allgemeine Erkenntnisse zum photoinduzierten Verhalten von D-A Biarylen gewonnen und zum anderen mögliche Anwendungen in Bereichen der Solarenergienutzung und der Sondierung von Mikroumgebungen mittels Fluoreszenz diskutiert.

Neben experimentellen Methoden der stationären und zeitaufgelösten (ps bis s) Lumineszenz, transienten Absorption (sub-ps), Polarisationsspektroskopie, Hochdruck- und Tieftemperaturtechnik kommen quantenchemische Rechnungen zum Einsatz.

Die elektronischen Zustände der D-A Biaryle können durch die Wechselwirkungen der Elektronenzustände der jeweiligen Arylhälften beschrieben und den beobachteten Absorptionsbanden zugeordnet werden. Elektronentransferwechselwirkungen (D→A) führen zu einer tiefliegenden, intramolekularen Charge Transfer (CT) Bande, die aufgrund der starken Kopplung des reinen Elektronentransferzustandes mit dem Grund- (S_0) und ^1L_a-Zustand sehr intensiv ist. Der photoinduzierte intramolekulare Ladungstransfer weist sich durch stark solvatochrome Fluoreszenz aus, wobei das daraus abgeleitete Dipolmoment mit dem Verdrillungswinkel zunimmt. Überraschenderweise wird nach optischer CT Anregung der D-A Biphenyle zunächst ein unpolarer $\pi\pi^*$ Zustand (mit ^{1}FC bezeichnet) besetzt, bevor der Elektronentransfer in wenigen Pikosekunden von der Dimethylanilin- zur Benzonitrileinheit stattfindet. Die im Verhältnis zur Solvensrelaxation langsame und durch die im Mutter-Tochter Verhältnis stehende duale Fluoreszenzverstärkung eindeutig verfolgbare Elektronentransferreaktion wird auf eine interne conversion zwischen zwei schwach miteinander wechselwirkenden Zuständen zurückgeführt, die in erster Näherung unterschiedliche Symmetrien besitzen.

Wie schon für das unsubstituierte Biphenyl bekannt, deuten die transienten Absorptionsspektren und quantenchemischen Rechnungen eine photoinduzierte initiale Relaxation zu einer stärker planaren Konformation als im S_0 an. Die Analyse der elektronischen Übergangsdipolmomente aus Absorption, Fluoreszenz und semiempirischen Berechnungen zeigen, daß das flexible D-A Biphenyl (**II**) im ^{1}CT Zustand seine planare Konformation unabhängig von der Lösungsmittelpolarität beibehält. Das durch zwei Methylgruppen in ortho Position substituierte stark verdrillte D-A Biphenyl (**III**) bleibt im ^{1}CT nur in unpolaren Lösungsmitteln planarer als in S_0, während es in stark polaren Lösungsmitteln stärker verdrillt in ^{1}CT als im S_0 vorliegt. Dafür gibt es weitere Anzeichen, wie z.B. die Tatsache, daß das ausrelaxierte transiente Absorptionsspektrum dem der Summe aus Dimethylanilin Kation und Benzonitril Anion entspricht als auch die Beobachtung, daß nur

Zusammenfassung

die ^{1}CT Fluoreszenz von **III** durch Wasserstoffbrückenbindungen gelöscht wird; beides ist mit einer biradikaloiden stark ladungsgetrennten Elektronenstruktur, induziert durch eine stark verdrillte Konformation, erklärbar. In mittelpolaren Lösungsmitteln liegt für **III** ein Gleichgewicht zwischen dem planareren (**CT**) und verdrillteren Konformer (**CTR**) vor, welches für das Lösungsmittel Diethylether mittels temperaturabhängiger Globalanalyse von Fluoreszenzabklingkurven thermodynamisch und kinetisch quantitativ charakterisiert wird. Eine hohe Aktivierungsenergie (E_a=14 kJ/mol) ist verantwortlich für eine im Verhältnis zum Dimethylaminobenzonitril (DMABN) relativ langsame adiabatische Photoreaktion von der planareren Spezies **CT** hin zu der verdrillteren, weiter relaxierten Spezies **CTR** mit größerem Elektronentransfer Charakter. Durch den Vergleich von druck- mit temperaturabhängigen Fluoreszenzmessungen kann geschlossen werden, daß tatsächlich eine thermische, d.h. intrinsische Barriere vorliegt und die Photoreaktion zusätzlich viskositätskontrolliert ist.

Die Eignung der D-A Biphenyle als Fluoreszenzsonden für Mikropolarität, sich stark ändernde Mikroviskosität oder Matrixordnung, für protische Lösungsmittel und pH wird untersucht. Die komplementären Sondeneigenschaften von **II** und **III** werden herausgestellt und die mit dem Sondenprozeß verbundenen Mechanismen diskutiert. Insbesondere die Fluoreszenzsondierung des pH erscheint als ein vielversprechendes Anwendungsfeld.

Die ungewöhnliche Eigenschaft von planaren oder mäßig verdrillten D-A Biarylen, hohe Fluoreszenzquantenausbeuten mit großen Dipolmomenten und daraus resultierenden großen Stokes-shifts zu kombinieren wird demonstriert und zur Nutzung in Fluoreszenzsolarkonzentratoren vorgeschlagen. Andererseits bieten stark verdrillte D-A Biaryle durch ihre nahezu ladungsgetrennte Elektronenstruktur im angeregten Zustand die Möglichkeit, über intermolekularen Elektronentransfer photokatalytisch zu wirken. Entsprechende Perspektiven für die photochemische Nutzung der Solarenergie werden kurz diskutiert und aufgezeigt.

Abstract

Photoinduced Intramolecular Charge Transfer in Donor-Acceptor Biaryls and Resulting Applicational Aspects Regarding Fluorescent Probes and Solar Energy Conversion

This study is focused on the effects of photoinduced intramolecular charge transfer (CT) in three differently twisted donor-acceptor (D-A) biaryls. Taking into account another pair of differently twisted D-A biaryls new universal insights into the photoinduced electronic and conformation dynamics of D-A biaryls are obtained. Furthermore, possible applications in fields of solar energy conversion and fluorescence sensing of microenvironments are demonstrated.

Experimental means of stationary and time-resolved (ps to s) luminescence, transient absorption (sub-ps), polarization spectroscopy, high pressure and low temperature techniques are employed in conjunction with quantum chemical calculations.

Twist angle and solvent dependent electron transfer (ET) interactions between the D and A aryl moieties are responsible for the low lying and solvatochromic intramolecular CT electron band which gains unusually high intensity through strong electronic coupling of the pure ^{1}ET with the ground (S_0) and ^1L_a state. As regards the class of biaryl compounds, for the first time, an excited state electron transfer from the D to the A could be monitored by dual spectrally separated stimulated fluorescence bands with precursor-successor relationship on a sub-ps timescale for the D-A biphenyls. It is concluded that, in additon to the electronic interaction of ^{1}ET with S_0 and ^1L_a, the electronic interaction with a close lying ^1L_b state plays a fundamental role in the ET dynamics and the ^{1}CT-S_0 transition probability in D-A biaryls.

The initial photoinduced conformational relaxation occurs towards planarity in all biaryls investigated. However, various results evidence that the highly twisted D-A biphenyl additionally performs a slow "excited state intramolecular back twist rotation" leading to a solvent polarity dependent conformational equilibrium between a more planar (**CT**) and a more twisted (**CTR**) conformer in $S_1(^1CT)$. Using global analysis of the biexponential fluorescence decays as a function of temperature and pressure in medium polar solvents, the kinetics, thermodynamics, viscosity control and decomposed emission spectra associated with this adiabatic photoreaction are determined.

The twist angle dependent ability of the D-A biphenyls to serve as fluorescent probes of micropolarity, changes of microviscosity or matrix order, protic solvents and pH is investigated. In particular, fluorescence sensing of pH seems to be promising.

The advantageous property of planar or moderately twisted D-A biaryls to combine high fluorescence quantum yields with large Stokes-shifts is proposed for the use in fluorescent solar concentrators. Alternatively, almost full charge separation in S_1 of strongly twisted D-A biaryls provides the possibility of electron transfer initiated photocatalysis.

Contents

Danksagungen

Zusammenfassung

Abstract

Chapter 1	**Introduction**	1
Goals, text guide and nomenclature		1
Chapter 2	**Absorption Spectra and Electronic Structure of D-A Biphenyls**	4
2.1 Introduction		4
2.2 Experimental		5
2.3 Results and Discussion		6
2.3.1 Composite molecule approach for donor-acceptor biphenyls I-III		6
2.3.2 UV/VIS absorption spectra and their derivatives		10
2.3.3 Linear dichroic absorption spectra		13
2.3.4 Comparison of spectral data with calculated results		14
2.4 Conclusions		16
2.5 References		16
Chapter 3	**Fluorescence and Photophysics of D-A Biphenyls at 298 K**	18
3.1 Introduction		18
3.2 Experimental Section		21
3.2.1 Synthesis of the compounds		21
3.2.2 Solvents		21
3.2.3 Steady state absorption		21
3.2.4 Steady state fluorescence		22
3.2.5 Fluorescence quantum yields		22
3.2.6 Fluorescence lifetimes		22
3.2.7 Quantum chemical calculations		23
3.3 Results and Discussion		24
3.3.1 Steady state spectra		24
3.3.2 Excited state dipole moments from solvatochromic plots		29
3.3.3 Photophysics		32
3.3.3.1 Photophysical evidence for ^1L_a-type fluorescence		32
3.3.3.2 Structural and solvent dependence of nonradiative rates		33
3.3.3.3 Ratios of radiative rate constants as indicators for angular relaxations		36
3.3.3.4 Excited state conformational equilibrium for III		38
3.3.4 Theoretical calculations		40
3.3.4.1 Twist potentials in S_0		40
3.3.4.2 Electronic nature of S_1		42
3.3.4.3 Molecular structure in S_1		43
3.4 Conclusions		47
3.5 References and Notes		49
Chapter 4	**Sub-picosecond Transient Absorption of D-A Biphenyls.**	
	Intramolecular Control of the Excited State Charge Transfer Processes.	52
4.1 Introduction		52
4.2 Experimental		55
4.2.1 Materials		55
4.2.2 Picosecond pump-probe experiments		55
4.2.3 Correction and fitting of the results		55
4.2.4 Quantum chemical calculation of the transient spectra		56
4.3 Results		57
4.3.1 Steady-state spectra and expectations		57
4.3.2 Transient absorption measurements		58

4.3.2.1 n-Hexane solutions	59
4.3.2.2 Acetonitrile solutions	60
4.3.2.3 Diethylether solutions	62
4.3.2.4 Triacetine solutions	64
4.4 Discussion	67
4.4.1 Excited state absorption and gain bands	67
4.4.1.1 Assignments of the absorbing and emitting species to two states	67
4.4.1.2 Effects of solvent polarity and twist angle	68
4.4.2 Structural relaxation	71
4.4.3 Kinetics of state interconversion $^1FC \rightarrow {}^1CT$	74
4.5 Conclusions	77
4.6 References	79

Chapter 5   Fluorescence Polarization Spectroscopy of D-A Biphenyls at 77K. The Electronic Relaxations. 81

5.1 Introduction	81
5.2 Experimental	82
5.2.1 Polarization and millisecond luminescence spectroscopy	82
5.3 Results and Discussion	82
5.4 References	88

Chapter 6   Temperature Dependent Study of Excited State Conformational Relaxations in D-A Biphenyls Using Steady-State and Time-Resolved Fluorescence 89

6.1 Introduction	89
6.2 Experimental	91
6.2.1 Low temperature measurements	92
6.2.2 Band shape analysis	92
6.2.3 Global analysis of emission decays	92
6.3 Results and Discussion	93
6.3.1 Temperature dependence of radiative back charge transfer and excited state relaxations for the D-A biphenyls I-III.	93
6.3.1.1 Energetics from fluorescence band shape analysis	93
6.3.1.2 Analysis of the fluorescence lifetimes and quantum yields.	97
6.3.2 Temperature dependence of the conformational equilibrium in the excited state ^1CT of the strongly pretwisted D-A biphenyl III in diethylether	101
6.3.2.1 The method to recover dual fluorescence bands and to derive the reaction rate constants by global analysis of emission decays	101
6.3.2.2 Quantitative characterization of the excited state conformational equilibrium between two charge transfer species CT and CTR	106
6.4 Conclusion	111
6.5 References and Notes	112

Chapter 7   Pressure and Temperature Dependent Fluorescence of D-A Biphenyls. The Separation of Viscosity and Thermal Control of the Conformational Photoreaction in the Highly Twisted Compound 114

7.1 Introduction	114
7.2 Experimental	116
7.2.1 High pressure equipment	116
7.2.2 Ti:Sapphire Laser	116
7.3 Results and Discussion	116
7.3.1 Viscosity and temperature influence on the fluorescence and excited state relaxations of I-III in triacetine (TAC).	116
7.3.2 Analysis of reaction rate constants and its boundary conditions	119
7.3.3 The method to distinguish between viscosity and thermal control of a photoreaction	123
7.3.4 Separated thermal activation energy and viscosity dependence of the conformational photoreaction in III.	125
7.4 Concluding Remarks	128
7.5 References	129

Chapter 8 **The Influence of Conformation and Energy Gaps on Optical Transition Moments in D-A Biphenyls..**	**130**
8.1 Introduction	131
8.2 Experimental and Semiempirical Calculations	132
8.3 Results and Discussion	132
8.3.1 Experimental transition dipole moments	132
8.3.2 Electronic coupling between ^1ET and ^1L_a in dependence of the twist-angle	133
8.3.3 Description of transition moments with reference states	135
8.3.4 Influence of structural relaxations other than φ_{D-A} twisting on M_{CT}	138
8.3.5 Influence of the ^1ET-^1L_a energy gap on M_{CT}	139
8.3.6 Evaluation of S_0 and S_1 twist angles in solvents	143
8.4 Concluding Remarks	143
8.5 References and Notes	144
Chapter 9 **Conformation and Energy Gap Dependent Electron Transfer Interactions in Flexible D-A Biaryls:** **The Case of Two Twisted 9-(dimethylanilino) phenanthrenes**	**146**
9.1 Introduction	146
9.2 Experimental and Calculations	147
9.3 Results	148
9.3.1 Spectroscopic transition moments and energies	148
9.3.2 Quantum chemical calculations	151
9.4 Discussion	154
9.4.1 Influence of electron transfer interactions on absorption and emission bands.	154
9.4.2 Squared transition moments as indicators for molecular and electronic structure	154
9.5 Conclusions	158
9.6 References and Notes	159
Chapter 10 **Possible Applications of D-A Biphenyls as Fluorescent Probes**	**160**
10.1 Introduction	160
10.2 Experimental	163
10.2.1 Measurements of pH	163
10.3 Results and Discussion	163
10.3.1 Micropolarity	163
10.3.2 Transition temperature from liquid to solid (Tg)	165
10.3.3 Protic solvents	170
10.3.4 Sensitive and self-calibrating sensing of pH in aqueous solution	174
10.4 Conclusion	182
10.5 References and Notes	182
Chapter 11 **Implications to Solar Energy Conversion**	**186**
11.1 Introduction	186
11.2 Experimental	188
11.2.1 Preparation of polymer films	188
11.3 Results and Discussion	189
11.3.1 Requirements for low lying and highly fluorescent ^1CT states in D-A biaryls	189
11.3.2 Which host environment is favourable for Fluorescent Solar Concentrators (FSC)	191
11.4 Solar Perspectives	196
11.4.1 Fluorescent Solar Concentrators using planar or moderately twisted D-A biaryls	196
11.4.2 Towards Electron Transfer Initiated Photocatalysis (ETIP) using twisted D-A biaryls	197
11.5 References	
Chapter 12 **Final Conclusions**	**200**

Appendix A
Glossary of Abbreviations
List of Publications
Lebenslauf

Chapter 1

Introduction

This introductory chapter aims to point out the general goal of the whole work and to serve as a guide through all chapters. It is NOT an introduction to the specific field investigated. Each chapter consists of its own introduction giving the important information of the method employed (usually in the results and discussion section) and a report of the state of the art. An overview regarding the class of donor-acceptor biphenyls is given in Ch. 3.1 with more photophysical details in 3.3.

Goals

In view of a common goal to exploit light using *photon-driven molecular devices* (PMD), e.g. solar energy converters or microsensors, science fulfills the task on the one hand, *to investigate and characterize possible candidates* of PMD's and, on the other hand, to analyze the applied basic principles and mechanisms utilizing model compounds in order *to provide a theoretical platform* necessary to improve existing and complex PMD's.

The current study intends to give a contribution to both but with a main stress on the second item, i.e. the investigation of a promising photoinduced mechanism with model compounds. The photon-induced process studied is *intramolecular charge transfer* between two well-defined molecular fragments of controllable electronic interaction. The model compounds are three donor-acceptor biphenyls where the electron pushing dimethylanilino groups (D) are attached by a single chemical bond to the electron pulling benzonitrile groups (A) in a different spatial arrangement modified by the twist angle (φ_{D-A}) between the planes of both moieties.

Since the model compounds were synthesized and investigated for the first time (except compound **II** which was simultaneously investigated by F.Lahmani/Orsay Cedex in a different way) a large part of this Ph.D. thesis deals with the electronic and photophysical characterization of the compounds in order to derive the photon-induced intramolecular electronic and conformational relaxations. Furthermore, using this knowledge advanced contemporary methods to analyze charge transfer processes are applied and evaluated. Finally, possible applications of photoinduced *intramolecular charge transfer* in donor-acceptor biphenyls, or more commonly in biaryls, as PMD's in the field of solar energy conversion and microenvironmental sensing are principally discussed and partially demonstrated.

Text Structure and Guide

Scheme 1.1 illustrates the structure of the text and the arrows denote the flow of information needed to reach the conclusions in the relevant chapter. In ch. 2, the absorption spectra are analyzed and the electronic structure is interpreted in terms of a composite-molecule model. In ch. 3, the fluorescence is investigated and the electronic nature of the emitting state is analyzed. It gives a survey about the photophysics at room temperature. The conclusions in ch. 3 are substantiated by more detailed results given in the following chapters. Ch 3 may be regarded as the heart of the text structure giving the references to the parts where more details can be found. For example, the transient absorption experiments in ch. 4 indicate an initial relaxation towards planarity, ch. 5 is focused on the triplet behaviour, ch. 6 and 7 provide a detailed description of the excited state equilibrium and ch. 8 analyzes the solvent and conformation dependent transition moments. To address the question whether the derived photoinduced properties can be transferred to larger biaryl systems, two differently twisted phenanthryl derivatives are examined in ch. 9 analogously to **I-III**. All these informations are finally used to discuss and demonstrate aspects of possible applications regarding fluorescence sensing (ch.10) and solar energy conversion (ch. 11).

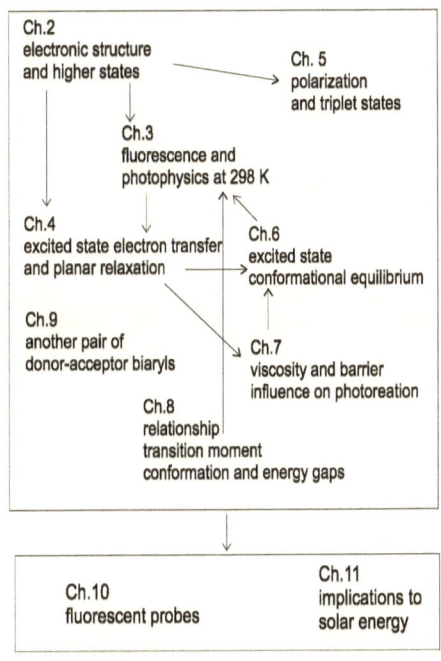

Scheme 1.1 *Flow diagram of text organization*

Comments to Nomenclature

The recommendations of the GLOSSARY OF TERMS IN PHOTOCHEMISTRY (GOTIP) [Pure & Appl. Chem. 1988, 60, 1055-1106.] are followed. The most important differentiation to be noted is the distinguished use of intramolecular charge transfer (CT) and intramolecular electron transfer (ET). Here and in the GOTIP, CT denotes the transfer of a FRACTION of electronic charge between localized sites in a molecule, while ET stands for the FULL transfer of one electron ($=1.602 \times 10^{-19}$C) between the relevant sites or units.

Note one exception of GOTIP notation: The preferred energy unit here is the wavenumber in cm^{-1} which is abbreviated here by ν without the usual queue.

Chapter 2

Absorption Spectra and Electronic Structure of D-A Biphenyls

Abstract
The electronic structure of 4-(N,N-Dimethylamino)-4'-cyano-biphenyl and its planar fluorene and twisted 2,6-dimethyl-substituted model compounds (**I-III**) is analyzed by experimental means of UV/VIS absorption spectroscopy including linear dichroic and derivative spectra. CNDO/S-CI calculations show that the electronic structure of the biphenyls investigated can be approximately described within a composite-molecule model based on the ^1L_b, ^1L_a states of the dimethylaniline and benzonitrile subunits. But in addition to unsubstituted biphenyl (BP), an intramolecular charge transfer (^1CT) state is active as the first excited singlet state and the twist angle dependent interaction with the higher lying, locally excited singlet states modifies the absorption spectra. The A, B, C and H absorption bands of unsubstituted biphenyl can be correlated with the absorption spectra of the donor-acceptor biphenyls **I-III** and the additional absorption band at fairly lower energy than the A band in biphenyl is assigned to a strong intramolecular CT band. This leads to a consistent and helpful interpretation of the electronic structure of donor-acceptor biphenyls including those (**IV-V**) investigated already in literature.

Keywords: biphenyl, UV/VIS absorption, linear dichroism, CNDO/S, electronic structure

2.1 Introduction

For the unsubstituted biphenyl (**BP**), several theoretical publications confirmed the correlation of the electronic states with states derived from the composite molecule model.[1-4] But in contrast to **BP**, in series of 4,4'-donor-acceptor substituted biphenyls charge transfer excitations cannot be neglected.

Until recently, a number of publications dealt with charge transfer properties of donor-acceptor (D-A) biphenyls[5-11] but none of them gave an in-depth and uniform interpretation of the electronic structure. Particularly the biphenyl compounds **II**, **IV** and **V** have already been investigated.[9-11] It has been reported for these compounds that the first strong absorption band is due to a ^1L_a-type state. HERBICH and WALUK[10] concluded that absorption and fluorescence of **IV** originate from the same state containing partial charge transfer character of similar magnitude. On the other hand, LAHMANI ET AL.[11] proposed for compound **II** primary excitation to a locally excited state of the ^1L_a type but radiative deactivation from a charge transfer state gaining intensity by mixing with locally excited states. Moreover, the involvement and assignments of the higher lying states was not investigated. BULGAREVICH ET AL.[9] gave assignments for the ^1L_b and ^1B_b states in the absorption spectrum of **V** but from the present investigation it follows that the ^1L_b assignment was not correct.

In order to improve and complete the interpretation often given for absorption spectra of D-A biphenyls and because we need a good knowledge of the electronic properties of **I-III** for the following studies including time-resolved transient absorption[12] in ch. 4 and fluorescence[13] in ch. 3, it is necessary to have a closer look on the electronic structure of the D-A biphenyl **II**. The biphenyl series is supplemented with the planar and more twisted biphenyl compounds **I** and **III** and especially the comparison of **II** with these sterically restricted biphenyls **I** and **III** should enable us to unambiguously assign the observed electronic transitions in D-A biphenyls. This study sets the basis for the following experimental studies on the D-A biphenyl series **I-III**. Therefore, in this chapter the assignments of the relevant electronic states are given and it is shown that the electronic structure is nearly independent of the methyl substitution pattern in **I** and **III** and that the observed changes can be satisfactorily described by the influence of the twist angle in the biphenyl **II**.

2.2 Experimental

CNDO/S-SCI calculations including 49 singly excited configurations have been performed with the QCPE program #333 modified to use the original CNDO/S parametrization[14] and to calculate the excited state dipole moments. All input geometries were fully optimized in the ground state by the Newton algorithm with the AM1 Hamiltonian within the AMPAC program.[15] All quantum chemical calculations were executed on a HP 735 workstation. The synthesis of the compounds **I-III** is described in ch. 3.[13]

All solvents used were of spectroscopic grade (Merck UVASOL) and the commercial polyethylene sheets have been checked for impurities by absorption and fluorescence. The absorption spectra presented were recorded on a ATI UNICAM UV4 UV/VIS spectrophotometer and the decadic molar extinction coefficients ε were repeatedly determined. The linear dichroic spectra have been obtained using a conventional arrangement[16,17] with a rotatable UV-Glan-Thomson polarizer and fixed stretched polyethylene sheets inside the ATI spectrophotometer.

2.3 Results and Discussion

2.3.1 Composite Molecule Approach for Donor-Acceptor Biphenyls I-III

In order to confidently treat the electronic structure of a super-molecule consisting of two different chromophoric parts in terms of a composite-molecule model[18] it is necessary to have a substantial localization of the molecular orbitals (MOs) on each subunit. Within the LHM method developed by LONGUET-HIGGINS and MURRELL,[19,20] this precondition is assumed per sé since it uses only localized MOs. The configuration interaction (CI) calculation then yields final wavefunctions for the electronic states which can readily be interpreted in terms of the composite fragments. An alternative procedure using delocalized MOs has been proposed by BABA, SUZUKI and TAKEMURA.[21] In their original work, they applied the Pariser-Parr-Pople (PPP) method[22] including a CI calculation. A configuration analysis was then performed to interpret the results in the language of the more comprehensible composite-molecule model.

In this work, the method of BABA, SUZUKI and TAKEMURA is principally followed but instead of the PPP method the more sophisticated CNDO/S[14] method is employed. A brief outline of our configuration analysis method is given in the Appendix A. In Fig. 2.1a, the most important MOs obtained for the separated phenyl-subunits dimethylaniline (donor part D) and benzonitrile (acceptor part A) as well as for the composite biphenyl molecule II (D-A) at three different twist angles are depicted. Using Platt's notation,[23] the one-electron configurations responsible for the ^1L_b and ^1L_a states of the free donor and acceptor molecules are indicated. As one can see, at 90° twist of the donor and acceptor phenyl part in the composite molecule, the MOs are effectively localized on the subchromophors and correspond to the MOs of the free donor and acceptor part. This is also substantiated by the retained MO energies and symmetries. As a consequence, in the composite molecule at 90°, the same local ^1L_b and ^1L_a transitions are active as in the separated D and A parts. Transitions between the subunits (electron transfer) are forbidden at 90° due to vanishing overlap of the donor and acceptor MOs. Deviations from perpendicularity reduce the symmetry of the molecule from C_{2v} to C_2 and thus lead to interactions between MOs of equal symmetry (Fig. 2.1a). Nevertheless, even at full planarity the MOs keep substantial localized character which gives the justification to continue the classification of localized ^1L_b and ^1L_a transitions. The strongest interactions occur between the occupied MOs 1 and 2 and the unoccupied MOs -1 and -4 (all of equal MO symmetry B) resulting in two important consequences for the HOMO-LUMO configuration:

1) The HOMO-LUMO transition is of pure electron transfer character at 90°, but at non-perpendicular twist angles the MO mixing dilutes the electron transfer character and introduces partial $L_a(D)$ and $L_a(A)$ character. Tab. 2.1 shows that the electron transfer (ET) character of the HOMO-LUMO transition decreases from 96% at 90° to 42% at 0° accompanied by an increase of 4% ^1L_a-type contributions at 90° up to 46% at 0°. Nevertheless, even at 0°, the ET contribution (HOMO(D)→LUMO(A) of the localized units) is by far largest for the state (2^1A)

Ch. 2 Absorption and Electronic Structure

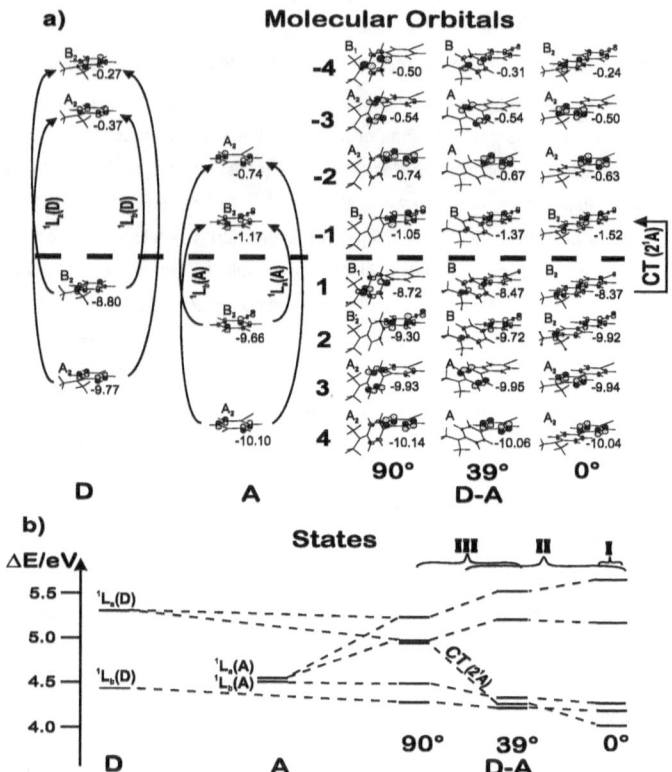

Fig. 2.1 a) *Calculated molecular orbitals of dimethylaniline (D), benzonitrile (A) and* **II** *(D-A) at 0°, 39° and 90°. MO energies (in eV) and MO symmetries (within molecular symmetry C_{2v} for 0° and 90° and C_2 for intermediate angle 39°) are also given.*
b) *Correlation diagram for the singlet state transitions of dimethylaniline (D) and benzonitrile (A) in regard to the composite molecule* **II**. *The braces indicate the available twist angle regions for each of the D-A biphenyls*

connected with around 90% (see below Tab. 2.2) of the HOMO-LUMO transition. Due to the strong ET character of 2^1A this state is called the intramolecular charge transfer state ^1CT to account for its sizeable charge separation between the phenyl units. This is also verified by the large dipole moments of the ^1CT state as compared to the other states calculated (see below Tab. 2.4).

2) The increased MO interaction with increasing planarity also leads to a reduced HOMO-LUMO gap and therefore to a stabilization of the ^1CT state. The ensuing twist angle dependent transition energy is well reflected by the correlation diagram shown in Fig. 2.1b. By changing the spatial arrangement of the donor and acceptor phenyl units from perpendicularity to planarity, the charge transfer state gets stabilized by 0.9 eV (7300 cm^{-1}). This means that in

contrast to the high lying ^{1}CT state in unsubstituted biphenyl **BP**, the ^{1}CT state of the D-A biphenyl **II** can become the lowest excited singlet state even in the gas phase for twist angles less than 39°.

Moreover, Fig. 2.1b shows that, similar to **BP**,[2] the interaction between the ^1L_b states is weak due to their small transition moments, so that they nearly retain their transition energies with respect to the free subunits and exhibit only a slight stabilization with increasing planarity. The fact that the stabilization of the ^{1}CT state is much stronger than that of the ^1L_b state suggests a quite stronger tendency of the D-A biphenyl **II** for a structural relaxation to planarity in the excited state than it was observed for **BP** itself.[24-27] In fact, an excited state relaxation towards planarity for **II** (and **III** only in non-polar solvents) in the excited state is concluded from time-resolved fluorescence measurements (ch. 3)[13,28,29] Similar as for **BP**,[2,4] the interaction between the more allowed local ^1L_a-states is much stronger. In **BP**, the enhanced interaction is reflected by a larger energy splitting between the out-of-phase and in-phase linear combination of the local ^1L_a excitations of equal energy. The strong interaction between the non-degenerate ^1L_a states in the D-A biphenyls is testified by a nearly 50:50 mixing of the donor and acceptor ^1L_a configurations in the present calculation. A comparable enhancement of the energy splitting between the ^1L_a states as in **BP**[2] is observed with decreasing twist angle (Fig. 2.1b). The most important differences to **BP** result from the additional interaction of the ^1L_a states with the low lying charge transfer state. All three states possess the same symmetry (^{1}A within C_2) with polarization in the long molecular axis. The interaction between these states increases with decreasing twist angle and partly accounts for the stabilization of the ^{1}CT state at the cost of destabilization of the ^1L_a states, as well as for the decreasing oscillator strengths of ^1L_a with decreasing twist angle. Thus, for the absorption spectra of **I-III** we have to expect an inverse behaviour of the ^1L_a band as it is observed for **BP**. For **BP**, it is well-known that with increasing twist angle the energy of the ^1L_a band shifts hypsochromically accompanied with a loss of intensity.[18,30-34] For the D-A biphenyls studied here, the behaviour of the ^1L_a band in **BP** is predicted to be transferred to the CT absorption band.

Tab. 2.1 Local (L_a or B_a) and Electron Transfer (ET and Reverse ET) Character of the Delocalized HOMO-LUMO (1→-1) Configuration. The Values for the Model Compounds **I** at 0° and **III** at 40° and 90° are Given in Brackets.

angle	L_a(D)	L_a(A)	ET(D$^+$A$^-$)	RET(D$^-$A$^+$)
90°	1.2% (1.2%)	3.2% (3.5%)	96% (95%)	0.04% (0.04%)
39°	20% (16%)	22% (24%)	50% (53%)	9% (7%)
0°	22% (21%)	24% (25%)	42% (41%)	13% (13%)

* The data are derived from the coefficients of the 2pπ atomic orbitals (eq. A-2) in Appendix A) involved in the 1→ -1 configuration. Because mixing between MOs of different symmetry is negligible, there is no L_b character in 1→ -1.

Let us now discuss whether the electronic structure of the planar fluorene **I** and the more twisted biphenyl model compound **III** can approximately be described by that of **II**. The configuration analysis shows that the MOs are only very weakly affected by the methyl substitution pattern in **I** and **III** yielding the same configurations as for **II**. As an example, Tab. 2.1 shows that the character of the HOMO-LUMO (1→-1) transition only depends on the twist angle but not on the methyl substitution pattern. In detail, the percentages of the character for the HOMO-LUMO configuration are nearly identical for **II** and its model compounds **I** and **III** the values of which are given in brackets in Tab. 2.1. Further, in Tab. 2.2 the contributions of the characteristic configurations (see Fig. 2.1) to the charge transfer and the first four locally excited singlet states are compared for **II** with **I** and **III**. Independent of the compound and the twist angle, the ^{1}CT state keeps the large 1→-1 contribution of about 90 % (with varying ET character, see Tab. 2.1). For **II** and **III**, the ^1L_b and ^1L_a assigned states are determined by similar weights of the contributions, too. Main differences are observed only between the contributions of **I** and **II**. The lower weights for **I** indicate more mixing between the states. This is not surprising since **I** is derived from unsubstituted fluorene which has C_{2v} symmetry with the twofold C_2 symmetry axis perpendicular to that in **BP** which is of D_2 symmetry with C_2 along the long-molecular axis as in **II**. Therefore, in fluorene the benzene ^1L_b (and ^1L_a) combinations split into a perpendicular and parallel polarized transition with respect to the long-molecular axis. The equal symmetry of the $^1L_b(+)$ and $^1L_a(-)$ states explains the enhanced mixing between these states in fluorene,[2] and can analogously explain the enhanced mixing behaviour of the D-A fluorene **I** (Tab. 2.2). This also leads to a relatively strong dependence of the calculated transition moment and its direction on the parametrization and geometry used for **I**.

Tab. 2.2 Contributions (%) of the Characteristic Configurations to the ^{1}CT, ^1L_b-type and ^1L_a-type States. The Contribution to the ^{1}CT State is Calculated by the Squared CI Coefficient of the 1→-1 Configuration and the Contributions to the ^1L_b (^1L_a) States are Calculated as the Sum of the Squared CI Coefficients of the Four ^1L_b (^1L_a)-type Configurations Denoted in Fig. 2.1a (see eq. A-4 in Appendix A).

state	D	A	II / 90°	III / 90°	II / 39°	III / 40°	II / 0°	I / 0°
^{1}CT($D^{\delta+}A^{\delta-}$)	-	-	90%	89%	94%	92%	95%	83%
^1L_b(D)-type[a]	98%	-	90%	98%	74%	66%	61%	39%
^1L_b(A)-type[a]	-	98%	83%	98%	49%	59%	54%	50%
^1L_a(D)-type[b]	96%	-	96%	97%	91%	92%	88%	84%
^1L_a(A)-type[b]	-	34%	93%	94%	80%	45%	80%	27%

[a] The contributions to the two ^1L_b states are mainly determined by the configurations of the respective subunit. Only for the ^1L_b(A) state at 0° for **I** and **II** the „wrong" ^1L_b(D) configuration contributions exceed 10%.
[b] Although the lower ^1L_a state has more donor character, the donor and acceptor localized contributions are strongly mixed up to 50:50.

However, previous results show that the electronic structure of fluorene is better comparable with biphenyl than with molecules of equal C_{2v} symmetry like carbazole or phenanthrene[35] which have the short molecular axis as the symmetry axis C_2. The similarity of the relative CI contributions for **I** and **II**, and even better for **II** and **III** leads us to conclude that the electronic structure of **I** and **III** can approximately be described by that of **II** at the corresponding twist angles. In this case, the absorption spectra of **I-III** (and **IV-V**) should be interpretable with the calculation results obtained for **II**.

2.3.2 UV/VIS Absorption Spectra and Their Derivatives

In Fig. 2.2, the absorption spectra of **I-III** in the solvents n-hexane and acetonitrile are shown together with the calculated electronic transitions for **II** at 0°, 39° and 70° which are indicated by the column bars of length proportional to the respective oscillator strength (for more details see caption of Fig.2.2). It is common knowledge that without environment correction CNDO/S-CI calculations overestimate the transition energies even for non-polar aromatics in low-temperature matrices[35,36] by up to 5000 cm^{-1}. Therefore, the positions of the column bars in Fig. 2.2 representing the calculated transitions are shifted 1700 cm^{-1} to the red in order to give a better representation of the experimental spectra. For comparison, the INDO/S calculations previously published for compound **IV** also overestimated the transition energies of the strongest bands by 4000 cm^{-1}.[10] Besides, for each of the ^1L_b and ^1L_a combinations the energies have been averaged (compare transition energy differences in Fig. 2.1b) and the corresponding oscillator strengths have been summed up.

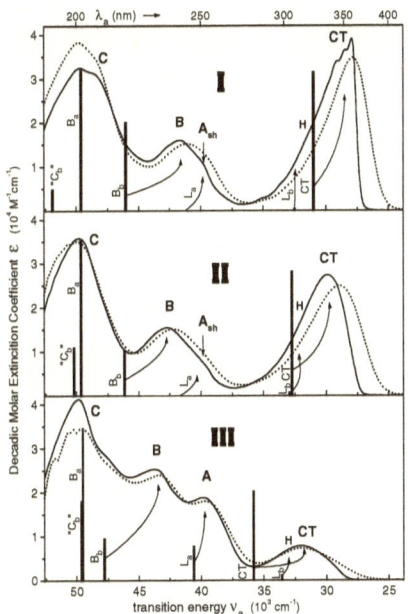

Fig. 2.2 Absorption spectra of **I-III** in n-hexane (———) and acetonitrile (- - -). Band notations are taken from SUZUKI.[18,33] The column bars symbolize the calculated transitions (gas phase) for **II** at 0°, 39° and 70°. The energy positions of the calculated transition bars are shifted by 1700 cm^{-1} to the red and the molar extinction coefficients ε are calculated from the oscillator strength for a Gaussian intensity distribution with a half width of 3400 cm^{-1}. The arrows indicate the proposed assignments.

Following increasing transition energy, the empirical band labels H, A, B, C have been introduced for **BP** absorption spectra in the literature.[18,34,37] Within PLATT's nomenclature and the symmetry corrections based on improved calculations including configuration interaction by HAM and RUEDENBERG,[38] the short-axis polarized H (hidden) and B bands in **BP** can be assigned to ^1L_b and ^1B_b benzenoid transitions and the long-axis polarized A and C bands belong to the corresponding ^1L_a and ^1B_a type transitions, respectively. Let us now localize these band series in the absorption spectra of **I-III** in n-hexane (Fig.2.2).

First of all, we unambiguously have to assign the lowest energy absorption band to the ^{1}CT state as the predicted hypsochromic shift and intensity loss with increasing twist angle from **I-III** is well observed. A similar hypsochromic shift by twisting is predicted for the strong $S_0 \rightarrow {}^1B_b$ transition which justifies our assignment of the band between 41-44000 cm^{-1} to the corresponding B band in **BP**. Consistently, the B band in **BP** exhibits a blue-shift upon methyl substitution in the ortho positions, too. Our assignment of this B band is in agreement with that of the strong absorption band around 42-44000 cm^{-1} in **IV** and **V** which has also been assigned to the ^1B_b transition in **V**.[9] At the low energy side of the B bands of **I, II, IV** and **V** a slight shoulder is apparent around 40000 cm^{-1}. The position of such band shoulders can better be extracted from the first and second derivatives of the spectra.[39-41] This procedure has been carried out for **I-III** and the results are shown in Fig. 2.3.

At band maxima or minima, the first derivative spectrum of the molar extinction coefficients $\varepsilon'(\nu_a)$ crosses zero providing an alternative possibility to localize the mean transition energies. Band maxima are usually also characterized by a corresponding minimum of the second derivative spectrum $\varepsilon''(\nu_a)$ at slightly higher energy coinciding with the zero crossing of $\varepsilon'(\nu_a)$. Additional minima of $\varepsilon''(\nu_a)$ can be used to locate absorption shoulders. It can therefore be deduced that an absorption band is located for **I** and **II** around 40000 cm^{-1} even though it is only barely visible in the spectrum $\varepsilon(\nu_a)$. Comparing the spectra of **I-II** (and **IV-V**) with those of **III** we can conclude that the shoulder is correlated with the clear band maximum of **III** also occurring around 40000 cm^{-1} but at somewhat lower energy. The behaviour of these correlated bands parallels the calculated prediction for the ^1L_a transition, i.e. increase of intensity and small shift to lower energy with increasing twist angle due to increased MO localization (Fig.2.1a). In **BP** the L_a

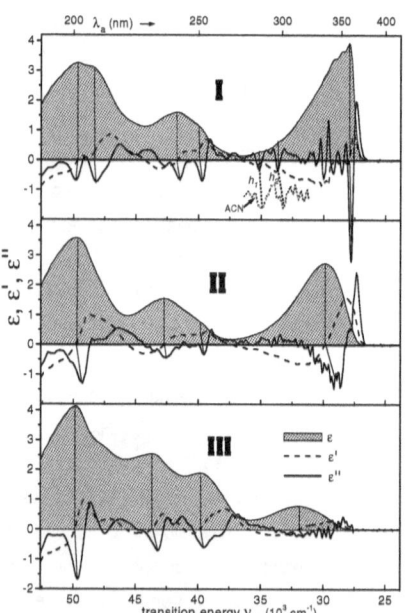

Fig. 2.3 *First and second derivative absorption spectra of **I-III** in n-hexane. For **I**, the second derivative spectrum in acetonitrile (ACN) is also shown at the high energy part of the CT band. Between 26500 and 29500 cm^{-1} the second derivative spectra of **I** have been divided by 8 in order to fit into the scale.*

band A can be regarded as originating from the intense out-of-phase and forbidden in-phase linear combinations of the individual benzene ^1L_a excitations without net electron transfer. Analogously, the A band of the D-A biphenyls is due to the mixing local ^1L_a excitations of the dimethylaniline and benzonitrile subunits, respectively. Furthermore, the similar transition energy of the A band of **BP** in cyclohexane (40490 cm^{-1}) encourages to assign the bands observed at around 40000 cm^{-1} to the A band produced by the $S_0 \rightarrow {}^1L_a$ transition. This contradicts the previous assignment of the shoulder in **V** to the ^1L_b state.[9]

The high energy band around 50000 cm^{-1} has not yet been studied for other D-A biphenyls like **IV** and **V**. It correlates with the longitudinally polarized C band in **BP**[18,27,34] at 49500 cm^{-1} and can thus be associated with the very strong ^1B_a-like transition. With increasing twist angle, we observe a narrowing and intensity gain of the C band in accordance with the calculation results. The main contribution to the C band is given by the ^1B_a-state but additionally, a perpendicular polarized transition contributes to this band which we may call „^1C_b" here to account for its nature of a higher series of the ^1L_b or ^1B_b type states. At full planarity, these two states ^1B_a and ^1C_b are distinctly split which may account for the appearance of two maxima in the C band of **I** (Fig. 2.2) and the two minima in the second derivative of the spectrum (Fig. 2.3), respectively. Clearly visible is the increase of the absorption intensity which according to the calculations is due to the increase of oscillator strength for the „^1C_b" like transition.

It remains to localize the so-called „hidden" because very weak H band[18] in the D-A biphenyls. At first sight, one could assume that the well pronounced vibrational structure of **I** around 28600 cm^{-1} (350 nm) indicates a vibronically coupled state. But it is quite more reasonable to explain this structure, only appearing in the planarity restricted D-A fluorene **I**, with a steeper twist potential in the ground and excited (^1CT) state providing a very narrow angular distribution and thus a more narrow energy distribution of the Franck-Condon transitions as compared to **II-V**. Furthermore, the 0-0 vibronic band is most intense which indicates a symmetry-allowed transition between $1^1A(S_0)$ to $2^1A(CT)$ if we consider **I** as a biphenyl derivative with the long molecular axis as the C_2 symmetry axis. Besides, in polar acetonitrile, the strong 0-0 band and the whole structure disappears (Fig. 2.2) because $2^1A(CT)$ is dipolar and thus strongly solvated giving rise to a broader energy distribution. Structure in the absorption spectrum of **I** also occurs around 33300 cm^{-1} (300 nm) with two minima in the $\epsilon''(\nu_a)$ spectrum at 35200 cm^{-1} and 33670 cm^{-1}. In contrast to the low energy region, this structure is well reproduced in other solvents and from the $\epsilon''(\nu_a)$ spectrum a slight red shift of the minima to 35000 cm^{-1} and 33220 cm^{-1} is found in polar acetonitrile. This structure, in fact, could be due to a symmetry forbidden ^1L_b state vibronically coupled to the long-axis polarized ^1CT (2^1A) state responsible for the red shift of the vibronic bands of 200 cm^{-1} (h_1) and 450 cm^{-1} (h_2) in acetonitrile (see Fig. 3.3). The absence of the structure around

33300 cm^{-1} for the biphenyls II-V can be explained by the overlap of the dominating CT band. Note, that the CT band for II-V is blue-shifted by the increased twist angle into the region of the supposed ^1L_b band.

2.3.3 Linear Dichroic Absorption Spectra

To obtain more reliable information about the hidden perpendicularly polarized H band, polarization experiments are quite useful. In particular the technique of linear dichroism (LD) in stretched polyethylene (PE) films allows to determine absolute transition moment directions if the molecule under study can be oriented.[16,17] For „rod-like" biphenyls, uniaxial orientation parallel to the stretching direction of the PE film preferentially is achieved in the long molecular axis.[37] This yields high LD ratios E_{par}/E_{ort} for long-axis polarized transitions of ^1A symmetry and correspondingly low ratios for transitions of ^1B symmetry (e.g. ^1L_b). The LD spectra of I-III and their derived dichroic ratios are shown in Fig. 2.4.

At the low energy side of the absorption spectra, the dichroic ratios reach their maximum values of 7, 10 and 1.4 for I-III, respectively. All ratios are larger than 1 which confirms that the first excited singlet state is of ^1A symmetry in a C_2 point group

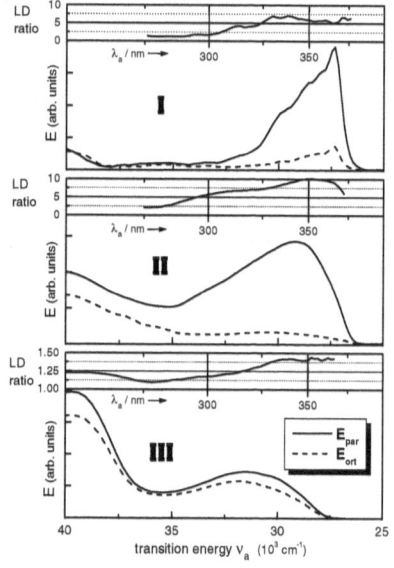

Fig. 2.4 *Linear dichroic (LD) spectra for* **I-III** *in stretched polyethylene (STPE). The curves of absorbance E_{par} (——) and E_{ort} (- - -) are obtained with light polarized parallel and perpendicular to the stretching direction of the fixed STPE film. The LD ratio E_{par}/E_{ort} is plotted in the upper box (——).*

with transition moment direction in the long molecular (C_2) axis. The lower LD ratio value for III may partially be due to the less efficient orientation in stretched PE connected with a more spherical structure of III as compared to I or II. Around 30000 cm^{-1} the dichroic ratio starts to decrease for all D-A biphenyls investigated. This means that at the high energy side, not far from the absorption maximum in I and II ($\Delta E \approx 2000$ cm^{-1} for I and $\Delta E \approx 1000$ cm^{-1} for II) there is evidently the expected contribution of a perpendicularly polarized transition assignable to the hidden ^1L_b band. For III, the relative ^1L_b contribution is much stronger than for I and II (compare E_{par} and E_{ort} for III with I and II in Fig. 2.4) and it already begins on the lower energy side of the absorption maximum. As the maxima of the dichroic spectra of II are situated at

29150 cm^{-1} (343 nm) for parallel (E$_{par}$) and at about 30300 cm^{-1} (330 nm) for orthogonal orientation (E$_{ort}$) and the respective maxima for **III** are at 31450 cm^{-1} (318 nm) and 31850 cm^{-1} (314 nm) we can conclude that the energy difference between the ^{1}L$_b$ (^{1}B) and ^{1}CT (^{1}A) state decreases with increasing twist angle (ΔE$_{max}$(**II**)=1150 cm^{-1} ΔE$_{max}$(**III**)=400 cm^{-1}) but does not lead to a level reversal in the Franck-Condon geometry. On the other hand, from the spectral analysis we cannot rule out the presence of a second ^{1}L$_b$ band at longer wavelength (as predicted from theory, see Fig. 1b). Such two hidden ^{1}L$_b$ bands have been reported for **BP** at 40320 cm^{-1} (248 nm) and 37170 cm^{-1} (269 nm) in PE of which the low energy one is very weak.[37] This corresponds to an experimental energy difference between the ^{1}L$_b$ states of 3150 cm^{-1} (0.39 eV) for **BP** in PE. Although the calculated energy difference is less than 1650 cm^{-1} (0.2 eV in Fig. 1b) for the D-A biphenyl **II** it can not be excluded that a lower lying, more forbidden ^{1}L$_b$ state is photophysically active (see also ch. 4,5).

2.3.4 Comparison of Spectral Data with Calculated Results

In Table 2.3, the absorption spectral data for **I-III** in n-hexane and acetonitrile are summarized together with the data of **IV**, **V** and **BP** as taken from literature. They can be compared with the CNDO/S calculated absorption spectral data of **II** at the angles 0°, 39° and 70° which correspond to the twist angle for **I**, the gas phase calculated twist angle for **II** and to the twist angle of 2,2' methyl substituted **BP**[33] shown in Tab. 2.4. The molar extinction coefficient ε_{max} is obtained for a Gaussian intensity distribution (eq. 2-1) for the absorption maximum $\nu = \nu_a^{max}$ from eq. 2-2[18] using a half width HW$_a$=3400 cm^{-1} and the calculated oscillator strength f.

$$\varepsilon_a(\nu_a) = HW_a^{-1}\sqrt{2/\pi} \ \exp(-2(\nu_a - \nu_a^{max})^2 / HW_a^2) \ \int \varepsilon_a(\nu_a) d\varepsilon_a \qquad (2\text{-}1)$$

$$\varepsilon_{max} = HW_a^{-1}\sqrt{2/\pi} \ \int \varepsilon_a(\nu_a) d\varepsilon_a = HW_a^{-1}\sqrt{2/\pi} \ f/4.319 \cdot 10^{-9} \qquad (2\text{-}2)$$

The first absorption band in the D-A biphenyls **I-V** has been attributed to the ^{1}CT state not existent in **BP**. The CT transition energies for the flexible biphenyls **II**, **IV** and **V** in n-hexane are nearly equal, and a bathochromic shift is induced with increasing polarity (Tab. 2.3). This points out the charge transfer character (i.e. partial ET) connected with a high dipole moment which is verified by the calculation results in Tab. 2.4. The ^{1}CT energy for **III** seems less affected by polarity because the absorption band is roughly halfway produced by the weakly polar ^{1}L$_b$ transition (H band). Thus only a part of the whole band is red-shifted as evidenced by the wider low-energy tail in polar acetonitrile On the other hand, the bathochromic shift as induced by decreasing twist angle supports the interpretation that a delocalized transition not associated with a full but partial electron transfer produces the first absorption band. In contrast to the gas phase calculations, the experimental data show that ^{1}CT is still situated below the ^{1}L$_b$ state in **III**, but strong interactions due to a small energy gap of about 400 cm^{-1} between ^1L_b and CT may explain the large absorption contribution of the H band in **III** as compared to **I** and **II** and as compared to the calculation results. The energy of the H band in **I-III** is in good agreement with the H band in **BP** and the calculated ^1L_b transition energies. The transition energy of the A band is quite insensitive to polarity because of negligible ET contribution and is similar for all biphenyls **I-IV** and **BP**. The calculated small dipole moments and nearly angle-independent transition energies of the ^1L_a state (Tab. 2.4) support our reassignment of this state to the A band. Similar to the CT band, the B band is of nearly equal energy for **II, IV** and **V** and exhibits a sizeable bathochromic shift with increasing polarity and decreasing twist angle. Red shift with decreasing twist angle is also known for the B band of differently twisted biphenyls.[34] On the other hand, the C band is insensitive to polarity and twist angle which is again in accord with the behaviour and energy of the C band in **BP**.[34] The comparison with the calculated results in Tab. 2.4 suggests that the C band is mainly due to the ^1B_a transition with the overlap of a second strong transition termed as „^1C_b" here.

Tab. 2.3 Experimental Absorption Spectral Data for (a) **I-III** in n-Hexane (HEX) and Acetonitrile (ACN)

band	I				II				III			
	v_{max} (10^3 cm^{-1})		ε_{max} (10^4 M^{-1}cm^{-1})		v_{max} (10^3 cm^{-1})		ε_{max} (10^4 M^{-1}cm^{-1})		v_{max} (10^3 cm^{-1})		ε_{max} (10^4 M^{-1}cm^{-1})	
	HEX	ACN	HEX	ACN	HEX	ACN	HEX	ACN	HEX	ACN	HEX	ACN
CT	28.34	27.8	3.92	3.48	29.85	29.07	2.73	2.49	32.05[a]	32.05[a]	0.778[a]	0.727[a]
H	30-35	30-35	-	-	30-35	30-35			30-35	30-35		
A	40.0	39.8	-	-	39.9	39.9	-	-	39.82	39.6	1.88	1.80
B	41.67	41.07	1.60	1.52	42.74	42.19	1.55	1.51	43.67	43.29	2.53	2.40
C	49.75	49.75	3.25	3.83	49.75	50	3.58	3.51	49.75	49.88	4.13	-

[a] due to overlap with the H band the values for the pure CT are somewhat smaller than presented

(b) **BP, IV** and **V** in Polyethylene (PE) or HEX.

band	BP[b]	IV[c]	V[d]
	v_{max} (10^3 cm^{-1})	v_{max} (10^3 cm^{-1})	v_{max} (10^3 cm^{-1})
	PE	HEX	HEX
CT	-	29.50	29.90
H	37.17	-	-
A	39.68	-	37-40
B	48.08	42.90	42-44
C	49.50	-	-

[b] from ref.18   [c] from ref 37   [c] from ref.10   [d] from ref.9

Tab. 2.4 CNDO/S Calculated Absorption Spectral Data for **II** in Dependence of the Phenyl-Phenyl Twist Angle φ in the Gas Phase.

state	φ = 0°			φ = 39°			φ = 70°		
	v (10^3 cm^{-1})	ε_{max}[a] (10^4 M^{-1}cm^{-1})	μ (D)	v (10^3 cm^{-1})	ε_{max}[a] (10^4 M^{-1}cm^{-1})	μ (D)	v (10^3 cm^{-1})	ε_{max}[a] (10^4 M^{-1}cm^{-1})	μ (D)
^{1}CT	32.7	3.16	13.1	34.4	2.82	15.7	37.5	2.03	21.7
^1L_b[b]	34.2	0.05	8.3	34.6	0.08	8.3	35.2	0.12	8.3
^1L_a[b]	41.4	0.01	5.2	43.2	0.02	8.6	42.3	0.78	8.1
^1B_b	47.8	2.02	7.0	47.9	1.03	14.9	49.5	0.96	10.2
^1B_a	51.3	3.24	8.2	51.3	3.57	6.7	51.2	3.47	9.4
"^1C_b"	53.7	0.49	3.2	51.9	1.10	6.5	51.3	1.81	6.2

[a] from eq. 1-2   [b] The energies ν and dipole moments μ are averaged for both possible ^1L_b (^1L_a) combinations and the calculated ε_{max} are the sum of both.

2.4. Conclusions

We may draw three main conclusions from this work:

1.) Using the notation of a composite-molecule model we are able to explain the properties and interactions of the electronic states in the donor-aceptor biphenyl **II**. The lowest excited singlet states are then characterized by two weakly interacting $^1L_b(D)$ and $^1L_b(A)$ states polarized perpendicularly to the long molecular axis, two strongly interacting $^1L_a(D)$ and $^1L_a(A)$ states and an intramolecular ^1CT state, the three latter ones being polarized parallel to the longitudinal axis.

2.) Both calculation results and the absorption spectral features of **I** and **III**, can be well understood with the calculated results for **II** at the respective twist angles of **I** and **III**. Thus, the rotationally restricted compounds **I** and **III** can be seen as model compounds for **II**. All three molecules can be regarded to belong to C_2 symmetry with the long molecular axis as the symmetry axis. This agrees with the conclusion elsewhere[35] that fluorene (C_{2v} with C_2 in the short molecular axis) in spite of equal symmetry to phenanthrene or carbazole (C_{2v}) is better comparable with biphenyl (D_2 with C_2 in the long molecular axis).

3.) The assignments of the absorption band series in **I-III** are derived from the parent biphenyl **BP** and can generally be applied to D-A biphenyls. These assignments possess the advantage to easily allow prediction of changes in the absorption spectra induced by different substitution patterns. Following increasing energy the main absorption bands in D-A biphenyls are:

CT   consisting of the main part of electron transfer character (40%-90% in **II** dependent on the twist angle),

H   associated with the forbidden ^1L_b state,

A   produced by the ^1L_a states (only visible as shoulder in donor-acceptor biphenyls of small twist angle),

B   assigned to the strong ^1B_b transition and

C   due to ^1B_a and higher lying states.

2.5 References

(1) Gamba, A.; Tantardini, G.F.; Simonetta, M. *Spectrochim. Acta* **1972**, 28A, 1877.
(2) Swiderek, P.; Michaud,M.; Hohlneicher, G.; Sanche, L. *Chem. Phys. Lett.* **1991**, 187, 583.
(3) Dick, B., Hohlneicher, G. *Chem. Phys.* **1985**, 94, 131.
(4) Rubio, M.; Merchán, M.; Ortí, E.;Roos, B.O. *Chem. Phys. Lett.* **1995**, 234, 373.
(5) Lippert E. *Z. Elektrochemie* **1957**, 61, 962 (b) Lippert E.; Lüder, W.; Moll, F. *Spectrochimica Acta* **1959**, 10, 858.
(6) P. Fromherz and A. Heilemann *J. Phys. Chem.* **1992**, 96, 6864; C. Röcker, A. Heilemann and P. Fromherz *J. Phys. Chem.* **1996**, 100, 12172.

Ch. 2   Absorption and Electronic Structure

(7) M. Yoon, D.W. Cho, J.Y. Lee, M. Lee and D. Kim *Bull. Korean Chem. Soc.* **1992**, 19(6), 613; D.W. Cho, Y.H. Kim, S.G. Kang, M. Yoon and D. Kim *J. Phys. Chem.* **1994**, 98, 558.

(8) P.T. Chou, C.P. Chang, J.H. Clements and K. Meng-Shin *J. Flu.* **1995**, 5(4) 369.

(9) D.S. Bulgarevich, O. Kajimoto and K. Hara *J. Phys. Chem.* **1994**, 98, 2278.

(10) J. Herbich and J. Waluk *Chem. Phys.* **1994**, 188, 247.

(11) F. Lahmani, E. Breheret, A. Zehnacker-Rentien, C. Amatore and A. Jutand *J. Photochem. Photobiol. A: Chem.* **1993**, 70 39.

(12) M. Maus, W. Rettig, G. Jonusauskas, R. Lapouyade, and C Rullière *J. Phys. Chem.*, **1998**, 102, 7393.

(13) M. Maus, W. Rettig, D. Bonafoux, Lapouyade, R. *to be published*.

(14) J. Del Bene and H.H. Jaffé *J. Chem. Phys.* **1968**, 48, 1807; **1968**, 48, 4050; **1968**, 49, 1221; **1969**, 50, 1126.

(15) AMPAC 5.0, Semichem, 7128 Summit, Shawnee, KS 66216 **1994**; M.J.S. Dewar, E.G. Zoebisch, E. F. Healy and J. P. Stewart *J. Am. Soc.* **1985**, 107, 3902.

(16) E.W. Thulstrup and J. Michl *J. Am. Chem. Soc.* **1982**, 104, 5594.

(17) J. Michl and E.W. Thulstrup, in *Spectroscopy with polarized Light*, VCH Publishers, New York **1986**; E.W. Thulstrup and J. Michl, in *Elementary Polarization Spectroscopy*, VCH Publishers, New York **1989**.

(18) H. Suzuki, in *Electronic Absorption Spectra and Geometry of Organic Molecules*, Academic Press, New York **1967**.

(19) H.C. Longuet-Higgins and J.N. Murrell *Proc. Phys. Soc.* **1955**, A68, 601.

(20) M. Godfrey and J.N. Murrell *Proc. R. Soc.* **1966**, A278, 60.

(21) H. Baba, S. Suzuki and T. Takemura *J. Chem. Phys.* **1969**, 50, 2078.

(22) R. Pariser and R.G. Parr, *J. Chem. Phys.* **1953**, 21, 466 and **1953**, 21, 767; R. Pariser, *J. Chem. Phys.* **1956**, 24, 250; J.A. Pople *Trans. Faraday Soc.* **1953**, 49, 1375.

(23) J.R. Platt, *J. Chem. Phys.* **1949**, 17, 484.

(24) I.B. Berlman, *J. Phys. Chem.* **1970**, 74, 3085.

(25) G. Swiatkowski, R. Menzel, W. Rapp *J. Lum.* **1987**, 37, 183.

(26) F.Momicchioli, M.C.Bruni and I.Baraldi *J. Phys. Chem.* **1972**, 76, 3983.

(27) Y. Takei, T. Yamaguchi, Y. Osamura, K. Fuke and K. Kaya *J. Phys. Chem.* **1988**, 92, 577.

(28) M. Maus, W. Rettig and R. Lapouyade *J. Inf. Recording* **1996**, 22, 451.

(29) W. Rettig and M. Maus *Ber. Bunsenges. Phys. Chem.* **1996**, 100, 2091.

(30) E.A. Braude and W.F Forbes *J. Chem. Soc.* **1955**, 3776.

(31) J.N. Murrell, in *The Theory of the Electronic Spectra of Organic Molecules*, Methuen & Co Ltd, London **1963**.

(32) J.N. Murrell *Proc. Phys. Soc.* **1956**, 69, 3779.

(33) H. Suzuki *Bull. Chem. Soc. Jpn.* **1959**, 32, 1340.

(34) G. Hohlneicher, F. Dörr, N. Mika and S. Schneider *Ber. Bunsenges. Physik. Chem.* **1968**, 72, 1144.

(35) M. S. Gudipati, J. Daverkausen, M. Maus and G. Hohlneicher *Chem. Phys.* **1994**, 186, 289.

(36) M. S. Gudipati, M. Maus, J. Daverkausen and G. Hohlneicher *Chem. Phys.* **1994**, 192, 37.

(37) J. Saciv, A. Yogev and Y. Mazur *J. Am. Chem. Soc.* **1977**, 99, 6861.

(38) N.S. Ham and K. Ruedenberg *J. Chem. Phys.* **1956**, 25, 1.

(39) G.Talsky, in *Derivative Spectrophotometry* VCH Verlagsgesellschaft, Weinheim **1994**.

(40) V.J. Hammond and W.C. Price *J. Opt. Soc. Am.* **1953**, 43, 924.

(41) J.D. Morrison *J. Chem. Phys.* **1953**, 21, 1767.

Chapter 3

Fluorescence and Photophysics of D-A Biphenyls at 298 K

Abstract:
*This photophysical study addresses to the general question, how electron transfer in bichromophoric molecules influences the conformational relaxation which can be either towards more or less π-conjugation. Therefore, the effects of photoinduced intramolecular charge transfer on the electronic and molecular properties of a series of differently twisted 4-N,N-Dimethylamino-4'-cyanobiphenyls are investigated by steady state and time-resolved fluorescence. The dipole moments, radiative rates and torsional relaxations in the excited state are analyzed by comparison with the absorption spectra and interannular twist angle (φ) dependent CNDO/S calculations. Independent of the twist angle φ and solvent polarity, the first excited singlet state of these donor-acceptor (D-A) biphenyls (**I-III**) is an emissive intramolecular ^1CT state of the ^1L_a-type transferring charge from the dimethylaminobenzene (D) to the cyanobenzene (A) subunit. Similar to the planar restricted D-A fluorene **I**, the flexible D-A biphenyl **II** shows only a weak dependence of the fluorescence radiative rate constants k_f (0.4-0.6 ns^{-1}) on the solvent polarity consistent with a planarization in the excited state of **II**. On the other hand, the strongly pretwisted biphenyl **III** behaves similarly to **I** and **II** only in nonpolar solvents ($<k_f> = 0.3$ ns^{-1} indicating partial excited state relaxation towards planarity), whereas with increasing polarity the mean radiative rate $<k_f>$ decreases down to 0.03 ns^{-1}. A fast equilibrium between a more planar and a more twisted rotamer distribution in the ^1CT state of **III** explains that only for **III** additional photophysical effects appear such as (a) strong decrease of the radiative rates with increasing polarity; (b) two long (> 200 ps) fluorescence lifetimes with precursor-successor relation; (c) excited state quenching by protic solvents.*

Keywords: *photophysics, biphenyls, photoinduced charge transfer, S_0/S_1 twist potentials, solvent polarity*

3.1 Introduction

One of the main challenges of applied research is the use of solar energy. The primary step to convert solar energy into chemical or electrical potential is photoinduced energy or charge transfer. In order to optimize the whole photoconversion process, the mechanisms of the primary steps, e.g. forward and back charge transfer, have to be understood. For this long-term goal, molecular model systems are needed which allow a convincing and clear analysis of the electronic and geometrical structure before and after the primary step of charge transfer. In the field of intramolecular charge transfer between strongly coupled organic donor and acceptor fragments (fast forward charge transfer), especially the essential conformational dynamics are usually not known. As an example, the photoinduced charge transfer in dimethylaminobenzonitrile (DMABN), one of the simplest donor-acceptor molecules, is still under controversial discussion for more than 20 years.[1-5] There is no agreement, whether the necessary conformational change is towards more or less π-conjugation. In biaryls and especially donor-acceptor biphenyls, donor (D) and acceptor (A) units are more clearly defined and they should therefore serve as less problematic model compounds for the study of photoinduced charge transfer.

In unsubstituted biphenyl (BP), charge transfer interactions can be neglected and the first excited singlet state (S_1) is a short-axis polarized ^1L_b-type state with low radiative rates ($k_f =$

0.01 ns^{-1}).$^{6\text{-}9}$ In numerous publications$^{8,10\text{-}13}$ about the conformational structure of BP in solution it has been worked out that in the ground state (S$_0$) the interannular twist angle φ amounts to between 15° to 40°, whereas in S$_1$ the planar geometry is preferred. The related 4-vinylbiphenyl (VBP) and p-terphenyl (TP) have the same tendency in S$_1$ to relax towards planarity^{8,13} but due to a state inversion in comparison to BP, a long-axis polarized state of the ^1L_a-type associated with a much faster radiative transition (VBP: k_f = 0.24 ns^{-1}, TP: k_f = 0.98 ns^{-1}) becomes the lowest lying excited state S$_1$.^{8} Now it is of interest to elucidate what consequences arise when charge transfer interactions are introduced. Several previous studies on D-A biphenyls are known,$^{14\text{-}25}$ but most of them show non-trivial photophysics strongly changing with solvent or solvent polarity, which complicates the assignment to well-defined electronic and conformational structure. The undesirable side-effects encountered in known D-A biphenyls containing considerable charge transfer character in the first excited state can be divided into three classes:

1) *electronic nature of mixed character, in between that of biphenyl (^1L_b-type) and 4-vinylbiphenyl (^1L_a-type)*

2) *strong and solvent-polarity dependent nonradiative channels due to intramolecular processes*

3) *strong nonradiative channels due to specific solvent interactions.*

Let us briefly cite some examples for these three categories.

category 1):

Para-amino substituted biphenyls like 4-diethylaminobiphenyl (EBA) show radiative rates which are in alkanes only slightly larger (k_f=0.02 ns^{-1}) than for BP whereas in polar solvents like acetonitrile (k_f= 0.05 ns^{-1}) they are in between the rates of BP and VBP.^{22} This indicates that the excited state electronic structure changes with increasing polarity from more ^1L_b-type to more ^1L_a-type.

category 2):

4-dimethylamino-4'-nitro-biphenyl (DNB), which is the first well-studied biphenyl with donor and acceptor substituents, is highly nonradiative in nonpolar solvents due to the nitro group.^{20} Similar effects have been observed for 4-dimethylamino-4'-formyl-biphenyl (FDBP) where the nitro group is replaced by the aldehyde group. Here, the non-radiative channel was found to be polarity dependent.^{18}

category 3):

FDBP as well as a D-A biphenyl with a pyrimidine ring as the acceptor phenyl unit undergo specific interactions with alcohols.^{16,18} The hydrogen bonding effects manifest themselves by increased nonradiative rates, shifted absorption and fluorescence maxima in ethanol as compared to equipolar but aprotic acetonitrile.

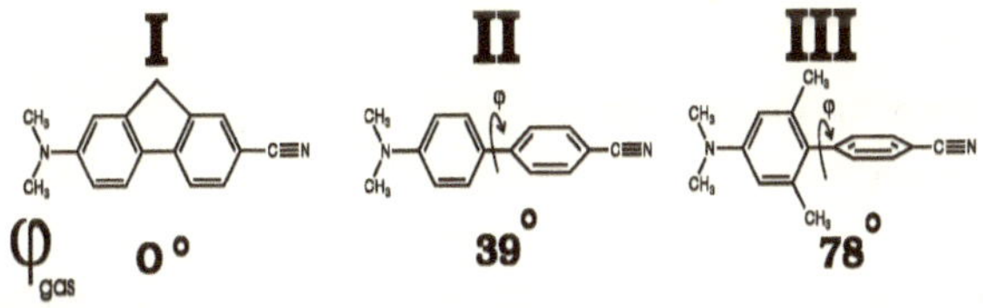

Scheme 3.1 *Molecular structures of the donor-acceptor biphenyls investigated (I-III) and their corresponding equilibrium ground state twist angles φ_{gas} as obtained by AM1 calculations.*

One purpose of this chapter is to show that the newly synthesized series of differently twisted donor-acceptor biphenyls in Scheme 3.1 do not exhibit the complications discussed above. The D-A biphenyl II has already been investigated by F.Lahmani et al.[14] but the additional photophysical study of the model compounds I and III in relation to the interannular twist angle φ should allow to handle the problems discussed above and to draw the desired conclusions about electronic and molecular structure.

In the previous chapter[26] the electronic structure of the D-A biphenyls I-III is interpreted in terms of a composite-molecule model [27,28] with dimethylaniline as the donor fragment (D) and benzonitrile as the acceptor fragment (A). CNDO/S-SCI calculations enabled an assignment of the absorption bands up to 52000 cm^{-1} and to correlate all observed bands with the typical band series A, B, C and H of BP. It was pointed out that in contrast to BP a charge transfer state (^{1}CT) becomes very low lying which is mainly responsible for the strong first absorption band. The quantitative configuration interaction (CI) analysis revealed that the ^{1}CT state consists of about 90% HOMO-LUMO configuration (scheme 3.2) which is a mixture of ^1L_a and electron transfer (^{1}ET)[29] character.[26] The (net) ^{1}ET character is due to the electron promotion from the HOMO(D) to the LUMO(A) orbitals and increases with the twist angle φ.

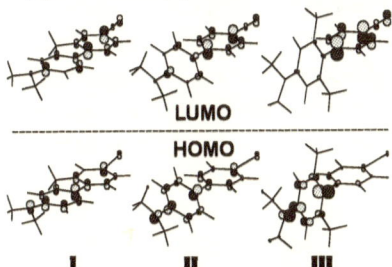

Scheme 3.2 *HOMO-LUMO configurations constituting the ^{1}CT state of I-III by ≈ 90%. Orbital localization, which is induced with increasing twist angle φ, enlarges the permanent dipole moment and decreases the transition dipole moment, respectively.*

Here, we investigate the fluorescence at room temperature and compare it to the absorption behaviour in order to get further insight into the photophysics and conformational behaviour of I-III involved in the photoinduced charge transfer.

3.2 Experimental Section

3.2.1 Synthesis of the Compounds

2-N,N-dimethylamino-7-cyanofluorene (**I**) and 4-N,N-dimethylamino-4'-cyanobiphenyl (**II**) were prepared from the commercial aminobromoderivatives in two steps: (1) reductive methylation of the aminogroup with sodium cyanoborohydride and formaldehyde[30] and (2) subsequent substitution of the bromo substituent by a cyano group.[31] 4-N,N-dimethylamino-2,6-dimethyl-4'-cyanobiphenyl (**III**) has been obtained in three steps: (**1a**) bromination of 1-N,N-dimethylamino-3,5-dimethylbenzene with 2,4,4,6 tetrabromo-1-cyclohexadienone[32] and (**1b**) monocyanation of 1,4-dibromobenzene[33] and (**3**) finally a cross-coupling reaction[34] of the organozinc derivative of 1-N,N-dimethylamino-3,5-dimethyl-4-bromo benzene with 4-bromobenzonitrile, catalyzed by Pd^0.

I: m.p. 226°C; MS m/z (%, fragment): 234 (100, M^+), 190 (34), 117 (11).

^{1}H-NMR(CDCl$_3$) δ: 2.93 (s,6H, N(CH$_3$)$_2$), 3.63 (s,2H, CH$_2$), 6.63 (s+d,2H), 7.33 (m,4H).

II: m.p. 217°C; MS m/z (%, fragment): 222(100,M^+), 206(13), 111(11).

^{1}H-NMR(CDCl$_3$) δ: 3.05 (s,6H, N(CH$_3$)$_2$), 6.88 (s,2H), 7.59 (d,2H), 7.7 (m,4H).

III: m.p. 132°C; MS m/z (%, fragment): 250(100,M+.), 234(16), 190(12).

^{1}H-NMR(CDCl$_3$) δ: 2 (s,6H, CH$_3$), 2.9 (s,6H, N(CH$_3$)$_2$), 6.5 (s,2H, ortho of N(CH$_3$)$_2$), 7.4 (m, 4H).

*This synthesis and characterization of **I-III** was kindly performed by Dominique Bonafoux and René Lapouyade (University Bordeaux/France) !*

3.2.2 Solvents

The solvents purchased from MERCK were of spectroscopic grade except triacetine (distilled twice) and iso-pentane (HPLC-grade). The absorption and fluorescence spectra of all solvents were checked for impurities and have been subtracted from the sample spectra to ensure spectra to be free from background effects, e.g., Raman and Rayleigh peaks. The abbreviations for the 12 solvents used in the text also collected in Tab. 3.1 are as follows: n-hexane (HEX), diethylether (EOE), acetonitrile (ACN), ethanol (EtOH), 4:1 mixture of iso-pentane/methylcyclohexane (IpM), 3:1 mixture of methylcyclohexane/iso-pentane (Mlp), di-n-butylether (BOB), triacetine (TAC), n-butylchloride (BCl), n-butyronitrile (BCN), N,N-dimethylformamide (DMF) and N-monomethylformamide (MMF).

3.2.3 Steady State Absorption

The absorption spectra presented were recorded on a ATI UNICAM UV4 UV/VIS spectrophotometer and the decadic molar extinction coefficients ε_{max} were repeatedly determined.

3.2.4 Steady State Fluorescence

All steady state fluorescence spectra were obtained on a Aminco Bowman 2 fluorimeter using a 150 W Xe lamp, 2 nm excitation and emission band pass and a photomultiplier tube in a right-angle geometry. All fluorescence spectra were corrected for detector response and time-drift and were additionally converted from the recorded wavelength scale $I_f(\lambda_f)$ to a linear energy scale according to $I_f(\nu_f) = I_f(\lambda_f) \cdot \lambda_f^2$.

3.2.5 Fluorescence Quantum Yields

The quantum yields were measured relative to quinine bisulfate in a 0.1N H_2SO_4 and calculated on the basis of[35]

$$\Phi_f = \Phi_f^0 \cdot \frac{n_0^2}{n^2} \frac{OD^0}{OD} \frac{\int I_f(\lambda_f) d\lambda_f}{\int I_f^0(\lambda_f) d\lambda_f} \qquad (3\text{-}1)$$

where n_0 and n are the refractive indices of the solvents, OD^0 and OD (≤ 0.1) are the optical densities, Φ_f^0 (= 52%)[36] and Φ_f are the quantum yields, and the integrals denote the (computed) area of the corrected fluorescence bands, each parameter for the standard and sample solution, respectively. All quantum yields were determined within one flash period using a single 1 cm quartz cuvette in order to minimize instrumental errors. Most values have been remeasured on the same fluorimeter as well as on a Perkin Elmer 650-60 and LS 50. The relative experimental error of the quantum yields is around ± 5%.

3.2.6 Fluorescence Lifetimes

Synchrotron radiation from the Berliner synchrotron facility BESSY was used as light source in conjunction with an excitation monochromator (Jobin Yvon, ≈ 20 nm bandpass). It delivers a 4.8 MHz pulse train with characteristic pulse widths of 600 ps. Emission was detected using a time-correlated single photon counting setup. It consists of a filter polarizer in magic angle position, emission monochromator (Jobin Yvon, ≈ 20 nm bandpass) and a microchannel plate photomultiplier (Hamamatsu R1564-U-01, 35 ps fwhm) cooled to -30°C. Using standard electronics from ORTEC, at most 0.1% of the signals (≈ 5kHz) were sampled in 1024 channels of a multichannel analyzer (ORTEC-Norland 5590) with a channel width of either 25 ps or 50 ps. The decays were analyzed by the „least square" and iterative deconvolution method on the basis of the Marquardt/Levenberg algorithm which is implemented in the homemade program „SP" as well as in the commercial global analysis program.[37] The quality of the exponential fits was evaluated by the reduced χ^2 (≤ 1.2) and the autocorrelation of the residuals quantified by the Durbin Watson parameter (1.9< DW <2.1).[38] This reconvolution technique allows an overall time-resolution down to 100-200 ps.

3.2.7 Quantum Chemical Calculations

CNDO/S-SCI calculations including 49 singly excited configurations have been performed with the QCPE program #333 modified to use the original CNDO/S parametrization[39] and to calculate the excited state dipole moments. All input geometries were fully optimized in the ground state by the Newton algorithm with the AM1 Hamiltonian within the AMPAC program.[40] In the experiments, we are dealing with solutions which shift the absorption and fluorescence transitions to lower energies v_a and v_f. For a rough theoretical estimation of the solvent shifts the classical Onsager model[41] is applied which treats the solvent as a dielectric continuum containing a spherical cavity within which the solute associated with a point dipole resides. With the usual assumption that the initial state is fully stabilized by solvent reorientation and electronic polarization, whereas the final state of the instantaneous electronic transition can only interact with the solvent by electronic polarization, the transition energies for the absorption and fluorescence process are calculated by[42]

$$hcv_a = hcv_{gas} - f(\varepsilon_r) \cdot (\mu_e \mu_g - \mu_g^2) - f(\varepsilon_\infty) \cdot (\mu_e^2 - \mu_e \mu_g) \qquad \text{(absorption)} \quad (3\text{-}2a)$$

$$hcv_f = hcv_{gas} - f(\varepsilon_r) \cdot (\mu_e^2 - \mu_e \mu_g) - f(\varepsilon_\infty) \cdot (\mu_e \mu_g - \mu_g^2) \qquad \text{(fluorescence)} \quad (3\text{-}2b)$$

using the well-known Onsager term[41] given in eq. 3.

$$f(\varepsilon) = \frac{\varepsilon - 1}{2\varepsilon + 1} \frac{1}{a^3} \qquad (3\text{-}3)$$

μ_g and μ_e represent the calculated ground and excited state dipole moments, h and c are the Planck constant and velocity of light, ε_r and ε_∞ denote the static and high frequency dielectric constant of the solvent. The Onsager cavity radius a is taken as $6 \cdot 10^{-10}$ m throughout the paper for the theoretical calculations and the solvatochromic plots. It is calculated by the half-length of the long-molecular axis using the PCMODEL program.[43] The potential energies of the ^1CT state $E(^1CT)$ are obtained by combining the ground state potential from AM1 with the gas phase excitation energies and adding the solvent stabilization energy for full solvent relaxation[44] according to

$$E(^1CT) = E_{gas}(S_0) + hcv_{gas}(S_0 \rightarrow {}^1CT) - f(\varepsilon_r)\mu_e^2 \qquad (3\text{-}4)$$

The ground state energies $E_{gas}(S_0)$ used are determined relative to the heat of formation (ΔH_f) of the fully optimized geometry (ΔH_f (**I**) = 395 kJ/mol, ΔH_f (**II**) = 366 kJ/mol, ΔH_f (**III**) = 319 kJ/mol). All quantum chemical computations were executed on a HP 735 workstation.

3.3. Results and Discussion

3.3.1 Steady State Spectra

In Fig. 3.1 and 3.2, the steady state absorption and fluorescence spectra are shown for I-III in solvents of different polarity. The detailed analysis of the absorption spectra in ch.2 assigns the B band to a ^1B_b-type state and the A band to the ^1L_a-type state analogously to the A and B bands in BP.[27] The maximum of the first absorption band (CT) of II is located at 4300 cm^{-1} and 8300 cm^{-1} to the red of the corresponding band in 4-dimethylaminobiphenyl[21] and 4-cyanobiphenyl,[19] respectively. This indicates a π-electron delocalization from the dimethylamino to the nitrile group which is consistent with the previous assignment to the long-axis polarized transition $1^1A(S_0) \rightarrow 2^1A(^1CT)$.

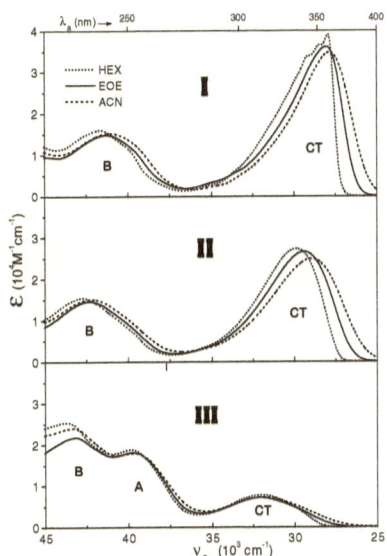

Fig. 3.1 *Absorption spectra of I-III in solvents of different polarity at 298 K.*

Now, let us come to the question regarding the electronic nature (^1A or ^1B symmetry) and geometrical structure of S_1 after vibrational relaxation. The fluorescence in n-hexane (HEX) already contains a wealth of information, since weak solute-solvent interactions in this nonpolar medium do not broaden the vibronic transitions too much, and the stabilization energy of the fluorescent state is more due to intramolecular than to solvation processes. The following information can be extracted from the steady-state fluorescence spectra in HEX (Fig. 3.2):

(i) The fluorescence spectra of all three biphenyl compounds I-III consist of a strong 0-0 vibronic transition which is not observed in BP.[7] Fluorescence in BP occurs from the forbidden ^1L_b state explaining the absence of the 0-0 transition. In addition, the degree of polarization (p) of I-III in EtOH at 77K (Fig. 3.2) is constant across the fluorescence band (ch.5) and is close to the maximum value of $p = +0.5$ (I: $p = +0.47$, II: $p = +0.44$ and III: $p = +0.48$). Under equal conditions, a value of only $p = +0.2$ was reported for BP and cyanobiphenyl derivatives.[19] Taking into account the long-axis polarization of the first absorption band,[26] this proves the long-axis polarization of S_1 emission in I-III, in contrast to BP. Thus, we can suppose that

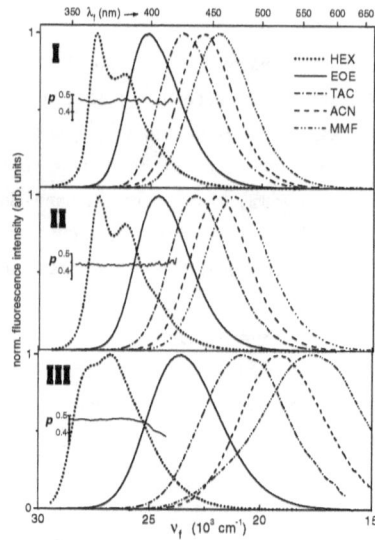

Fig. 3.2 *Normalized fluorescence spectra of* **I-III** *in solvents of different polarity at 298 K. The fluorescence polarization spectra* $p(v_f)$ *of* **I-III** *in EtOH at 77 K are also shown. All spectra are obtained by excitation at the absorption maxima. The* $p(v_f)$ *are shifted 2000 cm^{-1} to the blue side in order to match up the vibronic band positions in EtOH at 77 K with those in HEX at 298 K. The decrease of* $p(v_f)$ *for* **III** *below the energy of the 0-2 vibronic band is due to long-lived (*τ_p = 2.4 s*) phosphorescence (see Ch.5).*

fluorescence of **I-III** originates from the allowed 2^1A (^1CT) state similar to 4-vinylbiphenyl[8] and fluorene[7] where the observation of the 0-0 transition is also due to a lowest lying long-axis polarized state.

(ii) From a band-shape analysis (eq. 6-1) or similarly from fitting of the spectral profile to j=4 Gaussian functions (eq. 3-5), the average spacings of the progressions for **I** to **III** are found to amount to (1310±10) cm^{-1} or (1320±30) cm^{-1}, respectively. For BP, the spacing is 1000 cm^{-1} and is ascribable to the non-totally symmetric (C_2 axis) ring-breathing mode.[8] However, because the fluorescence of BP does not occur from a state with the same symmetry as **I-III**, we have to compare the spacings with those of biphenyls showing fluorescence of the long-axis polarized ^1L_a-type. In this case, the progression corresponds to the interannular bond stretching mode (totally symmetric to C_2). Such biphenyls are 4-vinylbiphenyl, p-terphenyl or 4-cyanobiphenyl. All these compounds possess a spacing[8,19] of about 1300 cm^{-1} which is in excellent agreement with the spacings of **I-III** and therewith, confirms the ^1L_a character of the emitting ^1CT state coupled to the ground state mainly by the totally-symmetric interannular bond-stretching vibrational mode.[45]

$$I_f(v_f) = \sum_j \frac{\int I_f(v_f)dv_f}{HW_f^j} \sqrt{\frac{2}{\pi}} \exp(-2\frac{(v_f - v_f^{j,max})^2}{HW_f^{j2}}) \quad (3-5)$$

(iii) In contrast to the absorption bands, the fluorescence bands of **II** and **III** exhibit vibrational structure and considerably narrower half widths. These observations indicate a narrower conformational distribution in the excited state.

(iv) The increasing Stokes shifts from **I** to **III** point to significant angular relaxations in the excited state of **II** and more extensively of **III**.

(v) The intensity of the 0-0 band relative to the second vibronic band is largest (ratio 1:0.74) for the compound **I** restricted to planarity, because the minima of the ground and emitting state are less displaced relative to each other. The high ratio 1:0.81 for the flexible biphenyl **II** also speaks in favour of an angular distribution close to 0°. The lower ratio of 0.93:1 for **III** indicates a larger difference between ground and excited state minima. This is most probably due to a relaxation from the initial conformational distribution towards an angular distribution with more planar angles $<\varphi>$ in S_1 than in S_0. In summary, the steady state fluorescence in HEX yields evidence that it originates from the allowed 2^1A (^1CT) state in which the conformer distribution of **II** and **III** is more planar than in the ground state 1^1A.

The fluorescence in dipolar solvents is structureless and exhibits a solvatochromic red shift with increasing solvent polarity for all D-A biphenyls investigated (Fig 3.2). Dipole-dipole interactions between solute and solvent are responsible for these large Stokes shifts and hence the strong solvatochromism proves the sizeable ^{1}ET character of the emitting state. Furthermore, the solvatochromism is similar for **I** and **II** which hints at a planar structure for excited **II** in the dipolar solvents, too. In contrast, the solvatochromic red shift for **III** is much more pronounced indicating a different molecular structure connected with a higher dipole moment in S_1. The quantitative determination of the excited state dipole moments μ_e is performed below in 3.3.2. The absorption and fluorescence spectroscopic data in different solvents are collected in Tab. 3.1. The Stokes shifts can be used to estimate the sum of the outer and inner reorganization energy by $\lambda_{sum} = \Delta v_{St}/2$ averaged for S_0 and S_1.[47] Such an evaluation leads to a difference between the reorganization energies λ_{sum} of **I** and **II** which remains approximately constant around 1000 cm^{-1} in all solvents. This amount of energy is most probably lost by the intramolecular rotation of **II** towards planarity. On the other hand, the Stokes shift of **III** immensely increases with solvent polarity ranging from 5000 cm^{-1} in non-polar to 14000 cm^{-1} in highly polar monomethylformamide (MMF) and the corresponding reorganization energies λ_{sum} are always larger than those of **I** and **II** (λ_{sum}(**III**)- λ_{sum}(**I**) = 2000 cm^{-1} in HEX and 4200 cm^{-1} in MMF). The half widths of the fluorescence bands HW$_f$ are also broader for **III** than for **I** and **II**. This points to the possibility that in **III** a single broad or even two different conformer distributions are fluorescent. Although dual fluorescence bands are not observed, it is not unusual that two sets of similar transition energies of two isomers result in the appearance of only a single band.[48-50] Here, the absence of vibrational structure in the steady state fluorescence prevents an obvious differentiation between two conformers (ch.6).

The differences $\Delta_{HEX-ACN}$ between the observed energies in HEX and acetonitrile (ACN) shown in the last lines of Tab 3.1 reveal further interesting aspects. As mentioned above, the differences of the fluorescence energies v_f, Stokes shifts Δv_{St} and half widths HW between nonpolar (HEX) and polar (ACN) solvents are similar for **I** and **II** but different to **III**. Different behaviour of **I** and **II** is only observed in the polarity dependence of the absorption energy. For

I, the absorption energy in ACN with respect to HEX is lower by only 80 cm^{-1} whereas for **II** this red shift is 10 times larger (780 cm^{-1}). This clearly shows that the ground state geometries of **I** and **II** are significantly different. **II** must be twisted in the ground state to such an extent that the excited state Franck Condon geometry possesses a sufficiently higher dipole moment than **I** to yield the observed polarity induced absorption red shift. Then the question arises why the absorption red shift for the strongly pretwisted biphenyl **III** is only 100 cm^{-1}, although a large red shift could be expected due to the higher excited state dipole moment. The reason is that in **III**, two absorption transitions of comparable intensity are responsible for the band, one of ^{1}CT and the other one of local partial ^1L_b character (ch.5).[26] The latter does not change its position with solvent polarity. Partially, this argumentation may also be valid for compound **I**. In this fluorene derivative, the ^1L_b-type and ^{1}CT transitions can mix which results in a better allowedness of the ^1L_b transition as compared to **II** (ch.2 and 8).[26] Thus, a larger part of the first absorption band might be produced by a nonpolar transition.

We can conclude at this point that **I** and **II** in polar solvents possess a similar conformation in the excited state analogously to the outcome for HEX, whereas in the ground state they differ in their twist angle φ. **III** in polar solvents has quite a different molecular structure than **I** and **II** in the excited state but the extent of twisting and its difference to the ground state is not yet obvious at this point.

Tab. 3.1 Spectroscopic Data of **I-III** at 298 K in Dependence on the Solvent Polarity: Absorption and Fluorescence Transition Energy Maxima v_a^{max} and v_f^{max}, Stokes Shifts Δv_{St} and Band Half Widths HW (all values in 10^3 cm^{-1}).

Solvent	abbrev	ε_r^c	n^c	v_a^{max}	v_f^{max}	Δv_{St}	HW$_a$	HW$_f$
I								
n-hexane	HEX[b]	1.9	1.372	27.93	26.89	1.0	3.5	2.5
i-pentane/MCH (4:1)[a]	IpM[b]	1.9	1.365	27.93	26.93	1.0	3.6	2.5
MCH/i-pentane (3:1)	MIp[b]	2.0	1.403	27.78	26.84	0.9	3.4	2.3
di-n-butylether	BOB	3.1	1.397	27.93	25.97	2.0	4.0	2.5
diethylether	EOE	4.2	1.350	28.09	25.13	3.0	3.8	2.8
triacetine	TAC	7.1	1.400	28.65	23.58	5.1	4.0	3.0
n-butylchloride	BCl	7.2	1.430	27.78	24.81	3.0	3.7	2.5
n-butyronitrile	BCN	24.1	1.382				3.8	
ethanol	EtOH	24.5	1.342	27.78	22.47	5.3	4.2	3.2
acetonitrile	ACN	36.2	1.359	27.86	22.47	5.4	4.0	3.0
N,N-dimethylformamide	DMF	37.0	1.428					
N-monomethylformamide	MMF	182.4	1.430	27.47	21.93	5.5	4.1	3.1
	$\Delta_{HEX-ACN}$			0.08	4.42	-4.3	-0.5	-0.5
II								
n-hexane	HEX[b]	1.9	1.372	29.85	26.64	3.2	4.3	2.7
i-pentane/MCH (4:1)[a]	IpM[b]	1.9	1.365	29.94	26.73	3.2		2.7
MCH/i-pentane (3:1)	MIp[b]	2.0	1.403	29.76	26.66	3.1	4.4	2.5
di-n-butylether	BOB	3.1	1.397	29.50	25.64	3.9	4.5	2.6
diethylether	EOE	4.2	1.350	29.41	24.51	4.9	4.6	3.0
triacetine	TAC	7.1	1.400	29.50	23.04	6.5	5.2	3.0
n-butylchloride	BCl	7.2	1.430	29.07	24.81	4.6	4.8	2.5
n-butyronitrile	BCN	24.1	1.382	28.90			4.9	
ethanol	EtOH	24.5	1.342	28.99	21.93	7.1	4.8	3.4
acetonitrile	ACN	36.2	1.359	29.07	21.79	7.3	5.1	3.1
N,N-dimethylformamide	DMF	37.0	1.428					
N-monomethylformamide	MMF	182.4	1.430	28.49	21.28	7.2	5.5	3.3
	$\Delta_{HEX-ACN}$			0.78	4.85	-4.1	-0.8	-0.4
III								
n-hexane	HEX[b]	1.9	1.372	32.05	26.97	5.1	5.0	3.6
i-pentane/MCH (4:1)[a]	IpM[b]	1.9	1.365	31.95	26.95	5.0	5.2	3.5
MCH/i-pentane (3:1)	MIp[b]	2.0	1.403	31.75	26.84	4.9	5.2	3.4
di-n-butylether	BOB	3.1	1.397	31.75	25.00	6.7	5.1	3.5
diethylether	EOE	4.2	1.350	32.05	23.42	8.6	5.4	4.0
triacetine	TAC	7.1	1.400	32.15	20.66	11.5	5.6	4.4
n-butylchloride	BCl	7.2	1.430	31.75	23.04	8.7	5.9	3.7
n-butyronitrile	BCN	24.1	1.382	31.85	19.69	12.2	6.0	4.4
ethanol	EtOH	24.5	1.342	32.05	18.69	13.4	6.2	5.1
acetonitrile	ACN	36.2	1.359	31.95	18.73	13.2	5.9	4.6
N,N-dimethylformamide	DMF	37.0	1.428	31.75	18.66	13.1		4.8
N-monomethylformamide	MMF	182.4	1.430	31.65	17.73	13.9	5.6	5.4
	$\Delta_{HEX-ACN}$			0.10	8.24	-8.1	-0.9	-1.0

[a] MCH = methylcyclohexane
[b] The maxima of the emission spectra with vibrational structure (in HEX, IpM and MIp) are derived from lognormal fits.
[c] Static dielectric constants ε_r and refractive indices n are taken from ref. 46.

3.3.2 Excited State Dipole Moments from Solvatochromic Plots

The general goal of the solvatochromic studies is to obtain the dipole moments of the emitting species and to derive whether different species are emissive. Different formulae have been developed by Bakshiev,[51] Kawski-Chamma-Viallet,[52] McRae,[53] and others. In this work the most frequently used formulae after Lippert[54] and Mataga[55] are employed and compared with a plot on a recently developed solvent polarity/polarizability scale (SPP).[56]

To obtain the excited state dipole moments by the Lippert eq. 3-6[54] the Stokes shift is plotted versus the solvent polarity function $F_1(\varepsilon_r,n)$:

$$\Delta v_{St} = v_{St}(0) - \frac{2(\mu_e - \mu_g)^2}{hca^3} F_1(\varepsilon_r,n) \quad (3\text{-}6a)$$

$$F_1(\varepsilon_r,n) = \frac{\varepsilon_r - 1}{2\varepsilon_r + 1} - \frac{n^2 - 1}{2n^2 + 1} \quad (3\text{-}6b)$$

with the static dielectric constant ε_r and the refractive index n of the solvent. In eq. 3-6, Δv_{St} and $\Delta v_{St}(0)$ denote the Stokes-shifts in cm^{-1} between the absorption and emission maximum in a given solvent and in the gas phase, respectively. h and c correspond to the Planck constant ($6.63 \cdot 10^{-34}$ J s) and the velocity of light in a vacuum ($3 \cdot 10^8$ m s^{-1}). The choice of the Onsager radius of the solvent cavity is somewhat arbitrary. Some authors prefer to calculate it from an estimated or determined molecular volume[42,57] and others follow the original method of Lippert[20,54] estimating it by 40% of the long-molecular axis including the van der Waals radii for elongated molecules.[15a] Here, a is calculated by different methods for **I-III** using the PCMODEL program,[43] such that among others Onsager radii of $4.4\text{-}4.7 \cdot 10^{-10}$ m for the first cited method and $6.4\text{-}6.7 \cdot 10^{-10}$ m for the second cited method are obtained. It seems that the first method always delivers too low values and the second method too high values. In view of this uncertainty, an intermediate Onsager radius a of $6.0 \cdot 10^{-10}$ m is chosen throughout the paper for the calculations and the solvatochromic plots. It equals the half-length of the long-molecular axis without van-der-waals radii and the same value has been used by different authors for similar donor-acceptor biphenyls.[14,58,59] Especially Lahmani et al.[14] who also investigated compound **II** used this value so that it is possible to compare the present results with theirs. The solvatochromic plots are shown in Fig. 3.3 and the obtained excited state dipole moments μ_e are collected in Tab. 3.2. The plots Δv_{st} vs. F_1 according to the Lippert eq. 3-6 (Fig. 3.3a) for compound **II** closely resemble those of Lahmani et al., and the dipole moments determined here are in full agreement with those (μ_g=6 D, $\Delta\mu_{eg}$=17 D, μ_e=23 D) of ref. 14. Reasonable correlation coefficients (≥ 0.98) of the linear regressions express the quality of the fits and indicate that in all compounds only one emitting species is observed. In line with the observations of the previous section, the excited state dipole moments μ_e observed for **I** and **II** are similar (around 23 D by the Lippert eq.3-6) but considerably smaller than those for

III (30 D). The Lippert eq. 3-6 is valid only, in the case of the same nature of the absorbing and emitting states. If this is questionable, the Mataga eq. 3-7 should be taken using the polarity function $F_2(\varepsilon_r,n)$:[55]

$$v_f = v_f(0) - \frac{2\mu_e(\mu_e - \mu_g)}{hca^3} F_2(\varepsilon_r,n) \tag{3-7a}$$

$$F_2(\varepsilon_r,n) = \frac{\varepsilon_r - 1}{2\varepsilon_r + 1} - 0.5 \frac{n^2 - 1}{2n^2 + 1} \tag{3-7b}$$

The dipole moments obtained with this method delivers similar excited state dipole moments as those from eq. 3-6. Therefore we can assume that the Franck-Condon state reached directly upon excitation is equal to the emitting state, namely the long-axis polarized ^{1}CT state 2^1A (cf. ch.5). For **I-III**, no transformation from a locally excited state to a charge transfer state is revealed like in other biaryl systems where bilinear solvatochromic plots can be found (cf. ch.9).[56,61,62]

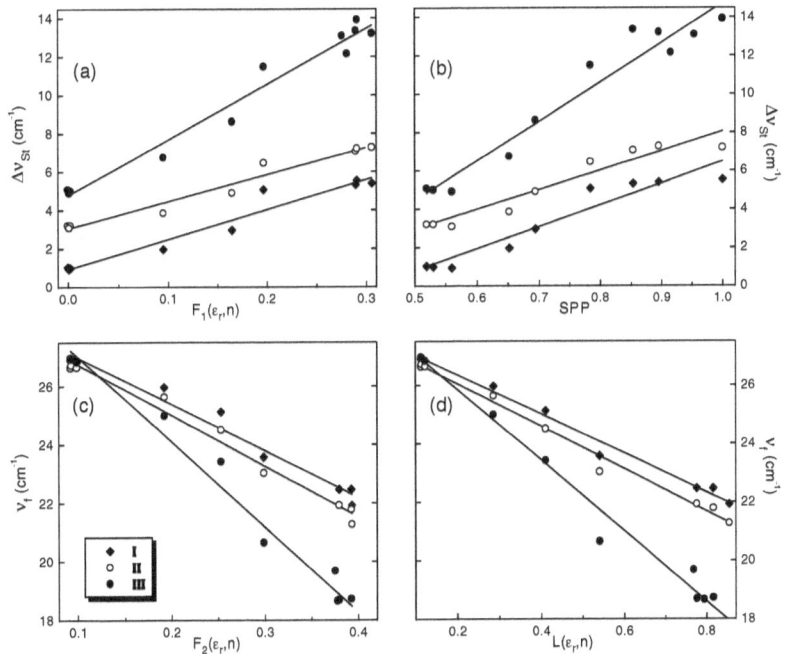

Fig. 3.3 Solvatochromic plots for **I** (◆), **II** (O) and **III** (●) dissolved in the solvents listed in Tab. 3.2: Stokes shift vs. (a) F_1 (eq. 3-6) and (b) vs. SPP values from ref.56; fluorescence maximum vs. (c) F_2 (eq. 3-7) and (d) vs. L (eq.3-8).

Tab. 3.2 Results of the Solvatochromic Plots Using an Onsager Radius of 6.0 Å

	I			II			III		
	$\Delta\nu_{St}(F_1)$	$\nu_f(F_2)$	$\nu_f(L)$	$\Delta\nu_{St}(F_1)$	$\nu_f(F_2)$	$\nu_f(L)$	$\Delta\nu_{St}(F_1)$	$\nu_f(F_2)$	$\nu_f(L)$
$\Delta\mu_{eg}$ (D)	18.0	18.2	11.8	17.1	19.0	12.3	24.5	24.6	15.9
μ_e (D)	23.7	21.3	15.0	22.7	22.0	15.5	29.5	27.2	18.6
μ_g (D)[a]	5.7	5.7	5.7	5.7	5.7	5.7	5.0	5.0	5.0
intercept (cm^{-1}):	0.95	28.55	27.70	3.05	28.50	27.50	4.80	29.90	28.30
slope (cm^{-1})	15.50	15.90	6.70	14.00	17.35	7.30	29.00	29.10	12.10
correl[b]	0.975	-0.988	-0.995	0.982	-0.991	-0.995	0.986	-0.990	-0.993

[a] obtained by AM1[40] for the optimized geometries in S_0 [b] BCl data are omitted

If we want to compare the experimentally determined dipole moments with calculated dipole moments for the gas phase (vide infra Fig. 3.8a and 3.9a), we have to exclude the solvent induced dipole moment involved with the polarizability of the solute. Assuming an average polarizability in S_0 and S_1 of about 0.5 a^3 as described elsewhere,[57] eq. 3-8 can be applied with the polarity function L(ε_r,n)[63]

$$\nu_f = \nu_f(0) - \frac{2\mu_e(\mu_e - \mu_g)}{hca^3} L(\varepsilon_r, n) \quad (3\text{-}8a)$$

$$L(\varepsilon_r, n) = \frac{\varepsilon_r - 1}{\varepsilon_r + 2} - 0.5 \frac{n^2 - 1}{n^2 + 2} \quad (3\text{-}8b)$$

As expected, these plots also yield linear correlations (R=0.99) as shown in Fig. 3.3d and the resulting dipole moments μ_e are 30% smaller than those from the above methods (Fig.3.3ac). Again, the dipole moments μ_e of **I** and **II** are practically equivalent (15 D) differing from the higher dipole moment for **III** (19 D). This demonstrates an increased ^{1}ET character in S_1 of **III** as compared to **I** and **II**. Latter both have similar μ_e and thus contain practically equal ^{1}ET character in S_1. In the framework of the minimum overlap rule,[1,64] these relations of μ_e can be understood by a planar structure of **I** and **II** and a highly twisted molecular structure of **III** associated with orbital localization and a higher dipole moment (cf. scheme 3.2).

One should keep in mind that beside the problem of arbitrary Onsager parameters, the solvatochromic methods detect only an average dipole moment value <μ_e>$_{obs}$ of an excited state distribution of rotamers $\eta_e(\varphi)$ which is, moreover, convoluted with the angle dependent fluorescence quantum yield $\Phi_f(\varphi)$ according to eq. 3-9. As a consequence, the detected weight of the higher dipole moments for the more twisted rotamers is less due to smaller quantum yields and can therefore lead to an underestimation of the real mean dipole moment <μ_e> defined by eq. 3-10, if more twisted rotamers are preferably stabilized in highly polar solvents.

$$\langle \mu_e \rangle_{obs} = \overline{\Phi}_f^{-1} \int \eta_e(\varphi) \Phi_f(\varphi) \mu_e(\varphi) d\varphi \quad (3\text{-}9)$$

$$\langle \mu_e \rangle = \int \eta_e(\varphi) \mu_e(\varphi) d\varphi \tag{3-10}$$

Recently, the empirical solvent polarity/polarizability (SPP) scale was proposed as a more sensitive tool for determining whether the solvent polarity can modify the emitting electronic state in such a way as to alter its dipole moments.[56] Such solvatochromic plots are shown in Fig. 3.3c for **I-III**. Although the data points in the extremely polar and nonpolar region are more stretched, a clear bilinear behaviour is not observed on this scale, either. At first sight, one might suppose a different slope for SPP values below 0.57. But the range between 0.52-0.57 only for non-polar and weakly polarizable solvents is too small to justify this assumption. The example compounds (e.g. bianthryl) with bilinear slopes given in the original publications[56] mostly exhibit the crossing of the two regression lines at around SPP=0.8. Moreover, the example compounds are known to emit from two states of locally excited and charge transfer character, respectively, with considerably different dipole moments. For **I-III**, the absence of emission from a locally excited state and the convolution of the dipole moments with rotamer dependent quantum yields (eq. 3-9) may prevent the observation of a bilinear solvatochromic plot.

3.3.3 Photophysics

Tab. 3.3 Photophysical Properties of **I-III** at 298 K:
(a) Integrated Area[65] of the CT Absorption Band $\int \varepsilon dv_a$, Fluorescence Quantum Yields Φ_f, Fluorescence Lifetimes τ_f.

solvent	$\int \varepsilon dv_a$ (M^{-1}cm^{-2})			Φ_f (%)			τ_f (ns)		
	I	II	III	I	II	III	I	II	III
HEX	1.51	1.28	0.38	57	71	35	1.4	1.2	1.2
EOE	1.56	1.35	0.38	83	84	55	1.7	1.7	4.6[a]
ACN	1.58	1.37	0.41	81	79	21	2.1	2.2	7.6
EtOH	-	-	-	84	81	4	2.0	2.2	1.4

(b) Nonradiative $k_{nr} = (1-\Phi_f)/\tau_f$ and Radiative Rate Constants $k_f = \Phi_f/\tau_f$ and Reduced Radiative Rates $k_f/v_f^3 n^3$.

solvent	k_{nr} (10^{7}s^{-1})			k_f (10^{7}s^{-1})			$k_f/v_f^3 n^3$ (10^{-7}s^{-1}cm^3)		
	I	II	III	I	II	III	I	II	III
HEX	32	25	55	42	61	30	83	124	60
EOE	10	9	10[a]	50	49	12[a]	128	134	37[a]
ACN	9	10	10	39	35	3	143	142	17
EtOH	8	9	66	42	37	3	146	140	19

[a] mean values from biexponential decay due to the equilibrium between two emitting species (see text).

3.3.3.1 Photophysical Evidence for ^1L_a-type Fluorescence
The photophysical data of **I-III** in n-hexane (HEX), diethylether (EOE), acetonitrile (ACN) and ethanol (EtOH) are collected in Tab. 3.3. The integrated absorption intensity $\int \varepsilon dv_a$, which is proportional to the oscillator

strength f (eq. 1-2), increases only slightly with solvent polarity. This indicates that the electronic and structural nature of the ground and Franck-Condon excited state responsible for the first absorption band does not vary dramatically with polarity. The same holds for the fluorescence of **I** and **II** in dipolar solvents. For these compounds, the fluorescence quantum yields Φ_f stay constant around 80% and the nonradiative rate constants k_{nr} are around 10^8 s^{-1} in all solvents except HEX. The slight variation of the radiative rate constant k_f of **I** and **II** can mainly be traced back to the inherent dependence on v_f^3 (eq.9)[66,67] as shown by almost constancy of the reduced radiative rates $k_f/v_f^3n^3$.

In apolar HEX, the fluorescence behaviour of all three compounds is similar, with the main feature that (i) Φ_f is decreased significantly as compared to medium polar EOE and (ii) all lifetimes are comparable, in particular for **II** and **III**. It seems astonishing that the radiative rate k_f for **I** is smaller than for **II** since in the dipolar solvents, **I** and **II** possess very similar k_f values. We can explain the decreased k_f for **I** by a slightly different electronic nature of S_1 in a nonpolar medium (ch. 8). As pointed out in ch. 2,[26] ^1L_b and ^1CT can mix more strongly in compound **I** which is derived from fluorene with parallel transition moments of its $^1L_b(+)$ and $^1L_a(-)$ states. As a result, the ^1CT state in **I** has to share a part of its oscillator strength with the ^1L_b states. In absorption, both ^1L_b and the ^1CT states all contribute to the first absorption band, whereas in fluorescence only the emission of the lowest state ^1CT can be observed. In polar solvents, on the other hand, the stabilization of the ^1CT state leads to a larger energy gap between ^1L_b and ^1CT which can account for reduced mixing between these states connected with an enhancement of the radiative rates for **I**. The polarity dependence of ^1L_b-^1CT mixing can further be substantiated by a comparison with the related biphenyl compound 4-diethylaminobiphenyl (EBA) possessing a ^1CT state somewhat higher lying than in **I-III**.[22] The observed larger radiative rate of EBA in ACN ($k_f = 5 \cdot 10^7$ s^{-1}) with respect to nonpolar cyclohexane ($2 \cdot 10^7$ s^{-1}) can be explained similarly to **I** with a stabilization of the fluorescent ^1CT state in polar solvents leading to less mixing between the forbidden ^1L_b (cf. k_f(BP)= 10^7 s^{-1}) and the allowed ^1L_a-type ^1CT transition to the ground state. In any case, the radiative rates k_f and, more meaningful, the emission energy corrected rates $k_f/v_f^3n^3$ for **I-III** are 20-150 times higher than for BP ($k_f/v_f^3n^3 = 10^{-7}$ s^{-1} cm^3)[8] and thus corroborate the assignment to ^1L_a-type fluorescence independent of the compound and solvent investigated.

3.3.3.2 Structural and Solvent Dependence of Nonradiative Rates In all dipolar and aprotic solvents, the nonradiative rates for **I-III** are close to $10 \cdot 10^7$ s^{-1}, but in HEX the increased rate constants of about $30 \cdot 10^7$ s^{-1} for **I** and **II** and $55 \cdot 10^7$ s^{-1} for **III** hint at an additional radiationless channel present only in nonpolar surrounding. Although phosphorescence is not observed in the liquid phase of the solvents, this channel is most probably due to intersystem crossing (ISC). A similar behaviour has recently been found for 9,9'-bianthryl where the ISC rate in HEX is ≈ 2 times larger than in EOE and ≈ 4 times larger than in ACN.[68] This was explained with a thermally activated ISC process to the second triplet state (T_2) which becomes relatively higher lying than the solvent stabilized S_1 with increasing polarity. The level reversal between S_1 and a triplet state may also take place in **I-III** when switching from HEX to polar solvents (ch.5). The suggestion of an additional ISC channel in HEX is further supported by the higher nonradiative rate for **III** than for **I** and **II**. Note also, that the emission polarization spectrum of **III** at 77 K, shown in Fig. 3.2, is contaminated by the perpendicular polarized phosphorescence contribution in the 0-2 vibronic band as evidenced by the long lifetime of 2.4 s (ch. 5). In contrast to the $^1\pi\pi^*$ singlet states, in the triplet $^3\pi\pi^*$ states the transition moments are usually polarized perpendicular to the π-system. Because the angle between the transition moment directions of the singlet transition in one phenyl ring is less perpendicular to that of the triplet transition in the other ring for **III** than for **I** and **II**, stronger excitonic interactions between singlet and triplet excitations should occur for the more twisted compound **III** with the consequence of a faster ISC rate. Such a twist angle dependent triplet yield has also been reported for BP.[7,69]

The behaviour of **III** in EtOH is rather outstanding. Whereas all photophysical data are equal for **I** and **II** in ACN and EtOH, only the radiative rate is the same in both solvents for **III**. Quantum yield and lifetime are strongly reduced which is the result of an additional radiationless channel resulting in a ≈ 7-fold increase of the k_{nr} rates.

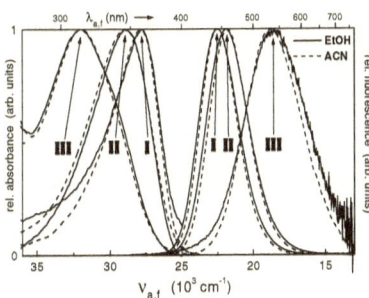

Fig. 3.4 *Normalized absorption and fluorescence spectra of **I-III** in protic ethanol and aprotic acetonitrile at 298 K. The coinciding maxima and similar shapes of the spectra in both (≈ equipolar) solvents indicate that hydrogen bonding occurs neither in the ground nor in the emitting state. The high noise in the fluorescence spectrum of **III** in EtOH (together with the shortened lifetime) reveals the deactivation to a nonradiative excited species due to protic solvent interaction.*

This can be further discussed with the aid of Fig. 3.4 which shows the comparison of the absorption and fluorescence spectra of **I-III** between protic EtOH and aprotic ACN. Both solvents are of comparable polarity ($F_2(\varepsilon_r,n)$ = 0.38 for EtOH and 0.39 for ACN). Absorption and fluorescence band shapes, maxima and their change with twist angle from **I** to **III** are similar in EtOH and ACN. From the similarity of the absorption spectra in the two solvents it can be deduced that there is no specific ground state complexation with EtOH and likewise the similarity of the fluorescence spectra indicates that the fluorescent species is the same in EtOH and ACN. However, a remarkable reduction of the fluorescence quantum yield of **III** in EtOH (Tab. 3.3) as testified by a lower S/N ratio (Fig. 3.4) is the result of a dynamic quenching process to a non-fluorescent product in the excited state of III. The effect of hydrogen bonding with the consequence of the formation of a non-fluorescent ^{1}CT state engaged in an excited hydrogen bonded pair is a well-known and frequently encountered phenomenon.[16,70-72] Although a 1:1 complex with the dimethylamino group has been reported leading to a nonfluorescent channel,[71] in our case, a ground state complexation between the dimethylaminonitrogen and EtOH can be ruled out since upon addition of EtOH to an solution of III in ACN, the absorption spectrum remains unchanged, whereas the fluorescence quantum yield strongly decreases by the dynamic fluorescence quenching.

Moreover, in the related D-A biphenyls 4-N,N-dimethylanilino-pyrimidine[16] and 4-N,N-dimethylamino-4'-formyl-biphenyl[18] where quenching hydrogen-bonding has been observed, too, complexation at the acceptor unit (more precisely at the ring nitrogen and the carbonyl oxygen) could be shown. Therefore, in the case of III, a specific hydrogen bonding interaction in S$_1$ with the negatively charged acceptor benzonitrile unit, in particular with the nitrile group is proposed. The question remains why the quenching does not appear in the D-A biphenyls I and II. In the carbonyl and pyrimidine biphenyl derivatives which are related to II, the hydrogen bonding interaction is stronger and takes place already in the ground state indicated by a shift of the absorption spectra in EtOH with respect to ACN. The difference of I and II as compared to III in S$_1$, however, can be understood by a strongly delocalized π-electronic structure for I and II with the central bond of partial double-bond character in the valence-bond description as denoted in eq. 3-11a, and on the other hand, by a different biradicaloid electronic structure for III (eq. 3-11b).

(a) $\text{D-A} \xrightarrow{h\nu} \text{D}^+\text{=A}^-$ (b) $\text{D-A} \xrightarrow{h\nu} \text{D}^{+\cdot}\text{-A}^{-\cdot}$ (3-11)

(I, II) (III)

This biradicaloid structure (eq. 3-11b) is reached only for a strongly twisted geometry and can explain the quenching process in the excited state of III by the following reaction cycle in EtOH (scheme 3.3).

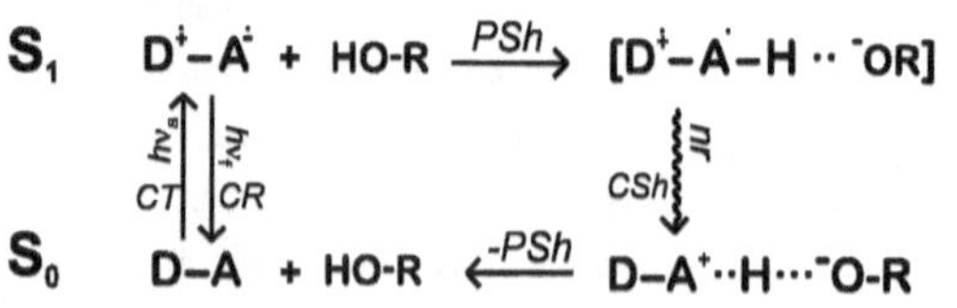

S_1 $\text{D}^{+\cdot}\text{-A}^{-\cdot}$ + HO-R $\xrightarrow{PSh}$ [$\text{D}^{+\cdot}\text{-A-H}\cdots{}^-\text{OR}$]

$\updownarrow h\nu \;\; \updownarrow h\nu'$ $\;\;\;\;\;\;\;\;\downarrow \pi$
CT   CR   CSh

S_0 D-A + HO-R $\xleftarrow{-PSh}$ D-A$^+\cdots$H$\cdots{}^-$O-R

Scheme 3.3

The proposed scheme involves a competition in the photoexcited ^{1}CT state between (i) radiative (and nonradiative) charge recombination (CR) and (ii) a proton shift (PSh) from ethanol to the acceptor side of III inside the solvent cage. Channel (ii) leads to a positively charged biradicaloid form of the donor-acceptor biphenyl III. Such cations are widely known for their fast radiationless decay to the ground state accompanied by an intramolecular charge shift (CSh).[17,73,74] In the ground state, the basicity of the acceptor unit is drastically reduced and hence, the proton captured in S_1 is released in S_0 (-PSh). The prototropic properties and the temperature dependence of the quenching in EtOH is analyzed in Ch. 10.

3.3.3.3 Ratios of Radiative Rate Constants as Indicators for Angular Relaxations

The central observation of this work is that the strong decrease of the radiative rate k_f (= Φ_f/τ_f) with increasing polarity occurs only for **III**. k_f is 10 times higher in HEX than in ACN or EtOH. To exclude the effect of the v_f^3 dependence on the radiative rate k_f according to eq. 3-12,[66,67] a decrease of the reduced radiative rates $k_f/n^3 v_f^3$ is also confirmed in Tab. 3.3.

$$k_f = \frac{64\pi^4}{3h} n^3 v_f^3 M_f^2 \qquad (3\text{-}12)$$

As compared to **I** and **II**, which show rather no polarity dependence of the reduced radiative rates $k_f/n^3 v_f^3$, a different and polarity dependent excited state process must take place in **III**. A change of the electronic or molecular structure between the absorption and fluorescence process can readily be evaluated by a comparison of the fluorescence rate constant k_f with the Strickler Berg rate constant k_f^{SB} obtained from the absorption spectra using eq. 3-13, where the integrated absorption spectrum is given by eq. 3-13:[66,75]

$$k_f^{SB} = \frac{8\pi c 10^3 \ln 10}{N_L} n^2 v_f^3 \int \varepsilon(v_a) d\ln v_a \qquad (3\text{-}13)$$

$$\int \varepsilon(v_a) d\ln v_a = \frac{8\pi^3 N_L n}{hc 10^3 \ln 10} M_a^2 \qquad (3\text{-}14)$$

After substitution of eq. 3-14 into eq. 3-13, it can be derived that the ratio k_f/k_f^{SB} is independent of v_f and n and equals to the squared ratio of the fluorescence to the absorption transition moment: $k_f/k_f^{SB} = M_f^2/M_a^2$. Hence, if no change of the electronic or molecular structure takes place in S_1 a ratio of unity must be obtained.

Tab. 3.4 Ratios of Fluorescence Rate to Strickler Berg Rate Constants k_f/k_f^{SB}

solvent	I	II	III
HEX	0.71	1.33	2.28
EOE	1.03	1.32	1.40
ACN	1.13	1.33	0.60

The ratios k_f/k_f^{SB} are shown in Tab. 3.4. As it is worked out, so far, that emission occurs only from the allowed ^1L_a-type ^{1}CT (2^1A) state, we can discuss the rate ratios k_f/k_f^{SB} mainly on the basis of conformational changes in S_1 (see also ch. 8). The ratios k_f/k_f^{SB} of **I** in EOE and ACN are both close to unity indicating similar conformers in S_0 and S_1. The deviation in HEX has been explained above by electronic coupling effects between close lying ^1L_b and ^1L_a type states occurring only in **I**. For **II**, all k_f/k_f^{SB} ratios amount to 1.33 which means that the conformer distribution in S_1 (a) is the same in all solvents, (b) is different to that in S_0 and (c) has higher transition dipole moments (M_f in eq. 3-12) than in S_0 (M_a in eq. 3-11). Two further observations confirm that mainly angular relaxations are detected by the ratios: (i) only the ratio of the planar model compound **I** is 1 and (ii) the energy corrected rates $k_f/v_f^3 n^3$ in Tab. 3.4 (except in HEX) are very similar for **I** and **II**. The ratio of 1.33 for **II** is then indicative of a fast relaxation of **II** towards full planarity in S_1.

In contrast to **I** and **II**, the pretwisted biphenyl **III** shows solvent dependent ratios k_f/k_f^{SB} due to varying k_f values. Consequently, the fluorescence of **III** is most probably connected with polarity dependent rotamer distributions. In the nonpolar solvent HEX, the large ratio k_f/k_f^{SB} of 2.3 points to the same process as for **II**, i.e. an angular relaxation in S_1 to conformations with higher transition probability than in S_0 consistent with a more planar geometry. In the highly polar ACN, conversely, an average conformation in S_1 is observed with a lower transition probability as compared to the conformational distribution in S_0. Let us denote the fluorescent species mainly observed in HEX as **CT** (with a photophysical behaviour similar as for **I** and **II**) and that observed at lower energies in ACN as the more relaxed species **CTR** (considerably less emissive than **CT**). The intermediate k_f/k_f^{SB} ratio in medium polar EOE ranging between HEX and ACN and can then either be the result of another single conformer distribution different from **CT** and **CTR**, or it can be due to an equilibrium of both species. In the equilibrium case of such differently emissive species, biexponential fluorescence decays should be observable. Indeed, all decays used for Tab. 3.3 are perfectly fitted with single exponential functions (χ^2 =1.05 ± 0.05, DW = 2 ± 0.1) without emission wavenumber dependence of the lifetimes within the error limits of the time-resolution, except the decay of **III** in EOE. Only in this case, a second short time component was necessary to improve the fitting parameters χ^2/DW from 1.29/1.62 to the acceptable values 1.1/2.01. The lifetime of this fast component at 298 K is near the limit of the time-resolution ($\geq$ 100 ps) of the experimental equipment.

3.3.3.4 Excited State Conformational Equilibrium for III

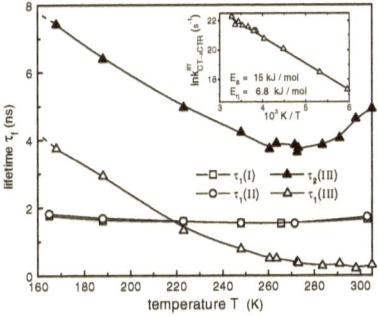
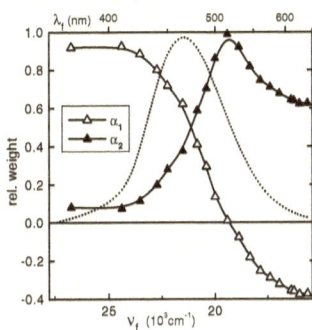

Fig. 3.5 *Fluorescence lifetimes in diethylether vs. temperature as obtained from monoexponential fits of the fluorescence decays for I (- -) and II (-O-) and biexponential fits for III (τ_1: -△-, τ_2: -▲-). The Arrhenius plot for III using the lifetime data of $\tau_1(T)$ is shown as an inset. τ_1 at 160 K is taken as the reference lifetime $\tau_o(CT)$ according to eq.3-16. The regression coefficient is 0.99 and the preexponential factor amounts to 1.5 ps^{-1}. The derived activation barrier E_a is more than twice as high as that for the solvent mobility E_η of EOE pointing to an intramolecular energy barrier between **CT** and **CTR**.*

Fig. 3.6 *Relative amplitudes $\alpha_1(\nu_f)$ and $\alpha_2(\nu_f)$ of the globally fitted fluorescence decay traces $I_f(\nu_f,t) = \alpha_1(\nu_f) \exp(-t/\tau_1) + \alpha_2(\nu_f) \exp(-t/\tau_2)$ of III in EOE at 188 K. Using the linked lifetimes τ_1 = 2.9 ns and τ_2 = 6.4 ns for 20 decay curves at different wavenumbers, a global χ_{glo}^2 =1.17 is achieved. The steady state fluorescence spectrum $I_f(\nu_f)$ is plotted as a dotted line.*

The lifetimes in EOE are plotted versus temperature in Fig. 3.5 to obtain a more reliable value for the fast time constant at room temperature by extrapolation. The short lifetime τ_1 derived by this way is around 200 ps. On the contrary, the fluorescence decays of I and II can be described by monoexponential fits over the whole temperature range from 168 K - 313 K.[76] Further, the lifetimes of I and II are equal and practically do not change across the whole temperature range supporting the view that emission of I and II occurs from a single and planar distribution of rotamers within the ^{1}CT potential surface. The biexponential behaviour of III, however, clearly demonstrates the emission from two different conformer distributions one of which is assigned to the **CT** species (τ_1(160 K) = 4.3 ns) and the other to the **CTR** species (τ_2(160 K) = 7.3 ns). This interpretation is confirmed by the strong and continuous wavenumber dependence of the amplitudes of both lifetimes as shown in Fig. 3.6. It is important to note, that in the low energy tail of the fluorescence spectra (at ν_f < 20·10^3 cm^{-1} in Fig. 3.6) the amplitudes α_1 of the short time component τ_1 become negative. Assuming primary population of the less relaxed charge transfer species **CT** (see ch. 4), it can be concluded that the **CT** distribution undergoes a net photoreaction to the **CTR** distribution and hence, the excited state (decay) behaviour has to be interpreted with a dynamic equilibrium as illustrated in scheme 3.4.

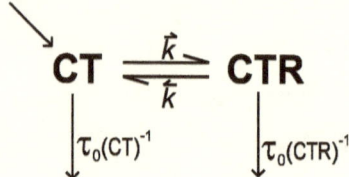

Scheme 3.4 *Excited state dynamic equilibrium between two conformationally different charge transfer species. The assumption that the more emissive **CT** species is primary populated instead of **CTR** is based on the initial relaxation towards planarity on a picosecond timescale as revealed by the transient absorption experiments in ch. 4 and the calculated S_1 twist potentials for small solvent stabilization (Fig.3.10b).*

Consequently, the observed two lifetimes τ_1 and τ_2 are not corresponding to the single decay times (τ_0) of **CT** and **CTR** but are a function of the forward and back reaction rate constants $k_{CT \rightarrow CTR}$ and $k_{CT \leftarrow CTR}$ and of the total deactivation rates X and Y of **CT** and **CTR** according to the well known Birks equations 3-15[66]

$$\tau_{1,2}^{-1} = \tfrac{1}{2}(X + Y \pm \sqrt{(X-Y)^2 + 4k_{CT \rightarrow CTR}k_{CT \leftarrow CTR}}) \qquad (3\text{-}15\ a,b)$$

$$X = \tau_0(CT)^{-1} + k_{CT \rightarrow CTR} \qquad Y = \tau_0(CTR)^{-1} + k_{CT \leftarrow CTR} \qquad (3\text{-}15\ c,d)$$

These equations can be used to derive that the decrease of the lifetimes τ_1 and τ_2 with increasing temperature (Fig. 3.5) is due to the acceleration of the equilibration rates $k_{CT \rightarrow CTR}$ and $k_{CT \leftarrow CTR}$, respectively.

If the possible back reaction **CT←CTR** is neglected in a first approximation, then τ_1^{-1} simplifies to X, i.e. to the sum of the photoreaction rate $k_{CT \rightarrow CTR}$ and the inverse of the intrinsic **CT** fluorescence lifetime $\tau_0(CT)^{-1}$. The latter can be replaced by τ_1 at the glass transition temperature ($T_g \approx 160$ K) where $k_{CT \rightarrow CTR}$ is negligibly small. Under these assumptions the activation barrier separating **CT** from **CTR** can directly be obtained from the Arrhenius plot (eq. 3-16) shown as an inset in Fig. 3.5.

$$\ln k_{CT \rightarrow CTR}^{irr} = \ln(\tau_1^{-1}(T) - \tau_0(CT)^{-1}) = -\frac{E_a^{irr}}{RT} + const. \qquad (3\text{-}16)$$

It has to be kept in mind that the derived activation barrier E_a^{irr} is representative for the case of an irreversible reaction from **CT** to **CTR** (the dominance of the irreversible reaction mechanism is proved by a „full kinetic analysis" in ch.6) and that it contains both the contributions from the intrinsic activation barrier $E_i^{\#}(CT \rightarrow CTR)$ and from a „dynamic barrier" E_η induced by the solvent viscosity.[77] Nevertheless, the obtained activation barrier E_a^{irr} of 15 kJ/mol (1250 cm^{-1}) is considerably higher than the barrier for the solvent mobility of EOE (E_η = 6.8 kJ/mol) and hence strongly supports the presence of an intrinsic barrier and the coexistence of two distinct conformer species denoted as **CT** and **CTR**. It is noteworthy that the barrier obtained is even around 4 times higher than that reported for DMABN in EOE (E_a = 4.0 kJ/mole).[78]

At room temperature in EOE, the two distributions are in fast equilibrium and in this case of thermodynamic control ($k_{CT \to CTR}$ >> $\tau_0(CT)^{-1}$, $k_{CT \leftarrow CTR}$ >> $\tau_0(CTR)^{-1}$) eq 3-15a and b can be simplified to

$$\tau_1^{-1} = k_{CT \to CTR} + k_{CT \leftarrow CTR} \tag{3-17a}$$

$$\tau_2^{-1} = f_{CT}\tau_0(CT)^{-1} + (1 - f_{CT})\tau_0(CTR)^{-1} \tag{3-17b}$$

$$f_{CT} = \frac{k_{CT \leftarrow CTR}}{k_{CT \to CTR} + k_{CT \leftarrow CTR}} \tag{3-17c}$$

where f_{CT} is the fraction of **CT** at equilibrium. The measured mean lifetime $<\tau_f>$ (= τ_2) in EOE (Tab. 3.3) at room temperature is therefore a weighted average of the intrinsic **CT** and **CTR** decay times $\tau_0(CT)$ and $\tau_0(CTR)$ yielding an intermediate value between that of HEX (mainly strongly emissive **CT** species) and that of ACN (mainly less emissive **CTR** species). Assuming a fraction of 100% **CT** in HEX and 100% **CTR** in ACN and correcting for the solvent dependence of k_{nr}, n^3 and v_f^3 to the conditions in EOE, the intrinsic lifetimes $\tau_0(CT)$= 3.5 ns and $\tau_0(CTR)$= 6.6 ns at 298 K can be derived.[79] With these values the fraction of **CT** in EOE is calculated f_{CT}=47% from eq. 3-17b, consistent with an equilibrium constant $K_{eq} \approx 1$ and with a reaction free energy change ΔG around zero in EOE at 298 K. This finding is well understandable with the similarity of the fluorescence energies of the **CT** and **CTR** species.

Kinetics, thermodynamics and activation barriers of the **CT/CTR** equilibrium are analyzed in detail by using time-resolved emission at low temperatures and at high pressures in ch. 6 and 7.

3.3.4 Theoretical Calculations

In this section, after some basic considerations regarding twist potentials, CNDO/S-CI calculations are employed in order to investigate whether different rotamers can be responsible for the fluorescence species **CT** and **CTR**. Moreover, the calculated properties of the two lowest lying states ^1L_b and ^1CT are compared to clarify that emission is due to the ^1L_a-type ^1CT state.

3.3.4.1 Twist Potentials in S_0

In general, twist potentials of biaryls, which are calculated here by AM1 for **II** and **III** in S_0 (Fig 3.7), result from an energetical compromise between the π-resonance energy (minimum at φ=0°) and the sterical potential (minimum at φ=90°) mainly induced by the substituents in the ortho-positions. Since this sterical hindrance of free intramolecular rotation is enhanced in **III** by the methyl groups, the calculated (intrinsic) barrier at $\varphi=0°$ is much higher for **III** ($E_i^{\#}(0°) = 66$ kJ mol^{-1}) than for **II** ($E_i^{\#}(0°) = 5.2$ kJ mol^{-1}), while at $\varphi=90°$, it is lower for **III** ($E_i^{\#}(90°) = 0.5$ kJ mol^{-1}) than for **II** ($E_i^{\#}(0°) = 7.7$ kJ mol^{-1}). Because both biphenyls possess C_2 symmetry,[80] the potentials and the corresponding Boltzmann rotamer distributions $\eta(\varphi)$ (eq. 3-18) are symmetric to $\varphi=0°$ and $\varphi=90°$. In the following, it is therefore referred only to the twist region $\Delta\varphi=0\text{-}90°$. This means that we have only a single distribution of equivalent rotamers around the equilibrium angle $\varphi_{eq}=39°$ for **II** and $\varphi_{eq}=78°$ for **III**. In addition, it should be noted that φ_{eq} differs from the mean twist angle $<\varphi_\eta>$ (eq. 3-19) which is convoluted with the distribution function $\eta(\varphi)$ and therefore gives a better representation for experimentally obsevable values ($<\varphi_{exp}>$). In S_0, $<\varphi_\eta>$ equals to 34.5° for **II** and 72° for **III** and depends only slightly on the solvent polarity (using $E_{ACN}(\varphi)=E_{gas}(\varphi)\text{-}f(\varepsilon_r)\mu_g^2$ for the solvent stabilized ground state energy in acetonitrile (cf. eq. 3-4) in eq. 3-18, $<\varphi_\eta>$ is obtained only 0.5° more planar for both **II** and **III**).

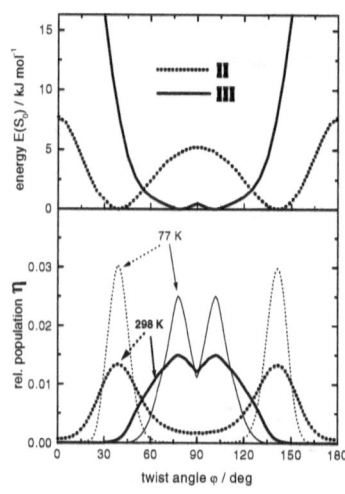

$$\eta(\varphi) = \frac{e^{-\frac{E(\varphi)-E(\varphi_{eq})}{RT}}}{\int_{0°}^{180°} e^{-\frac{E(\varphi)-E(\varphi_{eq})}{RT}} d\varphi} \quad (3\text{-}18)$$

$$\langle \varphi_\eta \rangle = \int_{0°}^{180°} \varphi \eta(\varphi) d\varphi \quad (3\text{-}19)$$

Tab. 3.5 Data from S_0 Twist Potentials[a]

	φ_{eq}	$E_i^{\#}(\varphi=0°)$[b]	$E_i^{\#}(\varphi=90°)$[b]	$\langle\varphi_\eta\rangle$
II	39°	5.2 kJ	7.7 kJ	34.5°
III	78°	66.3 kJ	0.5 kJ	72°

[a] obtained by AM1/gas phase/298K; see text for details [b] per mol

Fig. 3.7 Ground state twist potentials derived from AM1 optimizations (upper panel) and the corresponding rotational distributions $\eta(\varphi)$ according to the Boltzmann eq. 3-18 (lower panel) for **II** ($\cdots$) and **III** (———).

3.3.4.2 Electronic Nature of S_1

The calculated dipole moments μ, the absorption and fluorescence transition energies ν_a and ν_f in diethylether and the reduced radiative rates k_f/ν^3 of the ^1CT and ^1L_b states are plotted as a function of the twist angle φ for the flexible biphenyl **II** in Fig. 3.8a-c and for the sterically hindered derivative **III** in Fig.3.9a-c. The single symbols at 0° in Fig. 3.8a-c indicate the data for the D-A fluorene **I** which serves as the planar model compound.

Due to the calculated low dipole moments (Fig. 3.8a and 3.9a) of the ^1L_b state its stabilization by the solvent is negligible. Completely different to ^1L_b, the ^1CT state is strongly stabilized already in the calculated absorption process but distinctly more in the fluorescence transition (Fig. 3.8b and 3.9b). As a result, the ^1CT state of **II** becomes the lowest excited singlet state at twist angles φ=0°-50° in absorption and at φ=0°-70° in fluorescence (Fig. 3.8b). On the other hand, the twist angle region for the pretwisted biphenyl **III** is restricted to angles

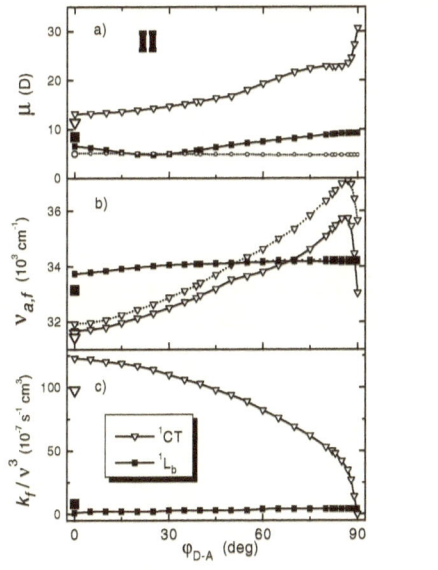
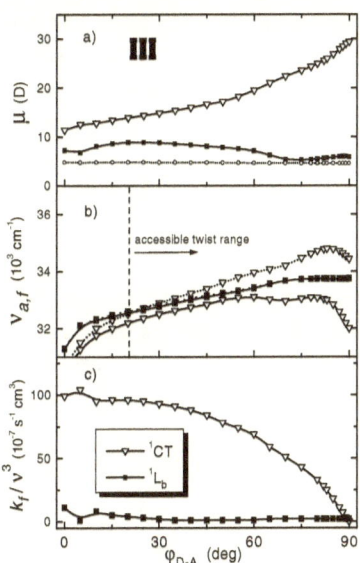

Fig. 3.8   **Fig. 3.9**
*Twist angle dependent results of the CNDO/S-SCI calculations using fully optimized (AM1) ground state geometries of **II** (and **I** at 0°) in Fig. 3.8 and of **III** in Fig. 3.9, respectively. The calculated properties of the two lowest lying excited states ^1CT (-∇-) and ^1L_b (-■-) are compared as a function of the interannular twist angle φ: (a) permanent excited state dipole moments μ, (b) absorption (ν_a symbols connected with dashed lines) and fluorescence (ν_f symbols connected with full lines) transition energies in diethylether calculated from eq. 3-2a and b (c) reduced radiative rates calculated from eq. 3-12. The ground state dipole moments for **II** and **III** (-o-) and all corresponding results for **I** (larger single symbols at φ=0° in Fig. 3.8) are also shown.*

φ above 20° in S_1 and S_0 due to the sterical strain (in S_0: E(20°)=30kJ mol^{-1}=2500 cm^{-1}). In the accessible angular region (φ > 20°), the calculated ^{1}CT state of III is not sufficiently stabilized to become the lowest lying state in absorption. ^1L_b and ^{1}CT are very close lying under these conditions in agreement with the large absorption contribution of a perpendicularly polarized transition within the first absorption band.[26] The situation changes when solvent relaxation in S_1 is switched on. The ^{1}CT transition is then the lowest one across the whole angular range and should be responsible for the fluorescence. Taking also into account that the experimental dipole moments (**I-III**: μ_e = 15-19 D using L(ε,n)) and reduced radiative rates $k_f/v_f^3 n^3$ (**I**: 80-150 10^{7}s^{-1}cm^3, **II**: 120-140 10^{7}s^{-1}cm^3, **III**: 15-60 10^{7}s^{-1}cm^3) given in Tab. 3.3 can be well correlated with those of the ^{1}CT state (Fig. 3.8bc and 3.9bc), but are considerably larger than those calculated for the ^1L_b states (μ_e < 10 D, k_f/v^3 < 10^{8}s^{-1}cm^3), irrespective of the compound or twist angle φ, the assignment of the observed emission to the ^{1}CT(2^{1}A) state remains unambiguously. Besides, a CI analysis of the calculations reveals that the comparably low fluorescence rate of **I** in HEX follows from strong mixing between the ^1L_b-type and ^{1}CT state (ch. 2 and 8). Therefore, the calculated fluorescence rate of **I** is smaller than that of **II** at 0° (Fig. 3.8c). In the polar solvents mixing between ^1L_b and ^{1}CT is decreased leading to higher emission rates similar for **I** and **II** at 0° (Tab 3.3, see ch. 8).

3.3.4.3 Molecular Structure in S_1

In the experimental part, it was suggested that the similar fluorescence properties (band half widths, dipole moments, radiative rates and transition energies) of **I** and **II** point to a planar structure of **II** in S_1. Indeed, Fig. 3.8 supports this view, since the calculated dipole moments and fluorescence energies are nearly identical for **I** and **II** at φ=0°. Furthermore, the calculated fluorescence energy difference between a twisted (φ ≈ 35°) and the planar rotamer (Figure 3.8b) is in excellent agreement with the additional reorganization energy ($\lambda_{sum} = \Delta v_{St}/2$) of 1000 cm^{-1} for **II** as compared to **I** (Tab. 3.1). This solvent-independent energy loss of **II** in S_1 can therefore be attributed to the intramolecular rotation.

Concerning the initial (Franck-Condon) geometry, we can explain the slightly stronger solvatochromic shift, larger transition energy and lower intensity of the first absorption band (Fig. 3.1) for **II** as compared to **I**, on the basis of Fig. 3.8, by a less planar molecular structure in S_0 associated with (a) a higher Franck-Condon S_1 dipole moment, (b) larger absorption energy and (c) smaller radiative rates (eq. 3-13). These comparisons between theoretical and experimental results underline that **II** relaxes in S_1 from a medium twisted towards a planar rotamer distribution.

For **III**, the polarity dependent radiative rates (Tab. 3.3 and 3.4) and the biexponential decay behaviour in EOE (Fig. 3.5 and 3.6) lead to the proposal of the coexistence of two electronically different conformer distributions, even though, only single, but broad, emission bands are observed (Fig. 3.2). To explain this discrepancy, we have to note, that the transition energies in the solvent stabilized case are modified by two counteracting energy contributions (eq. 3-2). With increasing twist angle φ, the pure electronic energy increases but the stabilizing solvation energy also increases due to the larger dipole moments for higher twist angles φ (Fig. 3.9b). As a consequence, the fluorescence energies of the rotamers in the accessible twist range of III are very similar in EOE which prevents the occurrence of two separated emission bands. To evaluate whether both counteracting energy contributions either result in a relatively flat potential of the ^{1}CT state connected with a broad angular distribution or a double minimum potential with two peaks of the conformer distribution, the ^{1}CT twist potentials of II and III are calculated in different solvents applying classical Onsager theory (eq. 3-4).

Indeed, only for the gas phase, the ^{1}CT twist potentials (Fig. 3.10) of II and III exhibit a single minimum at an angle φ_{eq} which is 30-40° smaller as compared to the calculated equilibrium angle in S_0 (Tab. 3.5). With increasing solvent polarity, this ^{1}CT energy minimum slightly shifts to higher twist angles φ_{eq} (^{1}CT) and a second energy minimum evolves for the perpendicular geometries. For II in EOE, the barrier $E_i^\#(90)$ from the minimum at planarity to the steep minimum at 90° is calculated very high (45 kJ) which can explain why a second twisted rotamer distribution is not populated and not observed in all experiments with II. In the case of III in EOE, the calculated double minimum potential is associated with a low barrier of 2 kJ (170 cm^{-1}). The barrier $E_i^\#(90)$ theoretically found in HEX (14 kJ) agrees well with the activation barrier E_a^{irr} (Fig. 3.5) found experimentally in EOE (15 kJ). One could suggest to use a larger Onsager cavity radius a in eq. 3-4 to fit the theoretical to the experimental barrier. However, some general comments on the shape of the potential curves have to be noted. All potential curves obtained are based on the wavefunction of the ^{1}CT state in the gas phase. But due to energy stabilization of the ^{1}CT state in solution, its electron transfer (^{1}ET) character[29] is enhanced at the expense of the weakly polar ^{1}L$_a$ character (see ch.8). As the ^{1}ET character for the perpendicular geometries is

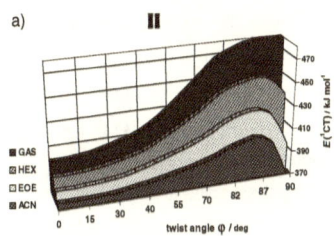

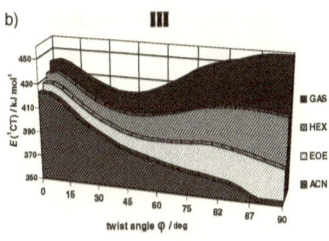

Fig. 3.10 ^{1}CT twist potentials of (a) II and (b) III in solvents of different polarity as obtained by eq. 3-4.

already nearly 100% (minimum overlap rule)[64] without solvent stabilization, it can easily be deduced that the solvent-induced increase of electron transfer character is stronger for medium twisted than for more perpendicular rotamers. This will lead to a flattening of the dipole moment dependence on the twist angle φ (Fig. 3.8a and 3.9a). As a result of the solvent interaction, the energy minima are expected to be broader and located at more planar angles φ than calculated with the gas phase dipole moments μ in Fig.3.8a and 3.9a. A shift and decrease of the barrier $E_i^\#(90)$ might also occur. In this sense, the potentials for **III** give already a more realistic representation, because the ^{1}ET weight at a given value of φ is intrinsically higher than for **II** due to the inductive effect of the two methyl groups (ch. 8).[26] A detailed analysis of the solvent effects on the ^{1}CT wavefunction indicates only minor effects on the torsional pathways in S_1 (ch.8). Although it has also been verified using idealized geometries as well as selected geometries optimized in the excited state that reasonable structural changes, e.g. planarization of the pyramidalization angle of the dimethylamino nitrogen or shortening of the interannular bond lengths, have negligible influence on the essential shape of the excited state twist potentials (e.g. the occurrence of two minima) as well as on the properties shown in Fig. 3.8 and 3.9, their possible influence has to be mentioned, too. Thus, the potential curves in Fig. 3.10 and the derived twist angles and barriers in Tab. 3.6 should mainly be regarded relative to each other to explain the excited state structural relaxations. Nevertheless, the calculated potentials in combination with the experimental results, obviously confirm that in the excited state (a) **II** relaxes to a planar structure irrespective of the solvent, due to the absence of a barrier towards planarity ($E_i^\#(0°) = 0$) but high barriers ($E_i^\#(90°)>30$ kJ) towards more twisted rotamers whereas (b) **III** can populate two rotamer minima, the population of which are modulated by their solvent polarity dependent energy differences and activation barrier ($E_i^\#(90°)=0-40$ kJ). In nonpolar solvents like HEX only the more planar rotamer distribution is populated which is experimentally observed as a more emissive species **CT**, and in highly polar solvents like ACN, only the more twisted distribution **CTR** is observed with smaller emission rates k_f. In medium polar solvents like EOE, a primary relaxation to the **CT** distribution with weak solvent stabilization can be assumed, followed by an equilibration between both rotamer distributions **CT** and **CTR**.

The above discussion can be further refined by a correlation of the experimental with the calculated reduced radiative rates, because they are monotonically decreasing with φ. If we employ the calculated $k_f/v^3(\varphi)$ curves of **II** and **III** in Figure 3.8c and 3.9c directly as calibration curves, ground and excited state mean twist angles $<\varphi_{kf}(S_0)>$ and $<\varphi_{kf}(S_1)>$ can be approximated from the experimental Strickler-Berg rates $k_f^{SB}/v_f^3 n^3$ and fluorescence rates $k_f/v_f^3 n^3$ (Tab. 3.3 and 3.4), respectively. The resulting ground and excited state twist angles $<\varphi_{kf}(S_0)>$ and $<\varphi_{kf}(S_1)>$ of **II** and **III** are collected in Table 3.6. In agreement with the gas phase AM1 optimizations, the obtained ground state angles $<\varphi_{kf}(S_0)>$ for **II** and **III** are around 40° and 80°, respectively. As expected, the radiative rates of **II** are consistent with a fully planar excited state geometry in all solvents. On the other hand, the S_1 conformation of **III** is obtained 20° more planar in HEX and 5° more twisted in ACN, as compared to the average S_0 conformation. For **III** in EOE, $<\varphi_{kf}(S_1)>$ is intermediate between that of HEX and ACN. In the light of the time-resolved experiments presented in Fig. 3.5 and 3.6, this is understandable by an equilibrium between two rotamer distributions, one associated with more planar twist angles than $<\varphi(S_0)>$ assignable to the precursor species **CT** and the other connected with more twisted angles than $<\varphi(S_0)>$ assignable to the more relaxed successor species **CTR**, respectively.

Tab. 3.6 S_1 Twist Angles and Energy Barriers as Derived from CNDO/S Calculations (eq. 2-4) Using the Obtained Gas Phase Energies in Combination with classical Onsager theory (eqs. 3-2 to 3-4).

	$\varphi_{eq}(^1CT)^a$	$E_i^{\#}(\varphi \to 0°)^b$	$E_i^{\#}(\varphi \to 90°)^b$	$<\varphi_{kf}(S_0)>^c$	$<\varphi_{kf}(S_1)>_c$
II					
GAS	0°	0 kJ	84 kJ	-	-
HEX	0°(+ 90°)	0 kJ	61 kJ	49	0
EOE	0°(+ 90°)	0 kJ	45 kJ	40	0
ACN	5°(+ 90°)	0.5 kJ	31 kJ	35	0
III					
GAS	40°	24 kJ	41 kJ	-	-
HEX	45° (+ 90°)	31 kJ	14 kJ	82	64
EOE	60° (+ 90°)	40 kJ	2 kJ	82	78
ACN	90°	79 kJ	0 kJ	82	85

[a] equilibrium twist angles of ^{1}CT potentials shown in Figure 3.10.
[b] intrinsic energy barrier for intramolecular rotation from the more planar minimum to the geometry with $\varphi = 0°$ and $\varphi = 90°$, respectively.
[c] mean angles obtained by a direct correlation of the experimental Strickler-Berg ($k_f^{SB}/n^3v_f^3$) and fluorescence ($k_f/n^3v_f^3$) rates with the calculation results in Figure 3.8c and 3.9c. All values should be considered only relative to each other, since solvent-induced changes of the wavefunction and other conformational reaction coordinates are not taken into account (see text and ch. 8)

3.4 Conclusions

This study provides a comprehensive survey of the fluorescence and photophysical behaviour of the differently twisted donor-acceptor biphenyls **I-III** and related biphenyls. It enables to draw the desired conclusions on the electronic and conformational structure in the ground state S_0 and the first excited singlet state S_1 reached after photoinduced intramolecular charge transfer.

The three problems complicating the photophysics of other known donor-acceptor biphenyls, as mentioned in the introductory section, can be excluded:

Concerning category 1 (S_1 state of mixed ^1L_b- and ^1L_a-type):
The electronic nature of S_1 in **I-III** observed in the fluorescence can definitely be assigned to a pure ^1L_a-type $^1CT(2^1A)$ state (except for a small deviation of **I** in n-hexane where ^1L_b mixing is also important) transferring charge from the HOMO(D) to the LUMO(A) orbital across the interannular bond. The three main cornerstones for this assignment are (i) the large radiative rates ($k_f > 0.3$ ns^{-1} except for **III** in polar solvents), (ii) a degree of fluorescence polarization close to 0.5 (p=0.44-0.48) and (iii) the comparison with CNDO/S calculated fluorescence energies, dipole moments and radiative rates, which is consistent only with $^1CT(2^1A)$.

Concerning category 2 (strongly polarity dependent nonradiative rates)
The influence of the (bulk) solvent polarity on the nonradiative rates k_{nr} of **I-III** is weak and can be explained. In polar and aprotic solvents, k_{nr} remains constant at 0.1 ns^{-1} for all compounds studied. The higher rates in HEX (Δk_{nr}/ns^{-1} = 0.2; 0.15; 0.45 for **I-III**) are understandable by an additional ISC channel, effective only in nonpolar solvents by a possible S_1-T_n level reversal (ch. 5). Thus, no strongly nonradiative channel due to intramolecular relaxations complicates the photophysics of **I-III**.

Concerning category 3 (specific solvent interactions in S_0 and S_1):
Specific solvent interactions are weak or even absent in **I** and **II**, and found only for **III** in EtOH by an enhancement of k_{nr} (Δk_{nr} = 0.56 ns^{-1}). Since the spectral and radiative properties of **III** are rather identical in EtOH and equipolar ACN, the nonradiative channel is not induced by polarity but by hydrogen bonding in S_1 only and can therefore be avoided, if only aprotic solvents are used (ch. 10).

Hence the knowledge of the electronic nature and photophysics, renders the donor-acceptor biphenyls **I-III** as excellent model compounds to investigate conformational changes with respect to the interannular twist angle φ in S_0 and S_1 associated with photoinduced charge transfer. The similar radiative rates k_f, excited state dipole moments μ_e and other fluorescence properties of **I** and **II** connected with a difference in their Strickler Berg rates k_f^{SB} (k_f^{SB}(**I**) >

k_f^{SB}(**II**) but k_f(**I**) = k_f(**II**)) show that the flexible compound **II** relaxes to a planar geometry after photoexcitation. On the other hand, the higher nonradiative rates k_{nr} in n-hexane, larger excited state dipole moments μ_e and smaller radiative rates k_f and k_f^{SB} of **III** confirm the twisted structure of this sterically hindered derivative in S_0 and S_1. In contrast to **I** and **II**, the energy corrected rates $k_f/v_f^3 n^3$ of **III** decline with increasing polarity. The latter behaviour can be explained with a solvent polarity independent mean twist angle $<\varphi>$ in S_0 whereas in S_1, the rotamer distribution is more planar in nonpolar solvents and more twisted in polar solvents as compared to S_0. The strongly twisted conformation in polar solvents leads to a biradicaloid electronic structure associated with electron localization on the benzonitrile acceptor unit. Therefore, the dynamic fluorescence quenching observed only for **III** in the protic solvent ethanol, which is interpreted by protic interaction of ethanol with the negatively charged benzonitrile unit, can be taken as further evidence of the highly twisted structure of **III** in polar solvents.

In the solvent of intermediate polarity, diethylether, the fluorescence decay is biexponential only for **III**. Together with the strong temperature dependence of the lifetimes and the spectral dependence of their amplitudes, this indicates the coexistence of two excited state rotamer distributions. One of them is more planar (**CT**) and the other one more twisted (**CTR**) than the original distribution in the ground state. At room temperature, both rotamer distributions are in a fast equilibrium. The risetimes observed at low temperature reveal a photoreaction with **CT** as precursor and the more relaxed **CTR** as successor species. The primary relaxation from the Franck-Condon geometry to the more planar geometry of the **CT** species, which takes place in all biphenyls during the solvation process, is not observed here, because it is either too fast or the emission properties of the more planar rotamers are too similar to be detected by the time-resolved experiments.[81] The quantum chemical calculations in combination with the Onsager model support the experimental conclusions regarding the barrierless excited state planarization of **II** in all solvents and of **III** in nonpolar solvents (**CT**), the occurrence of a double minimum potential for **III** in EOE (**CT** and **CTR**) and the preference of highly twisted conformers in more polar solvents (**CTR**).

In general, the results of these bichromophoric model molecules show that the introduction of strong coulombic interactions (here by photoinduced charge transfer) connected with a different interplay of solvent and mesomeric stabilization does not only determine the conformational structure, but can also lead to an equilibrium of conformers with different π-conjugation.

3.5 References and Notes

(1) Grabowski, Z.R.; Rotkiewicz, K.; Siemiarczuk, A.; Cowley, D.J.; Baumann, W. *Nouv. J. Chim.* **1979**, 3, 443.

(2) Rettig, W. *Angew. Chem. Int. Ed.* **1986**, 25, 971.

(3) Zachariasse, K. *Pure Applied Chem.* **1993**, 65, 1745.

(4) Visser, R.J.; Cyril, A.G.; Varma, G.O. *J. Chem. Soc. Faraday II* **1980**, 76, 453.

(5) Sobolewski, A.L.; Domcke, W. *Chem. Phys. Lett.* **1996**, 259, 119.

(6) (a) Berlman, I.B.; Steingraber, O.J. *J. Chem. Phys.* **1965**, 43, 2140; (b) Berlman, I.B. *J. Chem. Phys.* **1970**, 52, 5616.

(7) Hohlneicher, G.; Dörr, F.; Mika, N.; Schneider, S. *Ber. Bunsenges. Physik. Chem.* **1968**, 72, 1144.

(8) Momicchioli, F.; Bruni, M.C.; Baraldi, I. *J. Phys. Chem.* **1972**, 76, 3983. and ref. therein.

(9) (a) Dick, B.; Hohlneicher,G. *Chem. Phys.* **1985**, 94, 131; (b) Rubio,M.; Merchán,M.; Ortí, E.; Roos, B.O. *Chem. Phys. Lett.* **1995**, 234, 373; (b) Swiderek, P.; Michaud, M.; Hohlneicher, G.; Sanche, L. *Chem. Phys. Lett.* **1991**, 187(6), 583.

(10) (a) Kurland, R.; Wise, W.B. *J. Am. Chem. Soc.* **1964**, 86, 1877; (b) Eaton, V.J.; Steele, D. *J. Chem. Soc. Faraday Trans.* II **1973**, 69, 1601; (c) Le Gall, Suzuki, H.; *Chem. Phys. Lett.* **1977**, 46, 467; (d) Uchimura, H.; Tajiri, A.; Hatano, M. Bull. *Chem. Soc. Jpn.* **1981**, 54, 3279 and *Chem. Phys. Lett.* **1975**, 34, 34; (e) d'Amibale, A.; Lunazzi, L.; Boicelli, A.C.; Macciantell, D. *J. Chem. Soc. Perkins Trans. II* **1973**, 69, 1396; (f) Lim, E.; Li, Y.H. *J. Chem. Phys.* **1970**, 52, 6416; (g) Barrett, R.M.; Steele, D. *J. Mol. Struct.* **1972**, 11, 105.

(11) Takei, Y; Yamaguchi, T.; Osamura, Y.; Fuke, K.; Kaya, K. *J. Phys. Chem.* **1988**, 92, 577.

(12) Butler, R.M., Lynn, M.A., Gustafson, T.L. *J. Phys. Chem.* **1993**, 97, 2609.

(13) Swiatkowski, G.; Menzel, R.; Rapp, W. *J. Luminesc.* **1987**, 37, 183.

(14) (a) Lahmani, F.; Breheret, E., Zehnacker-Rentien, A.; Amatore C.; Jutand, A. *J. Photochem. Photobiol. A: Chem.* **1993**, 70, 39; (b) Lahmani, F.; Breheret, Benoist d'Azy, O.; Zehnacker-Rentien,A; Delouis, J.F. *J. Photochem. Photobiol. A: Chem.* **1995**, 89, 191.

(15) (a) Klock, A.M.; Rettig, W. *Pol. J. Chem.* **1993**, 67, 1375; (b) Maus, M.; Rettig, W.; Lapouyade, R. *J. Inf. Rec.* **1996**, 22, 451; (c) Rettig, W.; Maus, M. *Ber. Bunsenges. Phys. Chem.* **1996**, 100, 2091.

(16) Herbich, J.; Waluk, J. *Chem. Phys.* **1994**, 188, 247.

(17) (a) Ephardt, H.; Fromherz, P. *J. Phys. Chem.* **1991**, 95, 6792; (b) Fromherz, P., Heilemann, A. *J. Phys. Chem.* **1992**, 96, 6864; (c) Röcker, C., Heilemann A.; Fromherz, P. *J. Phys. Chem.* **1996**, 100, 12172.

(18) Chou, P. T.; Chang, C.P.; Clements, J.H.; Meng-Shin, K. *J. Fluoresc.* **1995**, 5, 369.

(19) David, C.; Baeyens-Volant, D. *Mol. Cryst. Liq. Cryst.* **1980**, 59, 181.

(20) Lippert, E. *Ber. Bunsenges. Phys. Chem. (Z. Elektrochem.)* **1957**, 61, 962.

(21) Zhu, Y; Schuster, G.B. *J. Am. Chem. Soc.* **1990**, 112, 8583.

(22) Foley, M.J.; Singer, L.A. *J. Phys. Chem.* **1994**, 98, 6430.

(23) Cowley, D.J.; O'Kane, E.; Todd, R.S.J. *J. Chem. Soc. Per. Trans.* **1991**, 2, 1495.

(24) Lequan, M.; Lequan, R.M.; Ching, K.C. *J. Mater. Chem.* **1991**, 1, 997.

(25) (a) Yoon, M.; Cho, D.W.; Lee, J.Y. *Bull. Korean Chem. Soc.* **1992**, 13, 613; (b) Kang, S.G.; Ahn, K.D.; Cho, D.W.; Yoon, M. *Bull. Korean Chem. Soc.* **1995**, 16, 972; (c)Cho, D.W.; Kim, Y.H.; Kang, S.G.; Yoon,M. *J. Phys. Chem.* **1994**, 98, 558; (d) Cho, D.W.; Kim, Y.H.; Kang, S.G.; Yoon, M.; Kim, D. *J. Chem. Far. Trans.* **1996**, 92, 29.

(26) Maus, M.; Rettig, W. *Chem. Phys.* **1997**, 218, 151.

(27) Suzuki, H. *Electronic Absorption Spectra and Geometry of Organic Molecules*, Academic Press, New York **1967**.

(28) (a) Longuet-Higgins, H. C.; Murrell, J. N. *Proc. Phys. Soc. A* **1955**, 68, 601; (b) Godfrey M.; Murrell, J.N. *Proc. R. Soc. A* **1966**,278, 60.

(29) In the present notation, it is distinguished between ^{1}ET and ^{1}CT. ^{1}ET represents the case for a complete transfer of one electron from D to A, whereas ^{1}CT denotes a state with partial electron transfer as resulting from mixed locally excited and ^{1}ET character (see Appendix A and ch.8)

(30) Gribble, G.W.; Nutaitis, C.F. *Synthesis* **1987**, 709.

(31) Ellis, G.P.; Romney-Alexander, T.M. *Chem. Rev.* **1987**, 87, 779.

(32) Fox, G.J.; Hallas, G.; Hepworth, J.D.; Paskins, K.N. *Org. Synth.* **1976**, 55, 20.

(33) Mc Namara, J.M.; Gleason, W.B. *J. Org. Chem.* **1976**, 41, 1071.

(34) Amatore, C.; Jutand, A.; Negri, S.; Fauvarque, J.-F. *J. Organomet. Chem.* **1990**, 390, 389.

(35) Parker, C.A. *Photoluminescence of Solutions*, Elsevier, Amsterdam, **1968**.

(36) Melhuish, W.H. *J. Phys. Chem.* **1961**, 65, 229.

(37) Globals Unlimited®, commercially available from the Laboratory of Fluorescence Dynamics at the University of Illinois, **1992**.

(38) O'Connor, D.V.; Phillips, D. *Time-correlated Single Photon Counting*, Academic Press, London, **1984**.

(39) Del Bene, J.; Jaffe, H.H. *J. Chem. Phys.* **1968**, 48, 1807; **1968**, 48, 4050; **1968**, 49, 1221; **1969**, 50, 1126.

(40) AMPAC 5.0, Semichem, 7128 Summit, Shawnee, KS 66216 (1994); Dewar, M.J.S.; Zoebisch, E.G.; Healy, E.F.; Stewart, J.P. *J. Am. Soc.* **1985**, 107, 3902.

(41) Onsager, L. *J. Am. Chem. Soc.* **1936**, 58, 1486.

(42) Zerner, M.C.; Karelson, M.M. *J. Phys. Chem.* **1992**, 96, 6949.

(43) commercially available from Serena Software, Box 3076, Bloomington, IN 47402-3076

(44) Böttcher, C. J. F. *Theory of Electronic Polarisation*, Elsevier, Amsterdam **1973**.

(45) The interannular bond lengths as obtained for the AM1 optimized structures of I-III are: 1.452 Å for I ($\varphi=0°$), 1.459 Å for II ($\varphi=39°$), 1.467 Å for III ($\varphi=78°$).

(46) Riddick, J.A.; Bunger, W.B.; Sakano, T.K. *Organic Solvents*, John Wiley & Sons **1986**.

(47) Marcus, R.A. *J. Phys. Chem.* **1989**, 93, 3078.

(48) Flom, S.R., Nagarajan, V.; Barbara, P.F. *J. Phys. Chem.* **1986**, 90, 2085.

(49) Ma, J.; Dutt, G.B.; Waldeck, D.H.; Zimmt, M.B. *J. Am. Chem. Soc.* **1994**, 116, 10619.

(50) Strehmel, B.; Seifert, H.; Rettig, W. *J. Phys. Chem.* **1997**, 101, 2232.

(51) Bakshiew, N.G. *Opt. Spectrosk.* **1964**, 16, 821; **1962**, 12, 473; **1961**, 10, 717.

(52) (a) Kawski, A.; Bilot, L. *Acta Phys. Polon.* **1964**, 26, 42; (b) Kawski, A. *Acta Phys. Polon.* **1966**, 29, 507; (c) Chamma, A.; Viallet, P. *C.R. Acad. Sci. Ser. C* **1970**, 270, 1901.

(53) McRae, E.G. *J. Phys. Chem.* **1957**, 61, 562.

(54) Lippert, E. *Z. Naturforsch.* **1955**, 10a, 541.

(55) Mataga, N.; Kaifu, Y; Koizumi, M. *Bull. Chem. Soc. Jpn.* **1956**, 29,465.

(56) (a) Catalán, J.; Díaz, C.; López, V.; Pérez, P.; Claramunt, R.M. *J. Phys. Chem.* **1996**, 100, 18392; (b) Catalán, J.; López, V.; Pérez, P. *J. Fluoresc.* **1996**, 6, 15; (c) Catalán, J.; López, V.; Pérez, P.; Martín-Villamil, R.; Rodriguez, J.G. *Liebigs. Ann.* **1995**, 241.

(57) Létard, J.F.; Lapouyade, R.; Rettig, W. *J. Am. Chem. Soc.* **1993**, 115, 2441.

(58) Klock, A.M.; Rettig, W.; Hofkens, J.; van Damme, M.; De Schryver, F.C. *J. Photochem. Photobiol. A, Chem.* **1995**, 85, 11.

(59) Van Damme, M.; Hofkens, J.; De Schryver, F.C. *Tetrahedron* **1989**, 45, 4693.

(60) Schneider, F.; Lippert, E. *Ber. Bunsenges. Phys. Chem.* **1970**, 74, 624.

(61) Hara, K.; Arase, T.; Osugi, J. *J. Am. Chem. Soc.* **1984**, 106, 1968.

(62) Onkelinx, A.; De Schryver, F.C.; Viane, L.; Van der Auweraer, M.; Iwai, K.; Yamamoto, M.; Ichikawa, M.; Masuhara, H.; Maus, M.; Rettig, W. *J. Am. Chem. Soc.* **1996**, 118, 2892.

(63) Bilot, L.; Kawski, A. *Z. Naturforsch.* **1962**, 17a, 621.

(64) Grabowski, Z.R.; Rotkiewicz, K.; Siemiarczuk, A. *J. Lumin.* **1979**, 18/19, 420.

(65) Possible ^1L_b contribution of the H band to $\int \varepsilon d\nu_a$ can be neglected because almost all of its intensity is borrowed from ^{1}CT. But the absorption spectrum $\varepsilon(\nu_a)$ of the CT band has been corrected for the contributions of the A and B bands by subtraction of these bands fitted to Gaussian functions (Fig.8.1).

(66) Birks, J.B. *Photophysics of Aromatic Molecules*, Wiley-Interscience, New York, **1970**.

(67) Förster, T. *Fluoreszenz organischer Verbindungen*, Vanderhoeck and Ruprecht, Göttingen, **1951**.

(68) Schütz, M.; Schmidt, R. *J. Phys. Chem.* **1996**, 100, 2012.

(69) Fujii, T.; Suzuki, S.; Komatsu, S. *Chem. Phys. Lett.* **1978**, 57, 175.

(70) Ikeda, N.; Miyasaka, H.; Okada, T.; Mataga, N. *J. Am. Chem. Soc.* **1983**, 105, 5206.

(71) Blias, J.; Gauthier, M. *Opt. Spectr.* **1978**, 9, 529.

(72) López Arbeloa, T.; López Arbeloa, F.; López Arbeloa, I. *J. Luminesc.* 1996, 68, 149; López Arbeloa, T.; López Arbeloa, F.; Hernández Bartolomé, P.; López Arbeloa, I. *Chem. Phys.* **1992**, 160, 123.

(73) Kakitani, T.; Mataga, N. *J. Phys. Chem.* **1986**, 90, 993 and **1987**, 91, 6277.

(74) Strehmel, B.; Rettig, W. *J. Biomed. Opt.* **1996**, 1, 98.

(75) Strickler, S.J.; Berg, R.A. *J. Chem. Phys.* **1962**, 37, 814.

(76) At temperatures below -90° an additional minor time component with small amplitudes only in the very blue wing of the fluorescence band has been used for the global analysis of the exponential fits of all three compounds. Because I also needs this time constant it is not related to twisting. This very short component can be attributed to solvent friction effects, the relaxation times of which fall into the time-resolution of the experimental setup under highly viscous conditions at low temperatures (ch. 6).

(77) Rettig, W.; Fritz, R.; Braun, D. *J. Phys. Chem.* **1997**, A101, 6830.

(78) Zachariasse, K.; Grobys, M.; von der Haar, T.; Hebecker, A.; Il'ichev, Yu.V.; Jiang, Y.-B.; Morawski, O.; Kühnle, W. *J. Photochem. Photobiol. A: Chem.* **1996**, 102, 59.

(79) To account for the $n^3\nu_f^3$ dependence and the k_{nr} difference, the intrinsic lifetimes in EOE can be calculated by

$$\tau_0(CT) = \frac{1}{k_{nr}(EOE) + k_f(HEX)\frac{\nu_f(EOE)^3 n_f(EOE)^3}{\nu_f(HEX)^3 n_f(HEX)^3}} \text{ and } \tau_0(CTR) = \frac{1}{k_{nr}(EOE) + k_f(ACN)\frac{\nu_f(EOE)^3 n_f(EOE)^3}{\nu_f(ACN)^3 n_f(ACN)^3}}$$

In addition, if we suppose the radiative rates as temperature independent, the nonradiative rates in EOE at 160 K can be obtained from $k_{nr}(160K) = k_{nr}(298K) \cdot \tau_0(298K)^{-1} + \tau_0(160K)^{-1}$. For the **CT** species $k_{nr}(160K) = 5 \cdot 10^7$ s^{-1} and for the **CTR** species $k_{nr}(160K) = 3 \cdot 10^7$ s^{-1} turns out. These values are rather similar to the room temperature value of $k_{nr}(298K) = 10 \cdot 10^7$ s^{-1} and the corresponding activation barrier is then less than RT. The neglection of $k_{nr}(T)$ in eq. 3-17 is therefore justified (cf. $\tau_f^{-1}(298K) - \tau_f^{-1}(160K) = 500 \cdot 10^7$ s^{-1}).

(80) The dimethylamino group is calculated to be coplanar with the linked phenyl unit in S_0, S_1 and S_2. A negligible deviation from C_2 symmetry is found only for S_0, where the pyramidalization angle is not exactly planar (Fig. 8.3).

(81) Recently a paper (Verbouwe, W.; Viaene, L.; Auweraer, V.d.; DeSchryver, F.C.; Masuhara, H.; Pansu, R.; Faure, J. *J. Phys. Chem. A* **1997**, 101, 8157.) appeared, where donor-substituted biphenyls were also found to relax to a planar geometry, in spite of monoexponential emission decays within their time-resolution of 20 ps. Furthermore, by comparison with **I** and **II**, we may interpret the published photophysical properties (lower $k_f/n^3\nu_f^3$ and higher k_{nr} rates than for **I** and **II**) and bilinear solvatochromic ν_f plots (smaller μ_e) as emission from a S_1 state with more ^1L_b and less ^{1}ET character (category 1).

Chapter 4

Sub-picosecond Transient Absorption of D-A Biphenyls. Intramolecular Control of the Excited State Charge Transfer Processes by a Weak Electronic Coupling.

Abstract:
The photoinduced intramolecular charge transfer processes of three differently twisted 4-dimethylamino-4'-cyanobiphenyl derivatives (**I-III**) have been investigated using time-resolved transient absorption and gain spectroscopy in the sub-picosecond range. Independent of twist angle and solvent polarity, the kinetics and spectral evolutions after excitation clearly reveal a precursor-successor relationship for the electron transfer from a less emissive state of mixed ^1L_b/CT character to a highly emissive charge transfer (^1CT) state. Beside the occurence of dual fluorescence gain, two transient absorption bands for the ^1CT state and one for the precursor state (^1FC) are observed. All bands are assigned to electronic transitions and correlated for all solvents and compounds. The band intensities are discussed with solvent polarity and twist angle controlled mixing between the charge transfer state ^1CT and the higher lying ^1L_b and ^1L_a states. In acetonitrile, the transient spectra of the pretwisted donor-acceptor biphenyl **III**, in contrast to the planar **I** and **II**, can be approximated by the sum of cation and anion spectra of the subunits demonstrating decoupled moieties. The kinetics of the CT processes are not dominated by solvation dynamics alone. As an example, in acetonitrile (τ_l=0.2ps, τ_s<1 ps) the kinetics are slower than 2.5 ps. The involvement of a weak electronic coupling matrix element is favored as a source for the intramolecular control of the CT reactions. Furthermore, for the strongly twisted biphenyl derivative **III** a secondary intramolecular process to a more relaxed species (CTR) occurs after the initial CT step, in agreement with fluorescence studies.

Keywords: electron transfer, picosecond transient absorption, dual fluorescence, electronic coupling.

4.1 Introduction

Photoinduced intramolecular electron transfer (ET) processes in biaryl systems have been extensively studied by a large number of groups.[1-15] ET dynamics in solution are determined by an interplay of solvation and intramolecular modes leading to two borderline cases:

    A) solvent control   B) intramolecular control

In the case of an adiabatic ET reaction with a negligible activation barrier (case A), it is generally accepted that the ET rate is strongly correlated to the solvent relaxation time τ_s.[5,6,16-21] In particular, for the biaryl system 9,9' bianthryl (BA) in polar aprotic solvents, though no agreement with the longitudinal relaxation times τ_l, quantitative agreement for the ET rate with the inverse solvent relaxation time τ_s was observed mainly by Barbara and coworkers.[5,6,20] On the other hand, some molecular systems are known to deviate strongly from the correlation of the solvent dynamics with the observed ET rates (case B) and the adiabacity of these reactions is always a point of discussion. As an example, the photoinduced ET in 4-(9-Anthryl)-N,N-dimethylaniline (ADMA) has been explained by the superposition of a very fast ET in the nonadiabatic regime with an adiabatic solvent-controlled ET.[8] An ET reaction much slower than solvent relaxation times has experimentally been shown for 4-N,N-dimethylamino-benzonitrile (DMABN) in aprotic solvents.[22-24] Interestingly, new results[7] of BA in nitriles suggest that the torsional relaxation towards a more planar structure is slower than τ_s, which is in accordance with molecular dynamics calculations.[25] In the second group of ET reactions (case B) intramolecular dynamics, such as large-amplitude motions, crossing of a thermal barrier or transitions connected with weak coupling matrix elements are rate determining instead of the solvent control (case A).

Different models have been applied to explain the dependence of such ET kinetics on intramolecular and solvent parameters. Early models from Sumi, Nadler and Marcus[26,27] did not yet take into account intramolecular quantum mechanical high-frequency modes, which have subsequently been treated in different extensions of these models.[28-29] In contrast to the purely classical model, the extended models were capable to explain observed rapid rates in slowly relaxing solvents with a change from solvent to vibrational control. Moreover, the importance of intramolecular vibrational modes of ET reactions especially in the Marcus inverted region has been substantiated by a number of recent experiments.[29-31] Theoretical approaches including large structural relaxations have been developed within two-dimensional stochastic treatments.[32-34] The model of Nordio predicts that the ratio of viscous to dielectric relaxation rate of the system is a key parameter and can lead either to solvent or to viscosity control.[34] Maroncelli[35] and Hynes[33] often accentuate that τ_s is only a simplified value for a distribution of solvent relaxation times and that fast inertial components of solvent motions are able to accelerate ET reactions. However, kinetic investigations on a particular molecular system are seldomly paired with a detailed analysis of the electronic structure.

ET reactions mainly occur on the femtosecond (fs) and picosecond (ps) time scale, and the progressive development of ultrafast spectroscopic experiments now allows very precise measurements of the rates. Thus a number of recent ultrafast time-resolved studies on classical laser dyes[36-40] which have frequently been used as probe molecules of solvation before,[35] suggest that the observed photodynamics are insufficiently described by a simple one-state kinetic model associated with ET occurring on the time scale of solvation (case A). A more appropriate description for these dyes seems to be that the ET reaction occurs in a precursor-successor relationship of two (or more) distinct species with different spectral and electronic properties evolving independently of solvation (case B). However, the spectral changes during the ET process are not very pronounced such that controversial discussions can be found regarding the question whether the temporal change is due to fast solvation components[35,41] (case A) or due to multiple excited states[39,41] (case B). A detailed electronic characterization of the precursor state is frequently missing in the literature.

In order to differentiate between the two cases A) and B) and to determine the states involved in the ET, time-resolved experiments on molecular systems with known electronic structure are necessary. A powerful tool is time-resolved transient absorption because it allows probing both ET kinetics as well as the electronic structure of the excited singlet states concerned. As intramolecular ET model systems the series of differently twisted donor-acceptor biphenyls **I-III** is applied possessing a structure with obvious donor and acceptor moieties which are directly linked in a varying spatial arrangement.

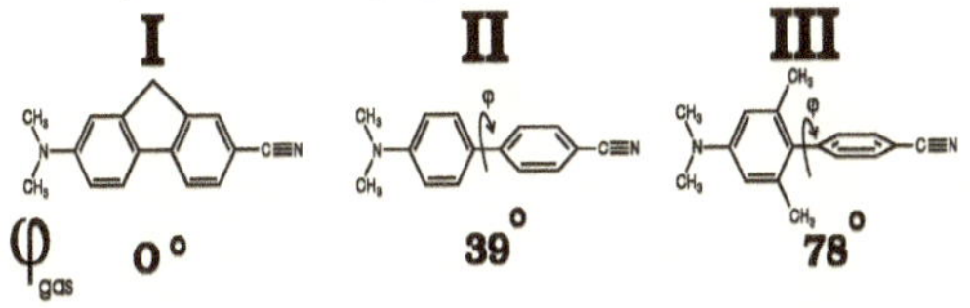

φ_{gas} **I** 0°   **II** 39°   **III** 78°

Scheme 4.1 *Molecular structures of donor-acceptor biphenyls investigated and corresponding ground state twist angles derived from AM1 calculations for the gas phase.*

From the previous investigations[3,4] in ch. 2 (and ch. 5) it is known that these biphenyl compounds possess a close lying ^1L_b-type state above the low lying intramolecular charge transfer S_1 state (^{1}CT). Fluorescence observed in the time range slower than 0.1 nanoseconds (ns) reveals only this intramolecular ^{1}CT state independently on excitation or emission wavelength as well as solvent and temperature (ch. 3 and 6).[1,2,4,9,10] Lahmani et al. recently reported that the charge transfer process in **II** is governed only by solvation[2] because the observed decay kinetics in alcohols occur on the time scale of solvent relaxation. But the resolution of their setup (30 ps) was too limited to observe an electron transfer that might be complete prior to solvent reorganization. With the sub-picosecond setup used here (time-resolution of 1 ps) we aimed to search for a possible initial relaxation.

The key question for the transient absorption experiments is, whether the initially populated state after vibrational relaxation, which will be termed as the (relaxed) Franck-Condon species **FC**, is of the same electronic nature as the ^{1}CT state or not. In other words, do we have to take into account only one state $S_1(2^1A)$ that changes its charge transfer chararacter gradually on the time scale of the solvent relaxation (case A) or are two distinct states involved (case B), the first one with locally excited character transforming into the second one with charge transfer character? In the one-state model, there should only be a continuous blue shift of the transient absorption bands on the time scale of solvation. However, if the two-state model is more appropriate, we should see different spectra of a first and a second species which are not necessarily transforming gradually on the time scale of the solvent relaxation (τ_s). Moreover, a decay and rise behaviour of the kinetics should be observable.

Ch. 4 Sub-ps Transient Absorption   55

4.2 Experimental

Technical details about the measurement of absorption and fluorescence spectra, lifetimes and quantum yields are reported in 3.2.[3,4,10]

4.2.1 Materials

The synthesis of the compounds **I-III** is described in 3.2.1.[4] The solvents n-hexane (HEX), diethylether (EOE) and acetonitrile (ACN) were of spectroscopic grade (Merck UVASOL). Triacetine (or glycerol triacetate, TAC), as purchased from Merck, has been washed with Na_2CO_3 and afterwards distilled twice under vacuum shortly before use. The optical density (OD) for all solutions was adjusted to OD=1.0 in a 1cm cuvette at the fixed pump energy of 33300 cm^{-1} (300 nm) yielding concentrations lower than 1.7×10^{-4} M (for **III** in triacetine and acetonitrile the optical density was slightly higher because of a too small transient signal under these conditions).

4.2.2 Picosecond Pump-Probe Experiments

The experimental set-up has been described in detail elsewhere.[42] Briefly, the laser system is based on a hybridly mode-locked dye laser (Coherent 702), synchronously pumped by a cw mode-locked Nd^{3+}/YAG laser (Coherent "ANTARES"), and is amplified in a three stage dye amplifier (Continuum PTA60) pumped by a regenerative amplifier (Continuum RGA 60). It delivers a typical output pulse of ≈1 ps duration, centered at 600 nm (16660 cm^{-1}) with an energy of 1.5 mJ at a 10 Hz repetition rate. The second harmonic of this pulse at 300 nm (33330 cm^{-1}) is used as pump pulse. The 600 nm pulse, after passing through an optical delay line (60 fs steps) is focused on a rotating quartz plate to generate a continuum of light which extends from 30000 cm^{-1} to 10000 cm^{-1}. This continuum is sent through the sample as the probe beam. Changing the position of the optical delay line allows observation of the transient spectra by use of an optical multichannel analyzer (spectral shape and optical density as a function of time) on a picosecond time-scale (sum of Excited State Absorption -ESA- and gain spectra). The angle between the polarizations of the probe and pump beams were adjusted to 54.7° (the "magic angle") to ensure kinetics free from reorientational effects.

4.2.3 Correction and Fitting of the Results

a/ Chirp correction of the transient spectra: Because of group velocity dispersion (GVD) and self-phase modulation (SPM), the spectral components of the probe beam do not all reach the sample at the same time. At short delays this effect, known as spectral "chirp", distorts the kinetics and deforms the recorded spectra. Using the optical Kerr effect in pure solvents, we measured this chirp from 10000 to 30000 cm^{-1}. All the spectra presented are corrected for this effect using a home-made algorithm.

b/ Fitting and modeling the kinetics: Kinetic modeling was based on the assumption that the population of excited molecules formed in the state directly reached after excitation (the Franck-Condon state) can be transformed into successive excited states until a final state is reached which decays to the ground state. If the number of possible excited states is N then the kinetic scheme can be described as follows:

$$S_0 \xrightarrow{h\nu} S_1 \xrightarrow{k_{12}} S_2 \xrightarrow{k_{2N}} \cdots S_N \xrightarrow{k_{N0}} S_0$$

where the S_1, S_i, S_N are the different excited states involved in the transformation process and the k_{ij} are the kinetic constants. Each state S_i has its own characteristic absorption and gain spectrum, with the molar absorption coefficient $\varepsilon_i(\nu)$ at wavenumber ν. If at time t_p, the population of excited state S_i is $P_i(t_p)$, the optical density OD (t_p,ν_n) at time t_p and wavenumber ν_n outside the $S_0 \rightarrow S_n$ absorption spectral range is given by eq. (1).

$$OD_{n,p}(t_p, \nu_n) \propto \sum_{i=1}^{i=N} P_i(t_p) \times \varepsilon_i(\nu_n) \qquad (4\text{-}1)$$

From the kinetic scheme described above, the populations P_i should follow the differential equation:

$$d/dt \begin{pmatrix} P_1 \\ P_2 \\ P_N \end{pmatrix} = (k_{ij}) \times \begin{pmatrix} P_1 \\ P_2 \\ P_N \end{pmatrix} + \begin{pmatrix} a\, I_{pump}(t) \\ 0 \\ 0 \\ 0 \end{pmatrix} \qquad (4\text{-}2)$$

From the measured optical density at different wavenumbers in the spectral range 10000-30000 cm^{-1}, the solution of the system of eqs. 4-1 and 4-2 by the „Runge-Kutta"-algorithm leads to the different kinetic constants k_{ij} and the absorption spectra $\varepsilon_i(n)$ of the states involved. The limited precision of the measurements enforced the number of states to be determined to a maximum of three (N=3). In the fitting procedure, the excitation pulse was modelled as a Gaussian shaped pulse the half-width of which was measured by the optical Kerr effect. All fits were performed by neglecting possible back reactions.

4.2.4 Quantum Chemical Calculation of the Transient Spectra

Transition energies (ν), oscillator strengths (f), symmetries (sym) and geometries of the first and second excited singlet states (S_1 and S_2) are obtained using the BFGS geometry optimization algorithm in combination with a complete active space configuration interaction calculation (CAS-CI) including the 5 highest occupied and 5 lowest virtual molecular orbitals.

Typically 10^5 configurations were generated from which around 100 are finally selected for the CI calculation. Increasing the orbital window up to 18 active MOs (10^8 configurations) does not significantly alter the spectral results. The semiempirical Austin Model 1 (AM1) Hamiltonian contained in the AMPAC 5.0 program was employed on a HP735 workstation.[43] The most characteristical differences of the geometries in S_1 and S_2 are the interannular twist angle (φ) and bond length (r_b) and the dihedral angle between the methyl groups of the dimethylamino substituent and the linked benzene unit (α_N), which is a measure for pyramidalization of the nitrogen. The twist angle of the dimethylamino group is 0° in S_0-S_2. These optimized structural data (S_0 geometries as obtained elsewhere[3,4]) for **I-III** are in increasing order (cf. Fig. 8.3):

$S_0(1^1A)$: φ = 0°, 39°, 78°; r_b = 1.452 Å, 1.459 Å, 1.467 Å; α_N = 18°, 18°, 19°

$S_1(2^1A = {}^1CT)$: φ = 0°, 15°, 46°; r_b = 1.407 Å, 1.413 Å, 1.427 Å; α_N = 2.0°, 1.6°, 1.4°

$S_2(1^1B = {}^1L_b)$: φ = 0°, 18°, 52°; r_b = 1.443 Å, 1.459 Å, 1.450 Å; α_N = 2.7°, 1.1°, 15°

4.3 Results

4.3.1 Steady-State Spectra and Expectations

The absorption spectra of the biphenyls **I-III** have been analyzed in detail in ch. 2.[3] Briefly, the four apparent bands denoted as CT, L_a, B and C in Fig. 4.1 mainly correspond to electronic transitions from the ground state $S_0(1^1A)$ into the terminal states ^{1}CT and ^1L_b (CT band, ^1L_b belongs to the H band in ch.2)[3], ^1L_a (L_a band), ^1B_b (B band) and ^1B_a, ^1C_b and ^1C_a (C band, the ^1C_a-type state is not listed in ch.2)[3] as specified in ch.2.[3] The lowest excited singlet state S_1 of the **I-III** investigated is identified as an intramolecular ^{1}CT state transferring charge from the dimethylanilino (donor) to the benzonitrile (acceptor) unit in any relaxed solvent surrounding.[3,4,9,10] The large solvatochromic shift of the fluorescence (Fig. 4.1) confirms the strong electron transfer character of S_1. It can be assigned to the long-axis polarized 2^1A

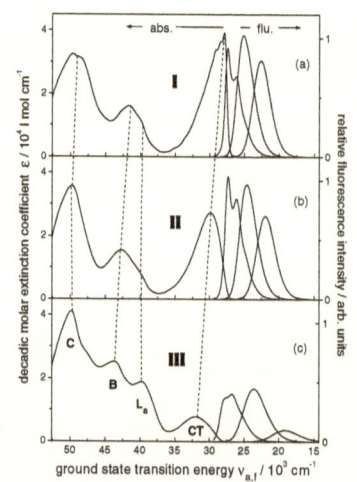

Fig. 4.1 *Steady-State absorption spectra in n-hexane and fluorescence spectra in n-hexane, diethylether and acetonitrile (from the left to the right) of* **I-III**.

state if we consider all three molecules to belong to the same symmetry point group C_2, disregarding the disturbing effect of the fluorene bridging in I, with the central biphenyl bond as the symmetry C_2-axis.

The short radiative lifetimes determined for I and II (see Tab. 4.1) demonstrate that the lowest singlet transition is of strongly allowed character and corroborates the above assignment for S_1 to the 2^1A state.[3,4] The polarity dependent and long radiative lifetimes for III in dipolar solvents have previously been interpreted by an adiabatic photoreaction towards a more decoupled geometry between the dimethylanilino and benzonitrile moiety induced by an increased twist of the interannular bond (ch.2,6 and 8).[4,9,10]

The fluorescence and absorption results (Fig. 4.1) lead us to predict the following main differences for the transient absorption experiments of the pretwisted biphenyl III as compared to the more planar biphenyls I and II:

1.) The stronger solvent stabilization of ^1CT in III should yield a larger blue shift of the main transient band as compared to I and II.

2.) An additional structural relaxation within the ^1CT state should be observed for III in polar solvents associated with different spectral features

3.) Similar to the L_a band observed only in the ground state absorption spectrum of III, the decoupling of the $^1CT(2^1A)$ and $^1L_a(3^1A)$ state in III may lead to an additional transient band not visible in I and II.

Tab. 4.1 Fluorescence radiative lifetimes $\tau_r = \tau_f/\Phi_f$ of I-III at room temperature in solvents of increasing polarity.

solvent	I	II	III
n-hexane (HEX)	2.4 ns	1.7 ns	3.8 ns
diethylether (EOE)	2.0 ns	2.1 ns	8.3 ns
triacetine (TAC)	2.4 ns	2.4 ns	10.2 ns[a]
acetonitrile (ACN)	2.6 ns	2.8 ns	36.5 ns

[a] average lifetime $<\tau_f>=(\alpha_1\tau_1+\alpha_2\tau_2)/(\alpha_1+\alpha_2)$ of the two long lifetime components at $\lambda_f = 410$ nm.

4.3.2 Transient Absorption Measurements

For these studies we used four different aprotic solvents in order to vary the polarity from non-polar n-hexane (HEX) to strongly polar acetonitrile (ACN), together with diethylether (EOE) of intermediate polarity. The viscosity of these three solvents is small and practically equivalent associated with comparable solvent relaxation times τ_S (see below Tab. 4.3). Triacetine (TAC) was additionally chosen because of its high viscosity at room temperature associated with intermediate polarity. This solvent should slow down the possible processes related to viscosity and allow a better understanding of the photophysical behaviour. The time-delays 1 ps, 5 ps and 40 ps are selected for the presented spectra in all fast relaxing solvents (HEX, EOE, ACN). This facilitates the comparison of spectral evolutions in different solvents. As a compromise, the kinetics can readily be followed by the time traces or contour plots. Let us now treat in detail these different experimental conditions.

4.3.2.1 N-Hexane Solutions

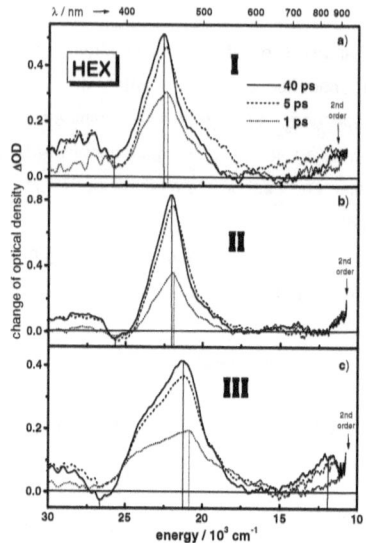

Fig. 4.2 *Transient absorption spectra in n-hexane (HEX) for* **I-III** *at 1ps (·····), 5ps (- - -) and 40 ps (——) time delays between pump and probe beam. Full vertical lines at maxima and minima indicate the transitions in the relaxed spectra.*

As shown in Fig. 4.2 the behaviour of the three compounds in n-hexane (HEX) is quite comparable. We mainly observed a broad absorption band around 22000 cm^{-1} (450 nm) and a possible gain band near 26000 cm^{-1} (385 nm) for **I** and **II** and 27000 cm^{-1} for **III** without strong spectral evolution as a function of time. Only a small narrowing of the 22000 cm^{-1} absorption band with time is slightly apparent. In compound **III**, also a weak absorption band near 11800 cm^{-1} (850 nm) appears at later times which is not observed, within experimental error, in compounds **I** and **II** (the positive signal in **I** and **II** at 11000 cm^{-1} is due to 2nd order of diffraction of the 22000 cm^{-1} absorption band). Whether or not the depression appearing in the spectra at 26-27000 cm^{-1} corresponds to gain is difficult to ascertain since negative optical density is weak at this energy due to the overlap with the strong absorption. This depression is more pronounced at later times and could be explained by a possible second gain species which is stronger than that of a first species. This would indicate that the transition to the ground state 1^{1}A of a second species is of more allowed character. From the steady state and nanosecond (ns) time-resolved fluorescence results, the relaxed S$_1$ state in HEX is indeed located around 27000 cm^{-1} and highly allowed (τ_r < 3.8 ns). Typical kinetics at the transient absorption maxima of **I-III** are shown in

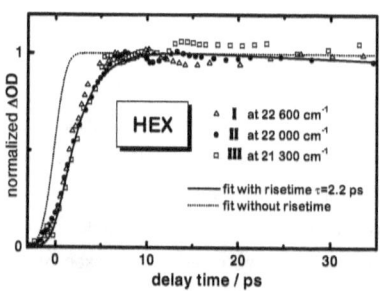

Fig. 4.3 *Normalized kinetics for* **I-III** *of the main transient absorption band in n-hexane. Simulation curves using a one state model (·····) and a two state model (——) indicate the existence of two excited state species, the precursor one being less visible than the successor one.*

Fig. 4.3. It is not possible to fit these kinetics within experimental error taking into account the presence of a single excited state only, formed during the excitation pulse and decaying on a ns time scale. The risetime observed (2.2 ps) is clearly longer than the limit imposed by the pulse duration as shown in Fig. 4.3. The presence of two excited species is therefore indicated. The initially populated species is denoted as the Franck-Condon species **FC**. (Note, that species are printed in bold in order to be distinguished from electronic states. In this respect, two species can belong to one state, but not vice versa.) In 2.2 ps a second excited species (charge transfer species **CT**) with stronger absorption signal is formed decaying on a ns time scale (1.2-1.4 ns). But on account of the fact that (only) in this solvent the spectral changes are very weak and a small narrowing is observed, at this point we cannot rule out the alternative explanation of vibrational cooling. Delayed absorption and band narrowing are typical features of such a process and may take place on the ps time scale.[44]

4.3.2.2 Acetonitrile Solutions

In the very polar solvent ACN with considerable stabilization of the ^{1}CT state (Fig. 4.1), strong evolution of the transient spectra (Fig. 4.4a-c) is observed for **I-III** as compared to HEX solutions. The behaviour of compounds **I** and **II** is practically identical in ACN as shown in Fig. 4.4 a and b. Following the excitation pulse profile in time, an absorption band appears near 24000 cm^{-1} (425 nm). As time evolves, in contrast to the behaviour in HEX, the intensity of this band begins to decrease after two picoseconds (Fig. 4.5a) accompanied by the simultaneous rise of a second absorption transition in the red part of the spectrum near 13-14000 cm^{-1} and a broad gain band at 21000 cm^{-1} (475 nm). The energy and strong intensity of the gain band in **I** and **II** (Fig. 4.4 a,b) is consistent with the static fluorescence maximum (Fig. 4.1) and the large radiative rates (Tab. 4.1) for the ^{1}CT state of **I** and **II**. Due to the obvious charge transfer properties of the successor species, it is termed as **CT**, whereas the less solvent stabilized precursor species is related to **FC** as above.

The kinetics at energies corresponding to the high-energy and low-energy absorption bands, respectively, are identical for **I** and **II** and are therefore shown in Fig. 4.5a only for **II**. They can obviously be rationalised by a precursor-successor relationship between the initially excited species **FC** only associated with the high-energy absorption band and the product excited species **CT** characterized by the gain band (21000 cm^{-1}) and the broad absorption at 14000 cm^{-1}. The **CT** species is formed from **FC** with a rate constant of (2.5 ps)$^{-1}$ and decays on a nanosecond time scale (Fig. 4.5a). A contribution from the **CT** species to the absorption around 24000 cm^{-1} has to be considered since the signal does not vanish after 2.5 ps.

Ch. 4 Sub-ps Transient Absorption

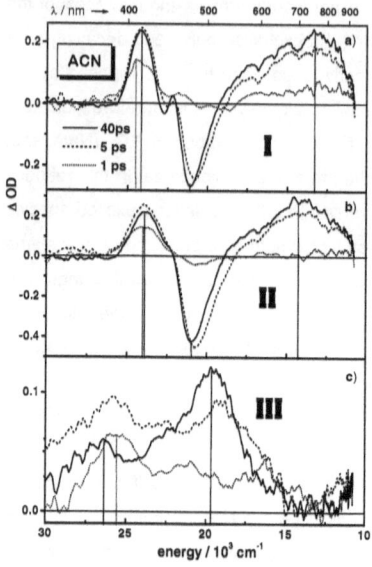

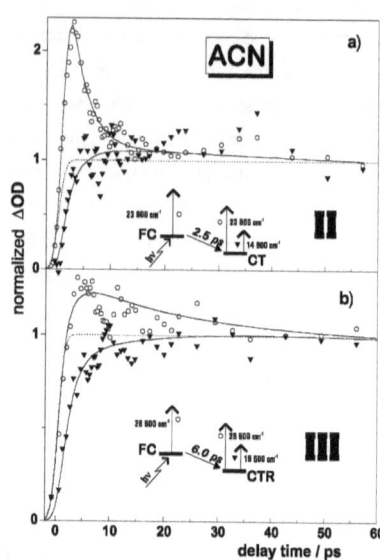

Fig. 4.4 *Transient absorption spectra in acetonitrile (ACN) for **I-III** at 1ps (·····), 5ps (- - -) and 40 ps (———) time delays between pump and probe beam. Full vertical lines at maxima and minima indicate the transitions in the relaxed spectra and the dotted vertical lines indicate the transitions in the first picoseconds.*

Fig. 4.5 *Normalized kinetics for **II** and **III** of the high energy and the low energy transient absorption bands in acetonitrile. The continuous lines are the fits obtained using the kinetic model of two states (FC and CT/CTR) shown as an inset. The dotted lines are the fits using a one-state model taking into account the pump pulse profile.*

The behaviour of **III** is completely different from that of compound **I** and **II** in this solvent (Fig. 4.4c and 4.5b). First of all, a gain band is absent for **III** in ACN which reflects the long radiative lifetimes of the S_1 state as compared to **I** and **II**. The excitation process is followed by the appearance of a weak absorption at 25600 cm^{-1} (390 nm) part of which decays with an approximate time constant of 6 ps. This transient band is strongly shifted to the higher energy region with respect to the corresponding bands of **I** and **II** indicating the relaxation from **FC** to a charge transfer species (denoted as **CTR** in Fig. 4.5b instead of **CT** in Fig. 4.5a) which is more relaxed than in **I** and **II**. A second and more intense absorption band near 19500 cm^{-1} (515 nm) appears with the same time constant as the high energy band decreases demonstrating their mother-daughter relation. The decay to the ground state occurs on a nanosecond time scale (7.6 ns).

Ch. 4 Sub-ps Transient Absorption

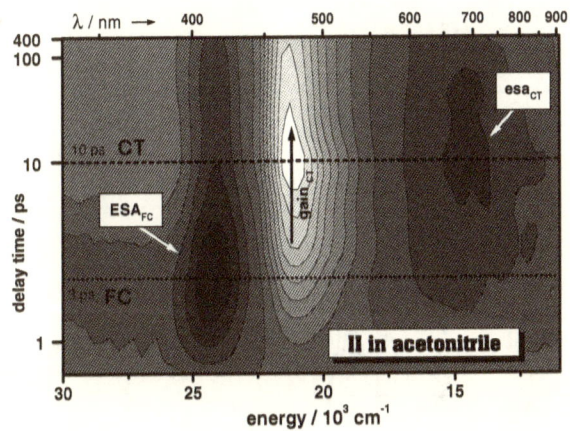

Fig. 4.6 *Contour plot of the transient absorption spectra of* **II** *in acetonitrile in the energy range of 30000 to 11000 cm^{-1} and in the delay time range of 0 to 400 ps. Broken lines indicate the time delays when the spectral characteristics of the* **FC** *and* **CT** *species, respectively, are best revealed.*

The spectral behaviour and existence of two species is well illustrated on the contour-map of **II** in ACN (Fig. 4.6). Here, black colour signifies positive absorption and white colour negative absorption or gain, respectively. The first species, the Franck-Condon excited species **FC**, is represented by one absorption region (**ESA$_{FC}$**) around 24000 cm^{-1}. With increasing time, part of this absorption decays quickly and a gain "valley" (**g$_{CT}$**) appears. Its temporal maximum exactly coincides with the low-energy absorption "mountain" (**esa$_{CT}$**). There, the maximum population of the **CT** species is reached.

4.3.2.3 Diethylether Solutions

In this solvent (EOE) of intermediate polarity between HEX and ACN, we expected intermediate behaviour. Because of almost equal features of **I** and **II** in HEX and ACN as well as in TAC (see 4.3.2.4), we did not investigate **I** in EOE supposing identical behaviour in this solvent, too. The spectral shapes of compound **II** are practically the same as in HEX (see Fig. 4.7a) with the transitions at similar energies. However, the assignment of the depression observed at 26000 cm^{-1} (385 nm) in HEX (Fig. 4.2b) to a gain band is more clearly demonstrated by the strong negative signal in EOE after 4 ps (Figure 7a). Furthermore, a significant temporal evolution of the bandshape is observed: the absorption band near

Ch. 4 Sub-ps Transient Absorption

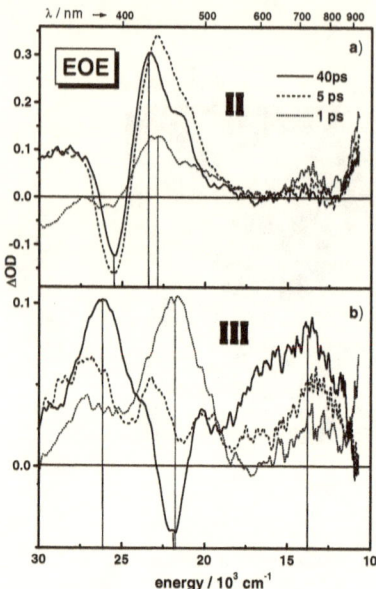

Fig. 4.7 *Transient absorption spectra in diethylether (EOE) for **II** and **III** at 1ps (·····), 5ps (- - -) and 40 ps (———) time delays between pump and probe beam. Full lines at maxima and minima indicate the transitions in the relaxed spectra and the dotted lines indicate the transitions in the first picoseconds of the Franck Condon species (**FC**).*

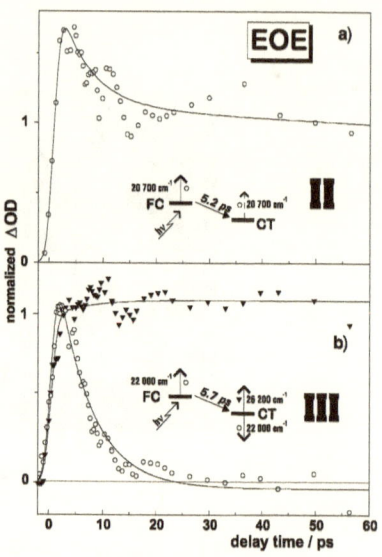

Fig. 4.8 a) *Normalized kinetics of the transient absorption band at 20 700 cm^{-1} for **II** in diethylether. b) Normalized kinetics of the rise of the high energy absorption for **III** at 26 200 cm^{-1} and the transformation of the Franck Condon absorption band into the gain band at 22 000 cm^{-1}. For a) and b) the applied kinetic schemes of two states (**FC** and **CT**) and the related simulation curves are also shown.*

23000 cm^{-1} (435 nm) narrows, shifts to the blue with a long wavelength shoulder similar but more pronounced as in HEX, and at the same time the gain band slightly shifts to the red.

Typical kinetics for **II** at an energy where the contribution of the **FC** species is relatively high are shown in Fig. 4.8a. The fit (τ = 5.2 ps) is based on a precursor (**FC**) - successor (**CT**) model with the spectral characteristics of both species being rather similar.

For compound **III** (Fig. 4.7b), the spectral changes are more evident. Just after excitation two maxima appear: one weak maximum near 28000 cm^{-1} and a stronger one near 21900 cm^{-1}. A minimum appears at about of 25000 cm^{-1}, between these two maxima. From comparison to the HEX spectra we can assign this minimum to a weak gain band of the **FC** species which is overlapped with a strong absorption band. As time evolves, the maximum at 21900 cm^{-1} transforms into a gain band with considerable negative optical density. At intermediate time delays (5 ps), the spectrum exhibits dual fluorescence gain of **FC** and **CT** (Figure 7). In the red part of the spectrum, a weak absorption band exists around 13800 cm^{-1} (750 nm), the kinetics of which could not be extracted due to a low signal-to-noise ratio. Within the first 40 ps, the kinetic behaviour of **III** can be described by the same model as above using a time constant of 5.7 ps (Fig. 4.8b). However, a blue shift of the low-energy absorption may indicate a further relaxation process.

4.3.2.4 Triacetine Solutions

In this solvent (TAC) of intermediate polarity but high viscosity, we expected a similar spectral behaviour as in EOE for all compounds. On the other hand, the slow solvation process should lengthen the observation time of the primary excited **FC** species.

In Fig. 4.9, typical transient spectra are shown for **I-III** in TAC. They are similar concerning the spectral shapes, but differ in time evolution, energy position and relative absorbance. All spectra of the initially populated species **FC** (3 ps) possess a strong absorption band (23-24000 cm^{-1} for **I** and **II**, 25000 cm^{-1} for **III**) modulated in **I** and **II** by a fluorescence gain band at 25400 cm^{-1} not far from the position of the gain (25500 cm^{-1} and 25900 cm^{-1}) and steady-state fluorescence band (26900 cm^{-1} and 26600 cm^{-1}) in HEX. Let us term this strong absorption band **ESA$_{FC}$** and the corresponding gain band **g$_{FC}$**. In **III**, only a depression in the **ESA$_{FC}$** band is apparent at 25000 cm^{-1}, but it is too weak for a firm assignment to a gain band. Thus we have to note that, comparing **III** with **I** and **II**, the gain **g$_{FC}$** is much stronger in **I** and **II**. At this short time delay, there is no indication for a low-energy absorption band neither for **I** and **II** nor for **III**. With proceeding time delay, **g$_{FC}$** weakens and a new fluorescence gain appears for all compounds at 23400 cm^{-1} (427 nm) in **I**, 23000 cm^{-1} (435 nm) in **II** and 22100 cm^{-1} (452 nm) in **III** which we assign to the gain **g$_{CT}$** of the charge transfer species. It is very important to stress that there is no gradual shift with time of the initial gain bands from around 25400 cm^{-1} (**g$_{FC}$**) to the position of the band **g$_{CT}$**, but a weakening of the gain **g$_{FC}$** accompanied by a growing of the **g$_{CT}$** gain. This means that all donor-acceptor biphenyls studied show dual fluorescence (here observed as dual gain) with precursor-successor relationship within a few picoseconds. The second gain band **g$_{CT}$** of **I** and **II** establishes exactly at the location of the steady-state emission at 23600 cm^{-1} (424 nm) for **I** and 23000 cm^{-1} (434 nm) for **II**.

In contrast to I and II, the centre of the gain g_{CT} in III after 22ps at 22100 cm^{-1} (452 nm) does not coincide exactly with the steady-state emission maximum at 20700 cm^{-1} (484 nm). But on a longer time scale (200 ps), the gain band g_{CT} of III broadens to the red and then its centre coincides with the steady-state emission maximum. The broadening of the gain band in III for longer times is not the only difference to I and II. III also exhibits an additional weak low-energy absorption band around 15-17000 cm^{-1} which can be assigned to the transition **esa$_{CT}$**, not seen in I and II. Similar to the behaviour in diethylether, this band seems to blue-shift with time, accompanied by the broadening of the gain g_{CT}. In Fig. 4.10, the kinetics at the energies of the gain bands g_{CT} for I-III are shown. To illustrate the confidence of the fits and to

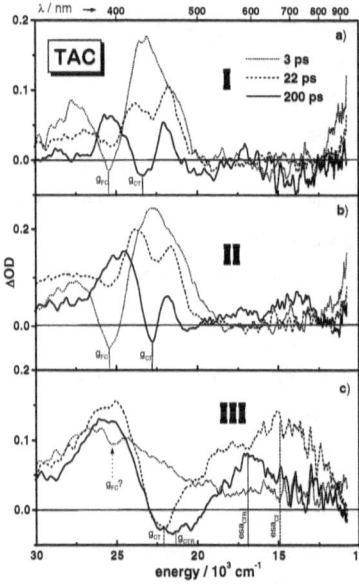

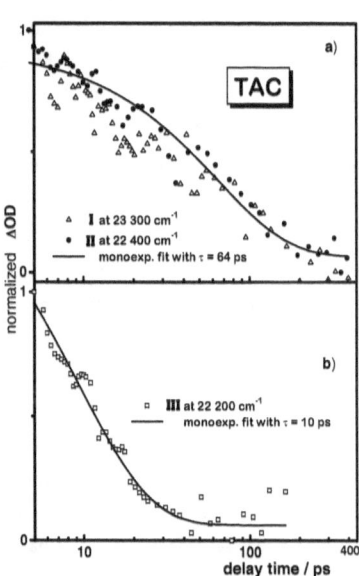

Fig. 4.9 Transient absorption spectra in triacetine (TAC) for I-III at 3ps (·····), 22ps (- - -) and 200 ps (——) time delays between pump and probe beam. For I and II dotted and full vertical lines at minima indicate the positions of the Franck Condon (g_{FC}) and charge transfer gain bands (g_{CT}). For III the Franck Condon gain is less developed (depression at 25 000 cm^{-1}). Broken and full vertical lines for III indicate the gain and low energy transient absorption bands for the **CT** species and the more relaxed **CTR** species, respectively.

Fig. 4.10 Normalized kinetics for I-III in triacetine of the transformation from the transient absorption bands of the **FC** species into the gain bands of the **CT** species plotted on a logarithmic time scale. Monoexponential decay simulation curves are plotted for II and III. The corresponding lifetimes are 69 (±20) ps for I, 64 (±15) ps for II, and 10 (±2) ps for III.

compare the different time scales of the kinetics in I and II with III, the time axis is on a logarithmic scale. For I and II, the kinetics of the CT formation are quite comparable (69 ps for I and 64 ps for II), but are clearly slower than for III (10ps).

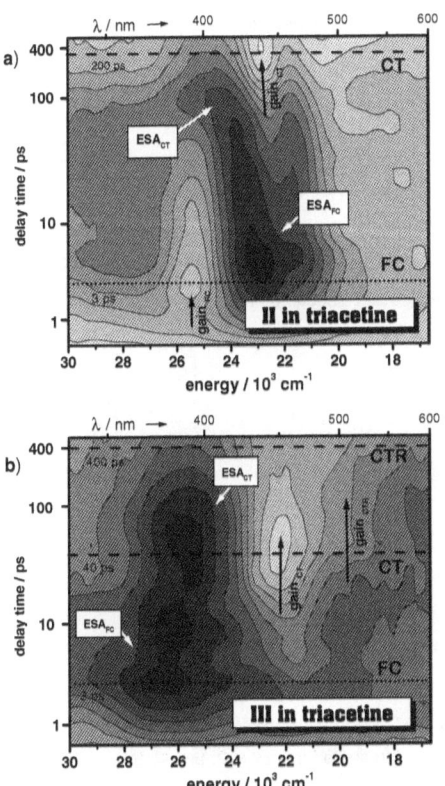

Fig. 4.11 Contourplots of the transient absorption spectra of II and III in triacetine in the energy range of 30000 to 16000 cm^{-1} and in the delay time range of a nonlinear time scale from 0 to 450 ps. Broken lines indicate the time delays when the spectral characteristics of the FC, CT and CTR species, respectively, are best revealed.

The time evolution of the spectra can vividly be followed on the contour maps of II and III in Fig. 4.11a and b. For II, the maximum population of the FC species is characterized by a strong absorption region (ESA_{FC}) and a gain valley (g_{FC}) at short time delays whereas for III only an absorption mountain of FC is well pronounced. The appearance of the gain valley (g_{CT}) at longer time delays indicates the population of the CT species in both maps. In accord with the gain time traces in Fig. 4.10, the gain valley of CT (i.e. maximum population of CT) is deepest at delay time 200 ps for II but already at 40 ps for III. The gain g_{CT} of the flexible biphenyl II exhibits a slight anomalous blue shift with increasing time which might be due to an overlapping red shifted absorption band of the CT species. However, the gain in III clearly broadens with time to the low-energy side indicating an additional relaxation to a more relaxed species CTR observable only in III.

4.4 Discussion

The discussion is divided into three main parts. In the first part, the observed transient transitions are assigned to electronic states and their polarity and twist angle dependence is interpreted. In the second section, evidences for structural relaxations are revealed and the last discussion section deals with the dependence of the charge transfer rates on solvent and molecular structure.

4.4.1 Excited State Absorption and Gain Bands

Five bands can be correlated for all compounds: Two gain bands, one of the precursur species **FC** (g_{FC}) and the other of the successor species **CT** or **CTR** (g_{CT} or g_{CTR}), and three transient absorption bands, the first one belonging to **FC** (**ESA$_{FC}$**) and the high and low energy successor bands associated with the **CT** or **CTR** species (**ESA$_{CT/CTR}$**) and **esa$_{CT/CTR}$**). The transient transitions from the **CT** and **CTR** species undoubtedly originate from the $^1CT(2^1A)$ state as evidenced by the solvatochromism of the gain bands. But the origin of the **FC** species is under discussion.

4.4.1.1 Assignments of the Absorbing and Emitting Species to Two States

In a first approximation, the experimental energy (ν_{exp}) of the excited state absorption $S_x \rightarrow S_n$ should resemble the difference ($\Delta\nu_{ss}$) of the photon energy for the ground state absorption $S_0 \rightarrow S_n$ and the emission $S_0 \leftarrow S_x$ process. Comparing the **ESA$_{FC}$** and **ESA$_{CT}$** transition energies in HEX (ν_{exp}=21-22000 cm^{-1}) with the energy differences between the absorption (Fig. 4.1) and gain bands ($\Delta\nu_{ss} = \nu_{abs}(S_0 \rightarrow S_n)-g_{FC/CT}$), a good correlation is only found for transitions into the C band ($\Delta\nu_{ss} \approx 23000$ cm^{-1}), since the alternative assignment of transitions into the B band ($\Delta\nu_{ss} \approx 16000$ cm^{-1}) is connected with a too low energy difference. The additional comparison with the quantum chemical (AM1/CI) calculations of the $S_1(^1CT) \rightarrow S_n$, as well as $S_2(^1L_b) \rightarrow S_n$ transitions, presented in Tab. 4.2, proves this assignment. In perfect agreement with experiment, the four ground state absorptions[3] $S_0 \rightarrow {}^1B_b$ (B band) and $S_0 \rightarrow {}^1B_a, {}^1C_a, {}^1C_b$ (C band) are calculated to be very intense ($f_{calc} \geq 0.20$; not shown), whereas from the excited states only the transitions $S_1(^1CT) \rightarrow {}^1C_a$ and $S_2(^1L_b) \rightarrow {}^1C_b$ carry large oscillator strengths ($f_{calc}(^1C_a, {}^1C_b) \geq 10 \times f_{calc}(^1B_a, {}^1B_b)$) and have suitable energy to account for the **ESA$_{CT}$** and **ESA$_{FC}$** bands. Further, the energy of the **esa$_{CT}$** band ($\nu_{exp} \approx 12000$ cm^{-1}) likewise is in nice agreement with both the energy difference between the gain g_{CT} and the L_a absorption band ($\Delta\nu_{ss} \approx 13000$ cm^{-1}) and, on the other hand, with the calculated transition energy for the $S_1(^1CT) \rightarrow {}^1L_a$ transition (Table 2). Up to $\nu_{calc} = 30000$ cm^{-1}, only long-axis polarized excited state transitions (symmetry transitions $^1A \rightarrow {}^1A$, $^1B \rightarrow {}^1B$ within C$_2$ point group) are found to be intense (cf. f_{calc} of $^1B \rightarrow {}^1A$ or $^1A \rightarrow {}^1B$ transitions are less than 0.03), which is exemplified in Tab. 4.2 by

Tab. 4.2 Assignments of Experimentally Observed Transient Bands (in n-hexane) to Calculated $S_1(^1CT) \rightarrow S_n$ and $S_2(^1L_b) \rightarrow S_n$ Transitions. The Calculated Twist Angles φ of I-III in S_1/S_2 are 0°/0°, 15°/32° and 46°/52°.

band	v_{exp} [10^3 cm^{-1}]			v_{calc} [10^3 cm^{-1}]			f_{calc}[a]			transition	sym
I-III	I	II	III	I	II	III	I	II	III	I-III	I-III
ESA$_{FC}$	22.5	21.9	20.9	24.0	22.4	20.2	0.28	0.28	0.22	$^1L_b \rightarrow {}^1C_b$	$^1B \rightarrow {}^1B$
ESA$_{CT}$	22.7	22.1	21.4	23.3	24.7	23.4	1.02	1.26	0.43	$^1CT \rightarrow {}^1C_a$	$^1A \rightarrow {}^1A$
esa$_{CT}$	-	-	11.8	7.8[b]	8.7[b]	10.3[b]	0.07[b]	0.11[b]	0.19[b]	$^1CT \rightarrow {}^1L_a$	$^1A \rightarrow {}^1A$
g$_{FC}$	-25.9°	-26.1°	-26.6°	-29.1	-32.0	-30.6	-0.02	-0.03	-0.02	$^1L_b \rightarrow S_0$	$^1B \rightarrow {}^1A$
g$_{CT}$	-25.9	-26.1	-26.6	-25.6	-27.2	-28.7	-0.57	-0.76	-0.58	$^1CT \rightarrow S_0$	$^1A \rightarrow {}^1A$

[a] The oscillator strength of all other calculated transitions up to 35000 cm^{-1} are less than 0.05 (except the $S_1(^1CT) \rightarrow {}^1B_a$ transition of I and II with $v_{calc} \approx 19000$ cm^{-1} and f_{calc}=0.15; cf. III: v_{calc} = 17000 cm^{-1} and f_{calc}=0.03).
[b] Due to strong coupling between the close lying ^1L_a states 3^1A and 4^1A (see ref. 3), v_{calc} corresponds to the more intense ^1L_a transition and the sum of both oscillator strength (less geometry dependent than single values) is given as f_{calc}.
[c] position of dips in early spectra

the low oscillator strength of the $S_0 \leftarrow S_2(^1L_b)$ emission transition. Hence, the short-axis polarization of the $S_2(^1L_b) \rightarrow {}^1L_a$ and the absence of another low-energy S_n state with ^1B symmetry explains that a low-energy transition is not found for the S_2 state, which in turn parallels the absence of a low-energy band **esa$_{FC}$** for the experimental precursor species **FC** in any solvent. It seems therefore very probable that the **FC** species is associated with the calculated $S_2(^1L_b)$ state. Two further correlations from Tab. 4.2 confirm this tentative assignment: (i) The calculated oscillator strengths for the proposed precursor state $S_2(^1FC/^1L_b)$ are less than those of the final $S_1(^1CT)$ state explaining that the kinetics of the main transient absorption band (Fig. 4.3) are observed as risetimes due to a transformation from a less allowed $S_2(^1FC)$ state to a more allowed $S_1(^1CT)$ state; (ii) the slight blue shift of the main absorption band with time (Fig. 4.2) is reproduced by the calculated higher energy of the final **ESA$_{CT}$** than of the **ESA$_{FC}$** band.

4.4.1.2 Effects of Solvent Polarity and Twist Angle

In order to obtain the experimental transition energies of the excited state absorption and gain bands, the individual spectra of **FC**, **CT** and/or **CTR** were extracted and decomposed into the absorption and gain contributions using Gaussian functions. The energetical positions of the lower states or species, respectively, relative to the ground state can be derived from the gain positions. In such cases where gain is not observed, e.g for **FC** in ACN, the energy is extrapolated taking into account the solvatochromic shift of the observable gain bands. Using all these informations, the solvent dependent transitions are plotted in an energy diagram (Scheme 4.2).

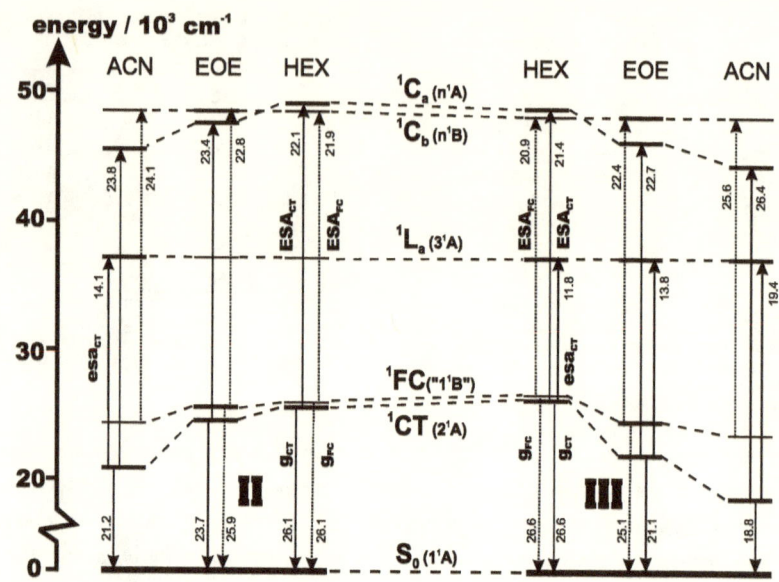

Scheme 4.2 *Energy level diagram of* **II** *and* **III** *in n-hexane, diethylether and acetonitrile (HEX, EOE and ACN) as derived from the transient absorption spectra. The energy of the ground state has been taken as zero. The energy levels of which the absolute energetical position could not be determined directly by the experiments are drawn with thinner lines. Transitions from the ^1FC state 1^1B (**ESA$_{FC}$** and **g$_{FC}$**) are indicated by dotted arrows and the transitions from the ^1CT state 2^1A (**ESA$_{CT}$**, **esa$_{CT}$** and **g$_{CT}$**) by full arrows.*

Gain bands. Gain bands (g_{FC} and g_{CT}) are produced by stimulated emission due to the probe beam. It has been shown[45] for organic dyes that gain is absent for radiative lifetimes longer than 10 ns, this property being inversely related to the fluorescence transition dipole moment M_f which controls the stimulated emission efficiency. Accordingly, gain of the ^1CT state is observed in all cases except for **III** in ACN, which has a radiative lifetime of more than 30 ns (Tab. 4.1) connected with a comparatively small M_f value (2.3 D).[4,9] On the other hand, the negative optical density (ΔOD) of the gain band g_{CT} of **I** and **II** increases relative to the positive ΔOD of the **ESA$_{CT}$** band with solvent polarity leading to a very intense gain band in ACN (Fig. 4.4ab). This is consistent with our previous observation that the transition dipole moment M_f of the $S_0(1^1A) \leftarrow S_1(^1CT)$ fluorescence increases with polarity in **I** and **II**[4,9] and, moreover, with the observation of Lahmani et al.[2] that the picosecond time decay of the integrated fluorescence intensity is not as fast as it would have been expected, if the solvation reduces M_f of the ^1CT emission. It can therefore be supposed that the electronic interaction of $S_1(^1CT)$ with the less allowed $S_2(^1L_b)$ state decreases with solvent polarity due to an increasing S_1-S_2 energy gap ($\Delta E^1_{CT \cdot ^1L_b}$). Furthermore, gain should be located at energies matching the steady state fluorescence maxima. In fact, the gain bands observed in HEX agree with the fluorescence maxima. However, only the gain g_{CT} exhibits a solvatochromism similar to the steady-state fluorescence (Fig. 4.1) confirming that the state responsible for g_{CT} is the ^{1}CT state.[1,4,9,10] Nevertheless, the slight red shift of the gain g_{FC} with increasing polarity reveals a nonnegligible ET character of the **FC** species, too.

ESA bands. The two ESA bands observed previously in 9,9'-bianthryl (BA) were assigned to a locally excited (^{1}LE) and ^{1}CT state.[7,11,12] The insensitivity of the ^{1}CT→S$_n$ transition energy on solvent polarity has been explained by a similar, near complete ET character in the lower and upper state. In contrast to BA, an increase of the energy for the **ESA**$_{CT}$ and **esa**$_{CT}$ bands with increasing solvent polarity is observed for **I-III** due to transitions from the strongly stabilized ^{1}CT state into the upper states ^{1}C$_a$ and ^{1}L$_a$ with smaller ET character. As expected from the larger dipole moment of the ^{1}CT state for **III** in comparison to **I** and **II**,[4] the increase of transition energy of **ESA**$_{CT}$ with polarity is strongest for **III**. We also note in scheme 4.2 that even the **ESA**$_{FC}$ bands are blue-shifted with increasing solvent polarity. In accord with the solvatochromism of the gain bands g_{FC}, this supports that the **FC** species possesses significant ET character, too. Regarding the twist angle effect in the case of weak solvent stabilization (HEX and EOE), the comparison of **I** and **II** with **III** reveals that the experimental as well as the calculated **ESA** transitions of the more twisted **III** are redshifted and less allowed as compared to those of the planar **I** and quasi planar **II**. The agreement between the experiments and calculations gives further confidence to the calculation results in Tab 4.2. Needless to say that in the highly polar solvent ACN, the stronger stabilization of the ^{1}CT state for **III** yields a higher **ESA**$_{CT}$ transition energy than for **I** and **II**.

***esa*$_{CT}$ bands.** The **esa**$_{CT}$ band is observed in nonpolar to medium polar solvents only for the highly pretwisted compound **III**. This reflects the observation of a maximum for the ground state absorption $S_0(1^1A) \rightarrow {}^1L_a(3^1A)$ only for **III**. Both long-axis polarized transitions to the 3^1A state, either from 1^1A or 2^1A, get more allowed with increasing twist angle φ as supported by the calculated oscillator strength of **esa**$_{CT}$ in Tab. 4.2. As discussed in section 4.2, the average twist angle $<\varphi>$ of **III** increases with solvent polarity, which may also explain the increasing intensity ratio $\Delta OD(\mathbf{esa}_{CT})/\Delta OD(\mathbf{ESA}_{CT})$ from HEX to ACN. In polar ACN, however, **I** and **II** also exhibit the **esa**$_{CT}$ band. One possibility is that the strong increase of the energy gap between the ^{1}CT(2^1A) and the $^1L_a(3^1A)$ state (ΔE_{La-CT} increase of 5000 cm^{-1} for **II** in ACN as compared to **II** in HEX) leads to a decoupling of these states, similar to the twist effect in **III**. Both decoupling effects, i.e. twisting of the benzene units and increased energy gap may enhance the intensity of the transient ^{1}CT(2^1A)→$^1L_a(3^1A)$ relative to the ^{1}CT(2^1A)→$^1C_a(n^1A)$ transition. On the other hand, the **esa**$_{CT}$ band of the planar **I** and **II** is located at lower energy than that of **III** in the solvent interaction case (Fig. 4.4) and gas phase case (v_{calc} in Tab. 4.2).

We can therefore not exclude the alternative possibility that the **esa$_{CT}$** band is simply concealed in the energy region (< 11000 cm^{-1}) where the detection sensitivity is too low. In any case, experiment and calculations agree that the intensity ratio ΔOD(**esa$_{CT}$**)/ΔOD(**ESA$_{CT}$**) is largest for the twisted **III**.

Mixed nature of ^{1}FC. The difference in the transition energies between the **FC** and **CT/CTR** species, especially of the gain bands (same final S$_0$ state) in combination with the precursor-successor relationship of their occurrence, the separate time scales in TAC and the additional low energy band **esa$_{CT}$** only for the ^{1}CT state leads to the conclusion that beside the ^{1}CT state, the **FC** species can be treated as a second excited state. Moreover, the appearance of the gain **g$_{FC}$** seems to necessitate a low energy gap to the highly allowed ^{1}CT(2^{1}A)$\rightarrow$S$_0$(1^{1}A) transition **g$_{CT}$** indicating that the ^{1}FC$\rightarrow$S$_0$ transition needs a vibronic coupling mechanism which in turn points to a different symmetry (^{1}B in C$_2$ point group) of the zeroth-order ^{1}FC state. These observations suggest that the ^{1}FC state is of substantial ^{1}L$_b$ character mixed with the ^{1}L$_a$-type ^{1}CT state.

Thus, the excited state electron transfer in the investigated donor-acceptor biphenyls is accompanied with more spectral features than in the extensively studied biaryls 9,9' Bianthryl (BA)[7,11,12] or 4-(9-Anthryl)-N,N-dimethylaniline (ADMA)[15] and derivatives as well as in the recently studied 9-dimethylanilino-phenanthrene (9DPhen)[13,14] biaryl compounds. For BA in triacetine, the ^{1}LE and ^{1}CT transient absorption bands could be separated,[12] but in ADMA,[15] 9DPhen[13] and related molecules, the time-resolved transient absorption spectra exhibit only a gradual shift with increasing time delay but not such clear spectral differences between the species before and after the electron transfer step as it is shown for **I-III**.

4.4.2 Structural Relaxation

In a previous fluorescence analysis it was concluded that ^{1}CT emission occurs from planar conformers of **II**, whereas the average twist angle <φ> in the ^{1}CT state of **III** increases with solvent polarity.[4,10] Because the twist angles in ^{1}CT are different to the ground state, photoinduced structural relaxations of **II** and **III** take place which are investigated in this section.

Independent of the solvent, the flexible biphenyl compound **II** shows rather the same spectral and kinetic excited state behaviour (fluorescence and transient absorption) like the planar model compound **I**. These results demonstrate that **II** performs a photoinduced relaxation towards planarity. For the pretwisted biphenyl **III**, similar behavior is observed in HEX. Thus, in agreement with the fluorescence studies, the **CT** species of **II** and **III** can be attributed to a more planar structure than in the ground state.

In the medium polar solvent EOE, the spectrum of the primary precursor species **FC** for **III** (1 ps in Fig. 4.7b) is quite similar to the one observed in HEX solvent (cf. Fig. 4.2c) and the spectrum of the successor species **CT** (40 ps in Fig. 4.7b), formed in 5.7 ps, is similar, while slightly shifted to the blue, to the spectrum observed for compounds I and II in ACN (see Fig. 4.4a,b). It seems then that compound III in this solvent of intermediate polarity exhibits a behaviour which starts, just after excitation, with the HEX situation and evolves to the situation observed in compound II and I in ACN for longer times. This indicates that in EOE the primary structural relaxation for III is the same as for II, namely to the more planar species **CT**. Hence for II and III, the **FC** species feeds the ^{1}CT state with a structure more planar than in the ground state. This also agrees with the observation that the relaxation times of II and III are similar in EOE (5-6 ps) but significantly different in ACN, where a different species **CTR** is populated (see below).

Using time-resolved fluorescence on a nanosecond timescale, an equilibrium between the initially populated more planar species **CT** and a more relaxed (conformationally and by solvent interaction) charge transfer species **CTR** has been deduced for III in EOE (Ch. 6).[4,9] Quantum chemical calculations in 3.3.4.3 supported the idea of a double minimum twist potential in the ^{1}CT state of III in EOE.[4,9,10] Searching for this additional slow relaxation in the transient spectra of III in the medium polar solvents EOE and TAC, we can note that (i) the low-energy absorption band **esa**$_{CT}$ in EOE (near 13800 cm^{-1} in Fig. 4.7b) reveals a slight blue shift from 5 ps to 40 ps and (ii) the **esa**$_{CT}$ band in TAC shifts to the blue with time, accompanied by a clear broadening of the gain **g**$_{CT}$ (Fig. 4.9c or 4.11b). These transient absorption results may also point to this later relaxation process from **CT** to **CTR**.

In analogy to the fluorescence results,[4,9] the **CTR** species should prevail for III in ACN. The observations (Fig. 4.4 and 4.5) that (i) the kinetics of III (6 ps) in ACN are distinctly slower than for I and II (2.5 ps), (ii) the absorption strongly shifts to the blue and (iii) that a gain band is absent, are indeed indications that a more stabilized charge transfer species **CTR** is populated in III than in I and II in ACN. In agreement with our fluorescence results, the more relaxed **CTR** species can then be characterized by a larger dipole moment and a strongly reduced fluorescence transition moment (M_f) as compared to the **CT** species.[4,9] Such properties of the ^{1}CT state are conceivable with a twisted structure of **CTR**. In this case, the transient spectrum of **CTR** should resemble a superposition of the dimethylaniline cation (DMA$^+$) and benzonitrile anion (BN$^-$) spectra.

DMA$^+$ and BN$^-$ ion spectra are characterized by $S_1 \rightarrow S_n$ absorption bands at 21300 cm^{-1} (470 nm)[46,47] and 24700 cm^{-1} (405 nm)[46,48] and both do not possess a gain contribution. The compounds I and II in ACN, however, show quite different spectra (Fig. 4.4ab) as compared to the sum of DMA$^+$ and BN$^-$ spectra.[46-48] This supports the idea of the strongly delocalized character of ^{1}CT in I and II connected with a planar structure. For III in ACN (Fig. 4.4c), the low-energy band **esa**$_{CT}$ is blue shifted (in comparison to less polar solvents) nearly to the energy that maches the DMA cation band and the high energy band **ESA**$_{CT}$ is close to the transient absorption of the BN anion. The close relation to DMA$^+$ and BN$^-$ indicates a nearly complete ET similar to BA and is therefore indeed understandable by the photoinduced strongly twisted structure of **III** in ACN as opposed to **III** in less polar solvents as well as compared to **I** and **II** in any solvent.

The photoinduced electronic and structural relaxation processes discussed above are summarized in Scheme 4.3. The rate determining step of the transition from the Franck-Condon region (^{1}FC) to the ^{1}CT state can be attributed to the ^{1}FC→^{1}CT interconversion (process 3 in scheme 4.3) which is associated with a partial electron transfer. The vibrational relaxation (process 1), on the one hand, can be excluded since both kinetic species **FC** and **CT** possess completely different spectral features and, on the other hand, the intramolecur rotation towards a more planar geometry (process 2) can be excluded, because the model compound **I** which is restricted to planarity, shows the same kinetic behavior as the flexible biphenyl **II**. In the following section, the possible rate limiting factors of the ^{1}FC→^{1}CT process are discussed.

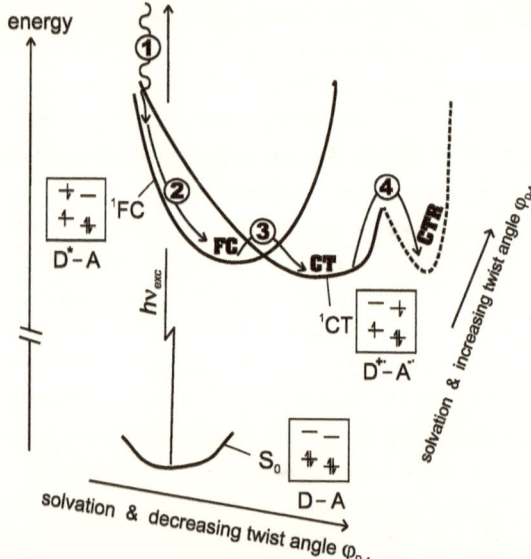

Scheme 4.3 *Schematic representation of the main relaxation pathes for the D-A biphenyls **II** and **III** after excitation into a manifold of vibronic states: (1) vibrational relaxation (2) initial solvation and geometry relaxation towards a more planar structure than in S_0 (3) ET interconversion (4) additional structural relaxation to a more twisted conformation than in S_0 occuring only for **III**. The characteristic one-electron configurations (within a composite-molecule description of localized HOMO and LUMO orbitals of the linked D and A phenyl subunits) of the involved states S_0 (1^1A), ^1FC (1^1B) and ^1CT (2^1A) are shown in boxes.*

4.4.3 Kinetics of State Interconversion $^1FC \rightarrow {}^1CT$

The observed relaxation times, $\tau_{FC \rightarrow CT}$, together with the relevant solvent parameters are listed in Tab. 4.3. They reveal that the $^1FC \rightarrow {}^1CT$ processes of **I-III** are slower than solvation dynamics (the longitudinal solvent relaxation times τ_l or the related solvation times τ_s) in the fast relaxing solvents EOE and ACN. Let us first test the possibility to describe the interconversion with a low barrier ET reaction in the purely adiabatic regime.

A numerical treatment of the experimental relaxation times within (semi-)classical ET models[57,58] needs parameters, such as reorganization energy for the solvent (λ_s) and low frequency intramolecular motions (λ_i'), the reaction free energy ΔG and activation barrier $\Delta G^\#$, as well as τ_l, which can only very crudely be determined. Best results can be obtained for ACN, since only for this solvent a reliable mean value (narrow distribution) of τ_l is known from various references.[17-19,53,56] The fact that $\tau_{FC \rightarrow CT} > 10 \times \tau_l > \tau_s$ in ACN (Tab. 4.3) may allow us to apply transition-state ET theory which requires a quasi-equilibrium between precursor and transition state.[20,21] In the case of a purely adiabatic and solvent diffusion controlled ET reaction, the theoretical rate constant can then be calculated by[57]

$$\tau_{et}^{-1} = \frac{1}{\tau_l} \cdot \sqrt{\frac{\lambda_s}{16 k T \pi}} \cdot e^{-\frac{\Delta G^\#}{k_B T}} \qquad (4\text{-}3)$$

This eq. is derived from transition state theory including Kramers diffusion treatment for a relaxing Debye solvent.[57,59] It is practically equivalent to the result of the classical two-dimensional Sumi-Marcus model[26] (adiabatic limit without high frequency vibrations) in the case of $\lambda_s > \lambda_i'$ and $\tau_{et} > \tau_s$ being appropiate for the conditions in the polar solvent ACN. The solvent reorganization energies λ_s may be determined from

$$\lambda_s = \frac{\Delta \mu^2}{a^3} \left(\frac{\varepsilon - 1}{2\varepsilon + 1} - \frac{n^2 - 1}{2n^2 + 1} \right) \qquad (4\text{-}4)$$

using the dipole moment differences $\Delta \mu = \mu_{CT} - \mu_{S0}$ and Onsager cavity radius $a = 6$ Å as obtained elsewhere[4] from Lippert-Mataga plots[51] (assuming $\mu_{S0} \approx \mu_{FC}$). Such estimates of λ_s for **I-III** in

Tab. 4.3 Parameters of the Solvents Used at 298 K and Time Constants $\tau_{FC \rightarrow CT}$ of the Charge Transfer Processes Observed by Time-Resolved Transient Absorption for **I-III**

solvent	polarity		viscosity	solvent relaxation		$\tau_{FC \rightarrow CT}$ [ps]		
	ε_r [a]	$f_2(\varepsilon_r)$ [b]	η [mPa s] [a]	τ_l [ps]	τ_s [ps]	**I**	**II**	**III**
HEX	1.9	0.09	0.30	-	-	(2.2 ± 0.2)	(2.2 ± 0.2)	(2.2 ± 0.2)
EOE	4.3	0.25	0.22	< 1.6[d]	< 2.4[d]	-	(5.2 ± 0.4)	(5.7 ± 0.5)[g]
TAC	7.0[c]	0.30	≈15	≈150[e]	≈690[e]	(69 ± 17)	(64 ± 12)	(10 ± 1)
ACN	35.9	0.39	0.33	0.2[f]	0.4 - 0.9[f]	(2.5 ± 0.1)	(2.5 ± 0.1)	(6.0 ± 0.5)[g]

[a] ref. 49, 50 [b] $f_2(\varepsilon_r) = (\varepsilon_r - 1)/(2\varepsilon_r + 1) - (n^2 - 1)/(4n^2 + 2)$ from ref. 51 [c] ref 52,53 [d] estimated from ethylacetate;[17,54] in EOE τ_s and τ_l are expected to be considerably faster because the solvent viscosity of ethylacetate ($\eta = 0.36$ mPa s) is larger than that of EOE ($\eta = 0.22$ mPa s) [e] τ_l is calculated in ref 55 by inverse Laplace transformation of Davidson-Cole data from ref. 52, τ_s is taken from ref 52, 56 [f] ref 17-19, 53, 56 [g] In **III** the CT process is superimposed by the structural relaxation to the **CTR** species.

ACN are (1800 ± 700) cm^{-1}, (2000 ± 700) cm^{-1} and (3600 ± 1500) cm^{-1}. The range of the resulting theoretical ET times of **I-III** according to eq. 4-3 for a low barrier $\Delta G^{\#}$ = (170 ± 90) cm^{-1} that has been given for the ET reaction in BA, are (1.1 ± 1) ps, (1.1 ± 1) ps, (0.8 ± 1) ps. For comparison, using λ_s=3100 cm^{-1} as reported for BA,[60] eq. 4-3 yields 1 ps which is close to the experimental ET time τ_{et} in BA obtained by Barbara et. al.[5] (0.75 ps) and close to the solvation time τ_s of ACN determined by time-dependent Stokes shift experiments.[17-19,53,56] By contrast, the observed rates $\tau_{FC \to CT}$ of **I-III** (Tab. 4.3) are even slower than the limits of the theoretical τ_{et} and also slower than τ_s. This indicates that the observed interconversion process $^1FC \to {}^1CT$ can not be described by a low-barrier crossing and solvent-controlled adiabatic ET reaction as proposed previously.[2] Thus, we have to ask for another rate limiting factor of the $^1FC \to {}^1CT$ process.

Within general ET theory, three reasons can be responsible for such a rate limitation: (i) a large amplitude motion (ii) a thermal activation barrier $\Delta G^{\#}$ larger than $k_B T$ or (iii) a weak electronic coupling matrix element $V^1_{FC \cdot {}^1CT}$ in the order of 100 cm^{-1} or less.

(i) The possibility of a rate limitation due to a intramolecular rotation around the interannular bond can be excluded since **II** and the planar fluorene **I** possess equal $\tau_{FC \to CT}$ (2.2 ps in HEX and 2.5 ps in ACN). However, the observed lower rate of **III** in ACN is consistent with a temporal superposition of the large amplitude motion towards a more twisted structure (see scheme 4.3). A strong influence of another motion including rotation of the dimethylaminogroup seems unlikely because the interconversion process is connected with a transfer of charge from the dimethylanilino donor to the benzonitrile acceptor group.

(ii) To conclude about the involvement of a thermal barrier, temperature dependent measurements are usually necessary. Rough estimates using classical ET theories (eq. 4-3) and solvent continuum models (eq. 4-4) indicate barriers at most in the order of $k_B T$ which could hardly explain such low rates. Furthermore, the influence of the solvent polarity on $\tau_{FC \to CT}$ of **I-III** (Tab. 4.3) is weak in comparison to known intramolecular ET reactions.[23,40] This also contradicts a significant activation barrier $\Delta G^{\#}$.

(iii) Following the explanations in the preceding discussion chapters, where the two kinetic species **FC** and **CT** are assigned to two states of different zeroth-order symmetry (scheme 3), an involvement of a weak electronic coupling matrix element $V^1_{FC \cdot {}^1CT}$ is favored. In order to test this idea, the one-electron matrix elements $<{}^1\Phi_{Lb}|H|{}^1\Phi_{CT}>$ between the leading delocalized SCF electron configurations of $S_2(^1L_b)$ and $S_1(^1CT)$ are calculated using the ZINDO/S method[61] with a CI expansion of 122 single excitations. The obtained $<{}^1\Phi_{Lb}|H|{}^1\Phi_{CT}>$ amounts to 185 cm^{-1} for **III** at the twist angle φ=70° and 30 cm^{-1} for **II** at φ=20°. By introducing an increasing self-consistent reaction field to simulate the solvent polarity influence,

$\langle{}^1\Phi_{Lb}|H|{}^1\Phi_{CT}\rangle$ of **III** continuously decreases down to 35 cm^{-1} (ACN conditions) due to an increasing energy gap between the leading configurations ^1L_b and ^1CT, while $\langle{}^1\Phi_{Lb}|H|{}^1\Phi_{CT}\rangle$ of **II** remains constant around (30 ± 5) cm^{-1}. Such low coupling matrix elements are indeed consistent with the observed slow interconversion rates from ^1FC to ^1CT. Moreover, the calculated larger coupling element of **III** could explain why the experimental time constant $\tau_{FC \to CT}$ in TAC is faster for **III**. Note that in the fast relaxing solvents EOE and ACN, $\tau_{FC \to CT}$ is slower for **III** than for **II** because the structural relaxation of **III** superimposes the electronic relaxation which is kinetically separated only in the viscous solvent TAC. The fact that the processes $^1FC \to {}^1CT$ of **I-III** in TAC are considerably faster than solvent dynamics ($\tau_{FC \to CT} < \tau_s$ and τ_l)[52,56] and likewise faster than the ET of BA in TAC (τ_{et} = 750 ps at 285 K)[5] further supports the view of a not purely adiabatic and not solvent-controlled reaction.

On the other hand, the comparison of the relaxation times $\tau_{FC \to CT}$ in the almost equipolar solvents EOE and TAC (Tab. 4.3) reveals that the rates are slowed down by the solvent viscosity. This is, of course, due to a nonnegligible influence of solvent diffusion processes which mainly control adiabatic ET reactions ($\tau_{et} \propto \tau_s$). Therefore, it appears reasonable to characterize the observed charge transfer processes as ET reactions in between the adiabatic and nonadiabatic limit. For such border cases, eq. 4-3 has been modified to[28,57]

$$\tau_{et}^{-1} = \frac{4\pi^2}{h} \cdot V^2 \cdot \frac{1}{\sqrt{4\pi\lambda_s k_B T}} \cdot e^{-\frac{\Delta G^*}{k_B T}} \cdot \frac{1}{1 + \frac{8\pi^2 V^2}{h\lambda_s} \cdot \tau_l} \qquad (4\text{-}5)$$

which considers the electronic coupling matrix element V as well as the solvent diffusional influence of τ_l. Using the same parameter for **II** in ACN as above and τ_{et} = 2.5 ps, a sensible value for $V^1_{FC\text{-}{}^1CT}$ of 70 cm^{-1} is obtained by eq. 4-5. However, it is important to stress, that the theoretical data derived for **I-III** in ACN, which rest on the continuum model, are just indicative and not expected to result in quantitative agreement. In addition, the fact that the observed rate constants in TAC are faster than the upper limit (= τ_l^{-1}) of eq. 4-5 indicates a possible acceleration by fast inertial solvent modes or intramolecular vibrational dynamics. Numerical simulations of time-dependent population distributions are necessary which include high-frequency intramolecular and inertial solvent modes within a quantum mechanical description[28-31,33] and which treat non-equilibrium solvation explicitly within a stochastic treatment.[8,26,27,32-34] However, because of strong spectral overlap and shift of bands, the present experimental results do not allow such a sophisticated analysis.

4.5 Conclusions

The intramolecular charge transfer processes in the investigated donor-acceptor biphenyls after photoexcitation can not be described using a single state which gradually relaxes from the Franck-Condon excited region to the charge transfer region on a non-perturbed potential surface governed only by solvation dynamics. On the time scale of a few picoseconds, two excited state species (**FC** and **CT**) with different transient absorption and stimulated emission (gain) spectra have been observed linked by a precursor-successor relationship. The different spectral features, in particular the occurence of dual gain bands, can also not be due to vibrational cooling, but are well consistent with an electron transfer from a primary populated $\pi\pi^*$ state ^{1}FC in the Franck-Condon region to the charge transfer state ^{1}CT that has been analyzed previously.[1,3,4,9] The short-lived ^{1}FC state is characterized by the ^1L_b-type $S_2(1^1B)$ state which interacts with the close lying ^1L_a-type $S_1(^1CT)$ state. The $S_2(1^1B)$ nature of the precursor state ^{1}FC follows from three main conclusions:

(i) In the less polar solvents, where the energy gap between the ^{1}FC ("1^1B") and lower lying ^{1}CT (2^1A) state is small, the gain of the primary species (**FC**) is weaker than that of the product species (**CT**). For high solvent polarity (acetonitrile, larger energy gap) there is even an absence of gain for the first species (**FC**) which can be explained by the reduced mixing with the allowed $1^1A \leftarrow {}^1CT(2^1A)$ transition.

(ii) By comparison with the steady state absorption as well as with quantum chemical (AM1/CI) calculations, the observed excited state absorption bands can be assigned to the first-order transitions $S_2(1^1B) \rightarrow {}^1C_b(n^1B)$, $S_1(2^1A) \rightarrow {}^1C_a(n^1A)$ and $S_1(2^1A) \rightarrow {}^1L_a(3^1A)$. Consequently, the observed gain bands are correlated with the stimulated fluorescence transitions $S_0(1^1A) \leftarrow S_2(1^1B)$ and $S_0(1^1A) \leftarrow S_1(2^1A)$. However, because of the high oscillator strength necessary for the observation of gain for this ^{1}FC state together with the polarity induced shifts of absorption and gain bands, ^{1}FC cannot be regarded as a pure ^1L_b state but better as a state with mixed $^1L_b/^1CT$ ($1^1B/2^1A$) character.

(iii) The observed rates for the charge transfer process $^1FC \rightarrow {}^1CT$ of **I-III** in the fast relaxing aprotic solvents are considerably slower than the characteristic solvent relaxation times τ_s and also slower as compared to the ET rates observed for other typical biaryl compounds, such as 9,9'-bianthryl,[5,7] ADMA,[8] 9DPhen[14] and their derivatives. This suggests that beside conformational dynamics of **III**, the different symmetry of the precursor and successor states, which holds only in the biphenyl series, is responsible for an additional rate reduction through a small electronic interaction matrix element.

Due to the strong spectral changes on the time scale of ET, **I-III** can be regarded as illustrative model systems for an intramolecular electron transfer of case B, i.e. ET between two distinct states with considerably less solvent than intramolecular control. Our observations are in line with recent reinvestigations on the well-known laser dyes DCM (4-dicyanomethylene-2-methyl-6-p-dimethylaminostyryl-4H-pyran),[36,37] C153 (Coumarin 153)[38,39] and DCS (4-Dimethylamino-4'-Cyanostilbene)[62,63] in polar solvents. On a femto- to subpicosecond-timescale, the ^{1}CT state is fed by different precursor states, the electronic structure of which could, however, not be characterized in such a detail as for **I-III**.

Concerning the conformational dynamics of **II** and **III**, this study gives additional support to the conclusions derived previously from time-resolved emission in ch. 3[4,9,10] and allows to go into more detail regarding the conformer species (**CT** and **CTR**) involved:

(i) A relaxation towards planar conformers (**CT**) is observed for **II** in all solvents and **III** in nonpolar n-hexane as demonstrated by the similarity of the spectral and kinetic behaviour as the compound **I** which is restricted to planarity.

(ii) In medium polar solvents, the initial relaxation of **III** towards planarity is followed by a viscosity controlled rearrangement towards a more twisted structure (**CTR**) as shown by further spectral evolutions after the ET process.

(iii) In strongly polar acetonitrile, the acceleration of the conformational relaxation to the **CTR** species leads to a temporal superposition with the ET interconversion ^{1}FC→^{1}CT such that the observable overall charge transfer rate is reduced and **CTR** is directly observed as the product species. The final transient spectrum resembles the sum of dimethylaniline (D) cation and benzonitrile (A) anion spectra which indicates the strong decoupling of the D and A submoieties in the **CTR** species of **III** most probably by a nearly perpendicular twist.

All spectral, kinetic and conformational aspects can be summarized in a 3 excited state species (2 electronic states) model (scheme3) for the investigated donor-acceptor biphenyls: **FC** („^{1}L$_b$/^{1}CT") → **CT** (^{1}CT) → **CTR** (^{1}CT). After photoinduced population of the ^{1}FC state the reaction to the ^{1}CT state takes place in all biphenyls **I-III** connected with a charge separation between the phenyl units and a more planar geometry than in the ground state. In addition, the subsequent adiabatic photoreaction to a more relaxed (more twisted and more polar) **CTR** species occurs only for the pretwisted compound **III** in dipolar solvents.

4.6 References

(1) Lahmani, F.; Breheret, E.; Zehnacker-Rentien, A; Amatore, C. and Jutand, A., *J. Photochem. Photobiol. A: Chem.* **1993**, *70, 39*.

(2) Lahmani, F.; Breheret, Benoist d'Azy, O.; Zehnacker-Rentien,A; Delouis, J.F. *J. Photochem. Photobiol. A: Chem.* **1995**, 89, 191.

(3) Maus, M., Rettig, W. *Chem. Phys.* **1997**, 218, 151.

(4) Maus, M., Rettig, W., Bonafoux, D.; Lapouyade, R. *submitted for publication*.

(5) (a) Kahlow, M.A.; Kang. T.J.; Barbara, P.F. *J. Phys. Chem.* **1987**, 91, 6452. (b) Barbara, P.F.; Jarzeba, W. *Acc. Chem. Res.* **1988**, 21, 195.

(6) Kang, T.J.; Jarzeba, W.; Barbara, P.F.; Fonseca, T. *Chem. Phys.* **1990**, 149, 81.

(7) Mataga, N.; Nishikawa, S.; Okada, T. *Chem. Phys. Lett.* **1996**, 257, 327.

(8) (a) Tominaga, K.; Walker, G.C.; Jarzeba, W.; Barbara, P.F. *J. Phys. Chem.* **1991**, 95, 10475. (b) Tominaga, K.; Walker, G.C.;Kang, T. J.; Barbara, P.F. *J. Phys. Chem.* **1991**, 95, 10485.

(9) Maus, M.; Rettig, W.; Lapouyade, R. *J. Inf. Rec.* **1996**, 22, 451.

(10) Rettig, W.; Maus, M.; Lapouyade, R. *Ber. Bunsenges. Phys. Chem.* **1996**,100,2091.

(11) Mataga, N.; Yao, H.;Okada, T; Rettig, W. *J. Phys. Chem.* **1989**, 93,3383.

(12) Lueck, H.; Windsor, M.W.; Rettig, W.; *J. Phys. Chem.* **1990**, 94, 4550.

(13) Onkelinx, A.; De Schryver, F.C.; Viane, L.; Van der Auweraer, M.; Iwai, K.; Yamamoto, M.; Ichikawa, M.; Masuhara, H.; Maus, M.; Rettig, W. *J. Am. Chem. Soc.* **1996**, 118, 2892.

(14) Onkelinx, A.; Schweitzer, G.; De Schryver, F.C.; Miyasaka, H.; Van der Auweraer, M.; Asahi, T.; Masuhara, H.; Fukumura, H.; Yashima, A.; Iwai, K. *J. Phys. Chem. A* **1997**, 101, 5054.

(15) Okada, T.; Mataga, N.; Baumann, W.; Siemarczuk, A. *J. Phys. Chem.* **1987**, 91, 4490.

(16) Weaver, M.J., *Chem. Rev.* **1992**, 92, 463.

(17) Rossky, P.J.; Simon, J.D. *Nature* **1994**, *370, 263*.

(18) Weaver, M.J.; McManis, G.E. *III.Acc. Chem. Res.* **1990**, 23, 294.

(19) Maroncelli, M.; MacInnis, J.; Fleming, G.R. *Science* **1988**, 243, 1674.

(20) Barbara, P.F.; Meyer, T.J.; Ratner, M.A. *J. Phys. Chem.* **1996**, 100, 13148.

(21) Yoshihara, K.; Tominaga, K.; Nagasawa, Y. *Bull. Chem. Soc. Jpn.* **1995**, 68, 696.

(22) Takagi, Y; Sumitani, M.; Yoshihara, K. *Rev. Sci. Instrum.* **1981**, 52, 1003.

(23) Schuddeboom, W.; Jonker, S.A.; Warman, J.M.; Leinhos, U.; Kühnle, W.; Zachariasse, K.A. *J. Phys. Chem.* **1992**, 96, 10809.

(24) Rotkiewicz, K.; Grabowski, Z.R.; Jasny, J. *Chem. Phys. Lett.* **1975**, 34, 55.

(25) Smith, M.J.; Krogh-Jespersen, K.; Levy, R.M. *Chem. Phys.* **1993**, 171,97.

(26) Sumi, H.; Marcus, R.A. *J. Chem. Phys.* **1986**, 84, 4894.

(27) Nadler, W.; Marcus, R.A.; *J. Chem. Phys.* **1987**, 86, 563.

(28) (a) Jortner, J.; Bixon, M. *J. Chem. Phys.* **1988**, 88, 167. (b) Bixon, M.; Jortner, *J. Chem. Phys.* **1993**, 176, 467.

(29) Walker, G.C.; Akesson, E.; Johnson, A.,E.; Levinger, N.E.; Barbara, P.F. *J. Phys. Chem.* **1992**, 96, 3728.

(30) Akesson, E.: Johnson, A.E.; Levinger, N.E.; Walker, G.C.; DuBruil, T.P.; Barbara, P.F. *J. Chem. Phys.* **1992**, 96, 7859.

(31) Kandori, H.; Kemnitz, K.; Yoshihara, K. *J. Phys. Chem.* **1992**, 96, 8042.

(32) Schenter, G.K.; Duke, C.B. *Chem. Phys. Lett.* **1991**, 176, 563.

(33) (a) Hynes, J.T., *J. Phys. Chem.* **1986**, 90, 3701. (b) Fonseca, T.; Kim, H.J.; Hynes, J.T.; J. Mol. Liq. **1994**, 60, 161.

(34) Polimeno, A.; Barbon, P.L.; Nordio, P.L.; Rettig, W. *J. Chem. Phys.* **1994**, 98, 12158.

(35) (a) Horng, M.L.; Gardecki, J.A.; Papazyan, A.; Maroncelli, M. *J. Phys. Chem.* **1995**, 99, 17311. (b) Reynolds, L.; Gardecki, J.A.; Frankland, S.J.V.; Horng, M.L.; Maroncelli, M. *J. Phys. Chem.* **1996**, 100, 10337.

(36) van der Meulen, P.; Zhang, H.; Jonkman, A.M.; Glasbeek, M. *J. Phys. Chem.* **1996**, 100, 5367-5373.

(37) Kovalenko, S. A.; Ernsting, N. P.; Ruthmann, J. *Chem. Phys. Lett.* **1996**, 258, 445.
(38) Kovalenko, S. A.; Ruthmann, J.; Ernsting, N. P. *Chem. Phys. Lett.* **1997**, 271, 40.
(39) Jiang, Y.; McCarthy, P.K.; Blanchard, G.J. *Chem. Phys.* **1994**, 183, 249
(40) Martin, M.M.; Plaza, P.; Changenet, P.; Meyer, Y.H. *J. Photochem. Photobiol. A: Chem.* **1997**, 105, 197.
(41) Agmon, C. *J. Phys. Chem.* **1990**, 94, 2959.
(42) Dumon, P.; Jonusauskas, G.; Dupuy, F.; Pée, P.; Rulliére, C.; Létard, J.F.; Lapouyade, R. *J. Phys. Chem.* **1994**, 98, 10391.
(43) AMPAC 5.0, Semichem, 7128 Summit, Shawnee, KS 66216 (1994); Dewar, M.J.S.; Zoebisch, E.G.; Healy, E.F.; Stewart, J.P. *J. Am. Soc.* **1985**, 107, 3902.
(44) (a) Elsaesser, T.; Kaiser, W. *Annu. Rev. Phys. Chem.* **1991**, 42, 83. (b) Foggi, P.; Pettini, L.; Santa, I.; Righini, R.; Califano, S. *J. Phys. Chem.* **1995**, 95, 7439.
(45) (a) Rulliére, C., *Can. J. Phys.* **1984**, 62, 73. (b) Rulliére, C; Joussot-Dubien, *J. Rev. Phys. Appl.* **1979**, 14, 303. (c) Rulliére, C. *PhD thesis* 1977, University of Bordeaux.
(46) Shida, T. *electronic absorption spectra of radical ions*; Elsevier Science Publishers B.V. 1988; Amsterdam.
(47) (a) Shida, T.; Iwata, S. *J.Am.Chem. Soc.* **1973**, 95, 3473. (b) Yoshihara, K.; Yartsev, A.; Nagasawa, Y.; Kandori, H.; Douhal, A.; Kemnitz, K. *Ultrafast Phenomena VIII*, Springer Series in Chemical Physics, Vol. 55, Ed. J.-L. Martin,; A. Migus; A.H. Zewail, Springer-Verlag Berlin Heidelberg 1993, pp. 571.
(48) (a) Shida, T.; Haselbach, E.; Bally, T. *Acc. Chem. Res.* **1984**, 17, 180. (b) Chutny, B.; Swallow, A. *Trans. Faraday Soc.* **1970**, 66, 2847.
(49) Riddick, J.A.; Bunger, W.B.; Sakano, T.K. *Organic Solvents*, John Wiley & Sons **1986**.
(50) *Landolt-Börnstein, Zahlenwerte und Funktionen aus Physik, Chemie, Astronomie, Geophysik und Technik*, 1969, Springer Verlag, Berlin. Vol. 2(5), ed. K.H.Schäfer; Landolt-Börnstein, New series IV/6
(51) Liptay, W. *Excited States*, Vol. 1, ed. E.C.Lim, Academic Press, New York 1974, pp. 129
(52) Ras, A.M.; Bordewijk, P. *Recl. Trav. Chim. Pay-Bas* **1971**, 90, 1055
(53) Pöllinger, F.; Heitele, H.; Michel-Beyerle, M.E; Anders, C.; Futscher, M.; Staab, H.A. *Chem. Phys. Lett.* **1992**, 198, 645.
(54) Simon, J.D. *Acc. Chem. Res.* **1988**, 21, 128.
(55) Braun, D.; *PhD Thesis*, Humboldt-University, Berlin, 1995, published as: Braun, D. *Fluoreszenzspektroskopische und theoretische Aspekte photoinduzierter intramolekularer Elektronentransferreaktionen und ihre Wechselwirkung zur Lösungsmittelumgebung* Wissenschaftliche Schriftenreihe Chemie 30; Verlag Dr. Köster, Berlin, 1995 (ISBN 3-89574-119-1).
(56) Barbara, P.F.; Jarzeba, W. *Ultrafast Photochemical Intramolecular Charge and Excited State Solvation*, Advances in Photochemistry, V.15, 1990, John Wiley & Sons
(57) Rips, I.; Jortner, J. *J. Chem. Phys.* **1987**, 133, 411.
(58) Marcus, R.A.; Sutin, N. *Biochim. Biophys. Acta* **1985**, 811, 265.
(59) Kramer, H.A. *Physica (Amsterdam)* **1940**, 7, 284.
(60) Schütz, M.; Schmidt, R. *J. Phys. Chem.* **1996**, 100, 2012.
(61) ZINDO/S SCF/CI program package available from M.C. Zerner; University of Florida, Gainesville, FL.; Zerner, M.C.; Loew, G.H.; Kirchner, R.F.; Mueller-Westerhoff, U.T. *J. Am. Chem. Soc.* **1980**, 102, 589.
(62) E. Abraham, E.; Oberle, J.; Jonusauskas, G.; Lapouyade, R.; Rulliere, C. *Chem. Phys.* **1997**, 214, 409.
(63) Eilers-König, N.; Kühne, T.; Schwarzer, D.; Vöhringer, P.; Schroeder, J. *Chem. Phys. Lett.* **1996**, 253, 69.

Chapter 5

Fluorescence Polarization Spectroscopy of D-A Biphenyls at 77K.
The Electronic Relaxations

Abstract:
Luminescence and excitation anisotropy spectra of donor-acceptor biphenyls in ethanol at 77K in conjunction with time-resolved emission are analyzed to derive the electronic relaxations in the absence of large structural and solvent relaxations. The excitation spectrum up to 40000 cm^{-1} is dominated by the parallel polarized $S_0 \rightarrow {}^1CT$ and $S_0 \rightarrow {}^1L_a$ transitions with a decrease of anisotropy where $S_0 \rightarrow {}^1L_b$ and $S_0 \rightarrow {}^1B_b$ are located. Short lifetimes and high anisotropy values of the fluorescence bands evidence that the lowest excited singlet state S_1 is the same oscillator strong ^1L_a-type ^1CT state responsible for the CT absorption band. For the twisted derivative, 7% of the emission occurs from a ^3L_a-type ^3CT state. With the aid of quantum chemical calculations, a polarity dependent energy level diagramm of the two relevant singlet and triplet states is proposed which can account for the polarity dependent nonradiative rate constants. The effect of a larger twist angle is discussed so as to result in stronger spin-orbit coupling and decreasing electron transfer character of the lowest triplet state.

5.1 Introduction

The electronic relaxations in photoexcited aromatic hydrocarbons are well studied.[1-7] Moreover, the conformational influence of the interannular twist angle (φ) on the photophysics of such flexible biaryls like biphenyl is well-known (e.g., the yield of ISC and phosphorescence increases with φ).[4-6] However, the electronic relaxations in donor-acceptor (D-A) biaromatic compounds are more complicated by a strong solvent influence due to low lying and polarity dependent charge transfer singlet (^1CT) and triplet (^3CT) states.[8-13] In particular, the role of ^3CT states is not yet clear. First profound experimental studies of triplet states in D-A biaromatic compounds[7-9] and in the extensively studied CT molecule p-dimethylamino-benzonitrile (DMABN)[10] appeared in the past few years. Briefly, the results suggest that in moderately twisted biaryls a ^3CT state can be lowered below a nonpolar locally excited triplet state (^3LE) by increasing the reduction potential of the acceptor unit ($E_{red}(A)$), while the delocalized ^3CT electronic structure is lost, if D and A moieties are decoupled by a high twist angle.

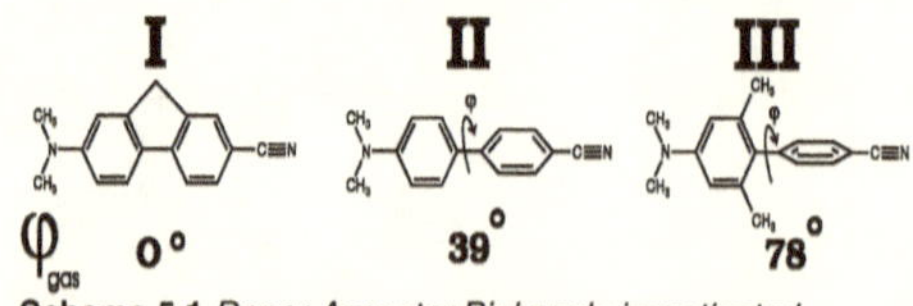

Taking into account the results of the preceding chapters,[11-14] this study focuses on the photoinduced electronic relaxations of the differently twisted donor-acceptor biphenyls **I-III** in the absence of large amplitude conformational and solvent reorganization. To characterize the electronic nature of the involved

Scheme 5.1 *Donor-Acceptor Biphenyls investigated*

I: φ_{gas} 0°
II: 39°
III: 78°

states polarized luminescence and excitation spectra at 77 K are employed in combination with time-resolved emission spectroscopy and gas phase semiempirical calculations.

5.2 Experimental

Details of the fluorescence steady-state and lifetime experiments and the quantum chemical calculations are given in ch. 3, 4 and 8.

5.2.1 Polarization and Millisecond Luminescence Spectroscopy

To obtain the relative polarizations of the electronic transitions the photoselection technique[3,15,16] at 77K is employed by polarizing the excitation beam perpendicular to the experimental plane with a Glan UV polarizer and detecting the emission intensity after passing through a filter polarizer in a parallel (I_{par}) and orthogonal (I_{ort}) arrangement to the incident light polarization. A polarization scrambler was mounted at the entrance slit of the emission monochromator which ensures the absence of polarization effects of the detection system and allows to apply the isotropic emission correction curves. The accuracy of the equipment was checked by measuring the polarization emission spectrum $p=(I_{par}-I_{ort})/(I_{par}+I_{ort})$ of purified triphenylene (Aldrich) which has a trifold symmetry axis and equally polarized one-photon $^1\pi\pi^*$ states[17] and should therefore have a constant theoretical value of p_f=1/7 for the fluorescence.[1,13,17] Fluorescence (p_f) and phosphorescence (p_p) polarization spectra from literature (p_f=0.12 and 0.14)[1,14] could excellently be reproduced (p_f=0.12) within an error of less than P=0.02 arising from usual „depolarizing effects". Here, the anisotropy parameter r=(I_{par}-I_{ort})/(I_{par}+2I_{ort}) is preferred because of its relevance for time-resolved experiments, too. Millisecond resolved luminescence was measured with a Xe-flash lamp as part of the Aminco Bowman 2 Luminescence Spectrometer. The sample solution contained in a small quartz pipe was immersed into a home-made quartz kryostat filled with liquid nitrogen.

5.3 Results and Discussion

Fig. 5.1 shows the polarized fluorescence excitation and luminescence spectra for **I-III** in EtOH at 77K together with the anisotropy values r which are given by

$$r = \frac{I_{par} - I_{ort}}{I_{par} - 2I_{ort}} = 0.2(3\cos^2\Theta - 1) \tag{5-1}$$

where I_{par} and I_{ort} are the relative intensities of luminescence polarized parallel and orthogonal to the electric vector of the incident radiation, respectively, and Θ is the angle between the directions of the absorption and luminescence electric transition dipole moment.[1,3,7,15,16]

Since the first absorption bands of **I-III** at around 28600 cm^{-1} (350 nm) have linear dichroism (LD) values larger than unity (ch.2)[14] and the ansiotropy excitation spectra are positive and exhibit the same shape as the LD spectra (Fig. 2.4), the limiting values of r=0.4 (r=-0.2) and Θ=0° (Θ=90°) for both emission and excitation are associated with pure transitions polarized parallel (perpendicular) to the long-molecular axis on the basis of C_2 symmetry for **I-III** (see more details in Ch. 2). Spectral overlap of bands with different polarization and transitions of mixed electronic character prevent the occurrence of pure anisotropy values. The better quality of the anisotropy spectra with regard to the LD spectra allows to confirm the long-axis polarization of the A band close to 40000 cm^{-1} assignable to the biphenyl-like $S_0 \to {}^1L_a$ transition.[5,6] The overlap with the sizeably stronger $S_0 \to {}^1B_b$ transition (B band) in **I** and **II** with a maximum at about 42000 cm^{-1} explains why the value of r decreases with higher transition energies for **I** and **II**. The decrease of r within the first absorption band (CT) towards higher energies reflects the overlap or electronic mixing of the lower lying and long-axis polarized ^{1}CT state with a weakly allowed ^1L_b-type transition (H band).[14] The higher lying H and the CT band of **III** are well separated for the orthogonal arrangement of the excitation and emission polarizers revealing their maxima at 32300 cm^{-1} and 29600 cm^{-1}, respectively. This separation of the H and CT band is also observed for the isotropic excitation spectra of **III** with decreasing temperature independent of the solvent. While **III** below 29400 cm^{-1} and **I** at the red edge of the excitation spectrum reach the limits r=0.4, the slightly lower value of r=0.35 for **II** indicates a small admixture of ^1L_b to ^{1}CT in the fluorescence or the long-wavelength absorption. Interestingly, an additional short wavelength emission

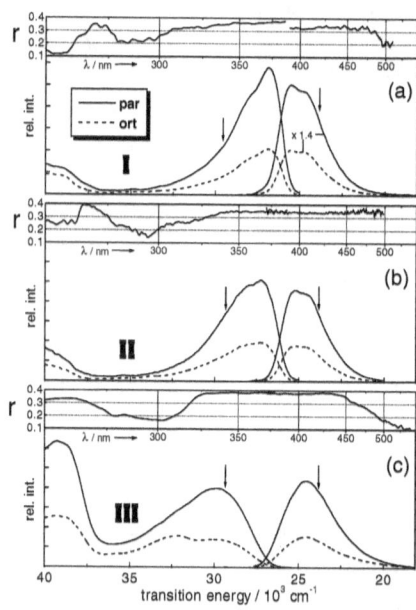

Fig. 5.1 Polarized fluorescence excitation (left) and emission (right) spectra of **I-III** in EtOH at 77K obtained by parallel (———) and orthogonal (- - -) orientation of the emission polarizator with respect to the polarization direction of the incident light. The anisotropy spectrum r = $(I_{par} - I_{ort})/(I_{par} + 2I_{ort})$ is plotted in the upper boxes. The arrows indicate the observation and excitation energies.

band assigned to a $^1L_b/^1CT$ mixed transition was found to be intense only for **I** and **II** within the first 100 ps (ch.4). It is therefore possible that the larger fluorescence anisotropy of **III** as compared to **II** is due to less mixing of ^1L_b with ^1CT in compound **III** which may then be traced back to a more twisted structure and/or a larger energy gap to the zeroth-order ^1L_b state. A superposition of two electronically different fluorescence bands as well as a vibronic coupling mechanism of S_1 is unlikely because the fluorescence anisotropy remains high and constant across the fluorescence band (behaviour at v_{em}<22000 cm^{-1} is discussed below) for **I-III** and the fluorescence lifetimes of **I** (1.7 ns) and **II** (1.6 ns) are monoexponential and do not vary with the emission energy. The wavenumber dependent fluorescence decay behaviour of **III** in polar and nonpolar solvents at 77 K is the consequence of emission from a single but broad distribution of frozen rotamers which have an extremely strong dependence of the radiative lifetime ($\tau_r = k_f^{-1}$) on the interannular twist angle as demonstrated by the theoretical calculations in ch. 3. The ground state rotamer distributions of **II** and **III** at 77 K can be compared in Fig. 3.7 and the large variation of k_f only in the twist angle range of **III** in Fig. 3.8 and 3.9). This outstanding decay behaviour has no influence on the present analysis of the electronic deactivation routes, if the obtained average lifetime <τ_f>= (4±1) ns is considered for **III** in EtOH.

The fluorescence lifetimes of **I-III** are then all shorter than the respective radiative lifetimes of the absorption bands (2.1 ns, 2.5 ns, 9.6 ns) calculated by the Strickler-Berg formula (eq. 3-13)[18] using the mean fluorescence energies 25100 cm^{-1}, 24900 cm^{-1} and 24600 cm^{-1} for **I**, **II** and **III**, respectively. Further, on the basis of the fluorescence quantum yields at 130 K (Φ_f=90% for **I** and **II** and Φ_f=60% for **III**) the resulting radiative lifetimes $\tau_r = \tau_f / \Phi_f$ are close to the Strickler-Berg lifetimes. This underlines that even for the 77 K conditions of almost frozen conformational and solvent relaxation, the fluorescence of **I-III** occurs from the same ^1L_a-type ^1CT state responsible for the first absorption band. As discussed already in Ch.3 this is clearly different to the behaviour of the unsubstituted biphenyl where long-lived emission (τ_f=16 ns, τ_r=89 ns) is observed from the ^1L_b state while the lowest energy absorption band corresponds to a radiative lifetime of 2.9 ns characteristic for the $S_0 \rightarrow {}^1L_a$ transition.[19]

Obviously, the anisotropy values of **III** decline considerably below 22000 cm^{-1}. This behaviour is absent for **II** and insignificant for **I**. Here, it should be noted that the measurement of monoexponential and short fluorescence lifetimes of **I** and **II** exclude the possibility of aggregation in EtOH, which has, however, been observed in Mlp at 77K for **I** and **II** by a loss of anisotropy and a large contribution of an additional long lifetime (ch.10). For **III** in Mlp, the excitation as well as the luminescence anisotropy values are only 0.02 units lower than in EtOH which is within the experimental accuracy of $\Delta r = \pm 0.02$. It is therefore reasonable to attribute the decrease of anisotropy at the low energy edge of the emission band to phosphorescence induced by spin-orbit coupling between a singlet and the lowest triplet state.

The phosphorescence spectrum can be derived from the polarized emission spectra as follows. For convenience, let us define the dichroic ratios

(a) $\quad y = \dfrac{F_{par}}{F_{ort}} = \dfrac{2r_f + 1}{1 - r_f} \quad$ (b) $\quad z = \dfrac{P_{par}}{P_{ort}} = \dfrac{2r_p + 1}{1 - r_p} \quad$ (5-2)

and (a) $I_{ort} = F_{ort} + P_{ort}$ (b) $I_{par} = F_{par} + P_{par} = yF_{ort} + zP_{ort}$ (5-3)

where I, F, P are the intensities of the total emission, fluorescence and phosphorescence of a given polarization, respectively, and r_f and r_p are the anisotropy values of fluorescence and phosphorescence. Making use of the relation for the intensity of isotropic phosphorescence

$$P = \frac{1}{3}(P_{par} + 2P_{ort}) = \frac{1}{3}(2+z)P_{ort} \quad (5\text{-}4)$$

after some algebraic transformations a simple expression is obtained

$$P = \frac{1}{3}\frac{2+z}{y-z}(yI_{ort} - I_{par}) \quad (5\text{-}5)$$

which allows to recover the phosphorescence P(ν) and fluorescence F(ν) = [I_{par}(ν)+2I_{ort}(ν)]/3 - P(ν) spectrum from the polarized spectra I_{par}(ν) and I_{ort}(ν), if the anisotropies r_p(ν) and r_f(ν) are known. Using the known and almost wavenumber independent value of r_f=0.38 (y=2.84) at the emission maximum and assuming a wavenumber independent value of r_p=0 (z=1), the obtained isotropic spectra I(ν), P(ν) and F(ν) are plotted in Fig. 5.2 for III in EtOH at 77 K. The quantum yield ratio of phosphorescence to the total emission

$$\frac{\Phi_p}{\Phi_{em}} = \frac{\int P(\nu_{em})d\nu_{em}}{\int I(\nu_{em})d\nu_{em}} \quad (5\text{-}6)$$

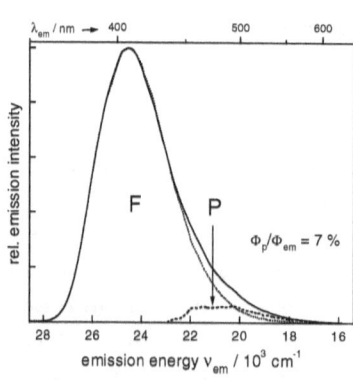

Fig. 5.2 *Isotropic emission spectrum of* **III** *in EtOH at 77 K. The fluorescence (F) and phosphorescence (P) bands are separated on the basis of eq. 5-5 using r_f=0.38 and r_p=0.*

yields a Φ_p/Φ_{em} value of 7% in EtOH and Mlp with an uncertainty of about 3% due to the assumption of r_p. A similar procedure for **I** in EtOH and **II** in Mlp suggests phosphorescence maxima at around 20000 cm^{-1}, but the derived values of Φ_p/Φ_{em}<1% are too small for a profound discussion. The assignment of the anisotropy decrease in the low energy emission tail for **III** to phosphorescence is confirmed by the agreement with time-resolved emission experiments as shown in Fig. 5.3. The emission band between 19-23000 cm^{-1} has a monoexponential lifetime of 2.4 s and by comparison of the time-gated spectrum after 5 ms with the time-integrated (=steady state) emission band recorded for equal conditions, a ratio of $\Phi_p/\Phi_{em}=(6 \pm 2)\%$ is found.

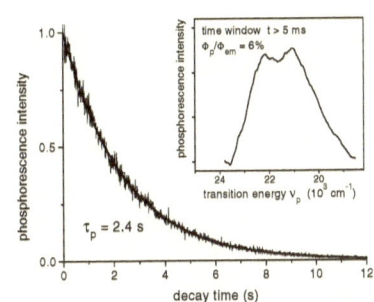

Fig. 5.3 *Phosphorescence time trace with the monoexponential fit and the corresponding spectrum of* III *in Mlp.*

According to these results, the comparably small fluorescence lifetimes of III in nonpolar solvents can be attributed to nonradiative ISC processes at higher temperature. Thus the temperature dependence of the fluorescence quantum yields and lifetimes in IpM shown in Fig. 5.4 derives from a small thermal activation barrier (E_a=2.2 kJ/mol) to the triplet state. Note, that the radiative rate constant is temperature independent because in the studied temperature range the average conformation does not change significantly.

In order to gain insight into the electronic nature of the triplet states involved, the gas phase transition energies of the lowest lying excited singlet and triplet states of both symmetry possibilities A and B within a C_2 molecular point group are calculated by the ZINDO/S and AM1 method for the optimized ground state geometries of I-III (Tab. 5.1). Even though both calculation results differ markedly and are not expected to achieve quantitative accordance with the experiments they agree that (i) the lowest lying triplet T_1 is of zeroth-order ^{3}A symmetry with ^3L_a-type ^{3}CT character, (ii) the energy difference to the higher lying triplet with ^{3}B symmetry of ^3L_b-type character is larger than between the respective singlet state pair 2^1A and 1^1B, (iii) in contrast to the singlet ^{1}CT(2^1A) state, the triplet ^{3}CT(1^3A) has decreasing electron transfer (^{3}ET) weight with increasing twist angle while an

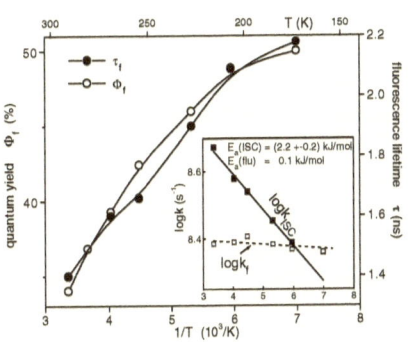

Fig. 5.4 *Temperature dependence of the fluorescence quantum yield and lifetime for* III *in IpM. The corresponding Arrhenius plots of the nonradiative and radiative rate constants are shown in the insert.*

electronically coupled higher lying $^3CT(^3A)$ state shows the opposite behaviour being degenerate with the singlet state $^1CT(2^1A)$ at $\varphi=90°$, (iv) the state 1^3B is the closest lower lying triplet with respect to the first or second excited singlet state.

Tab. 5.1
Calculated Singlet and Triplet Transition Energies in 10^3 cm^{-1} Using the ZINDO/S (AM1) Method.

state	III / 90°	III / 80°	II / 39°	III / 0°
1^1B (^1L_b)	33.3 (33.8)	33.3 (33.9)	33.5 (33.3)	32.9
2^1A ($^1CT/^1L_a$)	36.9 (33.7)	36.3 (33.7)	33.8 (32.0)	31.7
1^3B (^3L_b)	25.8 (27.8)	25.8 (27.7)	25.7 (28.8)	25.7
1^3A ($^3CT/^3L_a$)	17.1 (23.03)	17.1 (23.1)	17.8 (23.6)	16.7

^a The L_a character for the singlet state 2^1A decreases with increasing twist angle (ref. 14), whereas that of the triplet state 1^3A increases, such that at twist angles above 80° 1^3A is essentially ^3L_a.

These results lead to the conclusion that the emitting triplet of **I-III** is a ^3L_a-type ^3CT state and that the intersystem crossing process (ISC) occurs in a nonpolar environment from ^1CT to ^3L_b. Similarly to unsubstituted biphenyl, the twisted structure of **III** seems to enhance the magnitude of spin-orbit coupling between singlet and triplet states relative to **I** and **II** leading to higher nonradiative rates for **III** in nonpolar solvents at 298 K (Tab. 3.3b) and larger phosphorescence yields at low temperature.[4] In polar solvents, smaller nonradiative rate

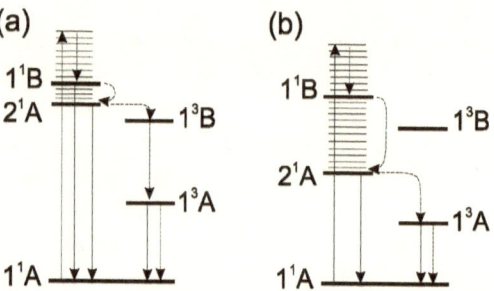

Scheme 5.2 *Jablonski diagram of main radiative (———) and nonradiative (······) electronic relaxations of D-A biphenyls (a) in a nonpolar or rigid medium and (b) in a polar solvent.*

constants are observed (ch.3). Taking into account the calculated electron transfer character of the low lying singlet 2^1A and triplet 1^3A state, the polarity dependent nonradiative rate constants may be explained with the proposed Jablonski diagram in scheme 5.2 which summarizes the main electronic relaxation processes of **I-III** in a nonpolar or rigid environment and in a polar solvent. After excitation into a manifold of vibronic states, the fluorescent ^1L_b-type 1^1B and the polar ^1L_a-type $2^1A(^1CT)$ singlet states are successively populated as worked out in ch.4.[13] In a weakly polar medium, the ^3B triplet state is close to $S_1(2^1A)$ enabling relatively fast ISC, while in a polar solvent, the charge transfer states 2^1A and 1^3A are stabilized, such that 1^3B is too high lying and ISC occurs directly to a ^1L_a-type ^3A state which may be slower than than the $1^1A \to 1^3B$ ISC because of a larger energy gap. The effect of the larger twist angle in III is (i) a faster interconversion rate $1^1B \to 2^1A$ connected with low $1^1B \to 1^1A$ fluorescence yields (ch. 4) and possibly higher fluorescence anisotropy values r_f (Fig. 5.1), (ii) a faster $2^1A \to 1^3B$ ISC rate and a faster $1^3A \to 1^1A$ phosphorescence rate constant probably due to more effective spin-orbit coupling induced by stronger excitonic interactions between the local triplet and singlet electron configurations and (iii) less ^{3}ET character (revealed by inspection of the CI coefficients to the calculated wavefunctions) which has experimentally been observed for a pair of differently twisted anilinophenanthrenes[8,9] and may therefore likewise explain the narrow and structured shape of the phosphorescence band of III (Fig. 5.3).

5.4 References

(1) Klessinger, M.; Michl, J. *Lichtabsorption und Photochemie organischer Moleküle*, VCH Verlag, Weinheim **1989**.

(2) Birks, J.B. *Photophysics of Aromatic Molecules*, Wiley-Interscience, New York, **1970**.

(3) Gudipati, M.S.; Daverkausen, J.; Hohlneicher, G. *Chem. Phys.* **1993**, 174, 143. (b) Gudipati, M.S.; Daverkausen, J.; Maus, M.; Hohlneicher, G. *Chem. Phys.* **1994**, 186, 289. (c) Gudipati, M.S.; Maus, M.; Daverkausen, J.; Hohlneicher, G. *Chem. Phys.* **1995**, 192, 37.

(4) Fujii, T.; Suzuki, S.; Komatsu, S. *Chem. Phys. Lett.* **1978**, 57, 175.

(5) Hohlneicher, G.; Dörr, F.; Mika, N.; Schneider, J. *Ber. Bunsenges. Physik. Chem.* **1968**, 72, 1144.

(6) Swiderek, P.; Michaud, M.; Hohlneicher, G.; Sanche, L. *Chem. Phys. Lett.* **1991**, 187(6), 583.

(7) Herbich, J.; Kapturkiewicz, A.; Nowacki, J. *Chem. Phys. Lett.* **1996**, 262, 633.

(8) Onkelinx, A.; De Schryver, F.C.; Viane, L.; Van der Auweraer, M.; Iwai, K.; Yamamoto, M.; Ichikawa, M.; Masuhara, H.; Maus, M.; Rettig, W. *J. Am. Chem. Soc.* **1996**, 118, 2892.

(9) Onkelinx, A.; Schweitzer, G.; De Schryver, F.C.; Miyasaka, H.; Van der Auweraer, M.; Asahi, T.; Masuhara, H.; Fukumura, H.; Yashima, A.; Iwai, K. *J. Phys. Chem. A* **1997**, 101, 5054.

(10) Chattopadhyay, N.; Rommens, J.; van der Auweraer, M.; De Schryver, F.C. *Chem. Phys. Lett.* **1997**, 264, 265.

(11) Maus, M., Rettig, W., Bonafoux, D.; Lapouyade, R. submitted.

(12) (a) Maus, M.; Rettig, W.; Lapouyade, R. *J. Inf. Rec.* **1996**, 22, 451; (b) Rettig, W.; Maus, M. *Ber. Bunsenges. Phys. Chem.* **1996**, 100, 2091. (c) Maus, M.; Rettig, W.; *J. Inf. Rec.* **1998**, 24(5/6), 461.

(13) Maus, M.; Rettig, W.; Jonusauskas, G.; Lapouyade, R. and Rullière, C. *J. Phys. Chem.* **1998**, 102, 7393.

(14) Maus, M.; Rettig, W. *Chem. Phys.* **1997**, 218, 151.

(15) Albrecht, A.C. *J. Mol. Spectrosc.* **1961**, 6, 84.

(16) Dörr, F.; Held, M. *Angew. Chem.* **1960**, 9, 287.

(17) Zimmermann, H.; Joop, N. *Z. Elektrochem. (Ber. Bunsenges. Physik. Chem.)* **1966**, 70, 803.

(18) Strickler, S.J.; Berg, R.A. *J. Chem. Phys.* **1962**, 37, 814.

(19) (a) Berlman, I.B.; Steingraber, O.J. *J. Chem. Phys.* **1965**, 43, 2140; (b) Berlman, I.B. *J. Chem. Phys.* **1970**, 52, 5616.

Chapter 6

Temperature Dependent Study of Excited State Conformational Relaxations in D-A Biphenyls Using Steady-State and Time-Resolved Fluorescence

Abstract:
*Steady-state and time-resolved fluorescence are performed to gain information about the temperature influence on excited state conformational relaxations of three differently twisted donor-acceptor biphenyls (**I-III**) in nonpolar and polar solvents. The band-shape analysis related to Marcus theory and the determination of the fluorescence rate constants both as a function of temperature indicate that the relaxation to a planar structure in **II** and **III** is not frozen in the liquid phase of the solvents. Only **III** in a medium polar solvent (diethylether) exhibits a nonlinear dependence of fluorescence quantum yields (Φ_f), reorganization ($\lambda_s+\lambda_i'$) and charge recombination (ΔG_{CT}) energies on temperature, which is found to be due to a temperature dependent population ratio of two conformationally different charge transfer species **CT** and **CTR**. Their thermodynamic equilibrium (reversible photoreaction) is reached above 265 K. Using global fluorescence analysis, the separated fluorescence bands of **CT** and **CTR** as well as kinetics, thermodynamics (ΔH=-2.5 kJ/mol, ΔS=-0.7 J/K/M) and activation barriers ($E_a(CT \rightarrow CTR)$=14.3 kJ/mol; $E_a(CT \leftarrow CTR)$=16.8 kJ/mol) associated with the adiabatic (forward and backward) photoreaction could be determined. The derived dipole moments (μ_{CT}=26D; μ_{CTR}=30D) and their k_f ratio ($k_f^{CTR}/k_f^{CT} \approx 0.7$) are consistent with a photoreaction from a more planar (**CT**) to a more twisted (**CTR**) conformer.*

Keywords: *dual fluorescence, kinetic global analysis, band shape analysis, adiabatic photoreaction*

6.1 Introduction

The phenomenon of dual fluorescence has attracted great attention in the past three decades.[1-11] From time to time, new molecules are presented which do not obey Kasha's rule[12] that „*emission always occurs from the lowest electronically excited state of a given multiplicity*". Small Franck-Condon factors and small electronic coupling matrix elements between the relevant higher (S_x) and lowest (S_1) excited diabatic states are frequently discussed as a source of dual emission from S_x and S_1.[1,2] Alternative possibilities in accord with Kasha's rule to observe dual fluorescence are adiabatic photoreactions.[3-11] In these cases, a nuclear rearrangement takes place on the adiabatic S_1 hypersurface, which can be slowed down by appropiate conditions of the environment to the nanosecond timescale of fluorescence. A typical example molecule is 4-dimethylamino-benzonitrile (DMABN) which is believed to exist in two stable molecular structures in the excited state.[3,5,7-9] The occurrence of well separated dual fluorescence bands in DMABN and derivatives allowed a detailed characterization of the excited state equilibrium parameters by a conventional analysis of single-photon counting and steady-state experiments.[4,5,7-9]

It is a further challenge to investigate the equilibrium conditions of photoreactive and emitting (conformer) species, if their fluorescence bands are not separated. Similar fluorescence energies (v_f) of two excited state isomers or species can result either from similar potential energies (E) or in case they are different from counteracting Franck-Condon ground state energies (ΔE_{FC}). Various methods, such as the maximum entropy method,[13] principal component analysis,[14] global compartmental analysis[15] or the concept of decay and species associated spectra[16-18] have recently been developed to encounter such phenomena.

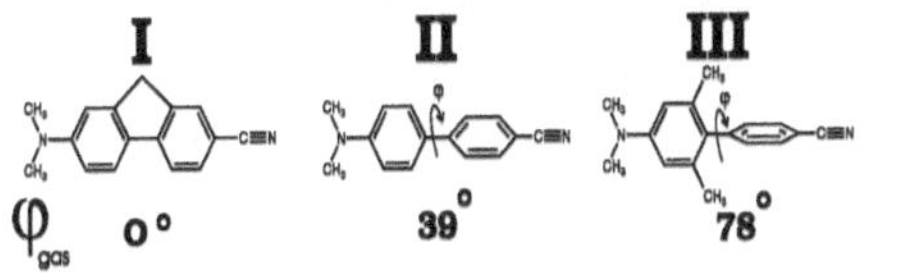

Scheme 6.1 Donor-acceptor biphenyls. Molecular structures and with AM1 calculated ground state twist angles φ.

In the series of the donor-acceptor biphenyls I-III (scheme 1), a biexponential fluorescence decay behaviour on a subnanosecond timescale was found for III in polar solvents (ch.3).[19,20] It has been interpreted with a dynamic excited state equilibrium between two charge transfer species denoted as **CT** and the more relaxed product species **CTR**.[19,20] It was suggested[19] in ch. 3 to analyze the excited state kinetics on the basis of a two-species Birks scheme where the observed two lifetimes τ_1 and τ_2 are described by[6,8,9,21]

$$\tau_{1,2}^{-1} = \tfrac{1}{2}(X + Y \pm \sqrt{(X-Y)^2 + 4k_{CT \to CTR}k_{CT \leftarrow CTR}}) \quad (3\text{-}15\,a,b)$$

$$X = \tau_0(CT)^{-1} + k_{CT \to CTR} \qquad Y = \tau_0(CTR)^{-1} + k_{CT \leftarrow CTR} \quad (3\text{-}15\,c,d)$$

On the other hand, time-resolved transient absorption revealed that the Franck-Condon excited state (^{1}FC) should be included as a third and primary kinetic species (**FC** in scheme 2) to describe the photoprocesses adequately (ch.4).[2] Excitation to a ^{1}FC state which is different to the emitting state, of course, always occurs. It is just a question of how fast is the ^{1}FC relaxation in relation to the photodynamics afterwards. If it is fast enough, which is, e.g., the case in the well known dual fluorescing compound DMABN, the model of two excited species can still be applied to analyze the kinetics. A further requirement to solve the Birks eq. 3-15 for an intramolecular photoreaction by well-known methods[8,9,15-18] analyzing the amplitudes of biexponential fluorecence decays is that only the precursor species, here **CT**, is initially populated by **FC**.[22]

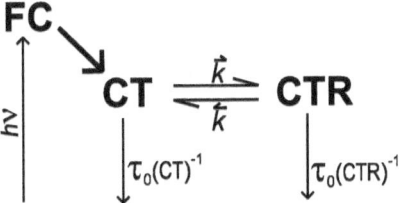

Scheme 6.2 Kinetic model for I-III. The FC→CT interconversion is more than two magnitudes of order faster than the photoreaction ($k_{CT \to CTR}$, $k_{CT \leftarrow CTR}$) or ground state deactivation ($\tau_0(CT)^{-1}$, $\tau_0(CTR)^{-1}$) rate constants. The CT→CTR photoreaction occurs only for III.

The time-resolved transient absorption experiments in ch. 4 supported this idea for room temperature conditions. Taking into account the fluorescence results of ch. 3,[19,20] it was concluded that within a few picoseconds, a solvent-independent, structural relaxation to a conformation with improved π-electron delocalization and connected with a more (overall) planar structure takes places for **I-III**. This process is regarded to be mainly responsible for the subsequent population of only the more emissive and more planar **CT** species instead of the less emissive **CTR** species. Here and especially for **III**, we are obliged to check, whether this fast structural and mainly torsional relaxation is frozen at lower temperatures, because in this case, the precondition of **CT** population only would not be fulfilled.

Swiatkowski, Menzel and Rapp[23] have shown for the parent unsubstituted biphenyl (BP), that this fast relaxation is not stopped by lowering the temperature in liquid solvent phases. Their conclusion is based on the observation that with decreasing temperature (i) the emission energy remains constant and (ii) their self-defined parameter 1-S monitoring the extent of vibrational structure in the emission spectrum linearly increases. This is consistent with an unhindered torsional relaxation and a narrowing of the excited state rotamer distribution around the twist angle $\varphi=0°$ with lowering the temperature down to the glass transition temperature (T_g).

For the donor-acceptor biphenyls, however, a more refined method is needed to assure the free torsional relaxation at low temperatures, because electron transfer interactions are involved causing an additional and different temperature influence on the emission maximum and vibrational structure. Here, the temperature dependent band-shape analysis related to Marcus electron transfer (ET) theory[24] is proposed as a possible way. Further, the comparison with the planar model compound **I** and an analysis of the radiative and nonradiative properties as a function of temperature should strengthen the conclusions. On the basis of these results, the possible application of the simplified two-species Birks scheme is verified and the excited state equilibrium between the two charge transfer species **CT** and **CTR** is quantitatively analyzed by a special global analysis method adapted from Löfroth[17] to yield the separated fluorescence bands as well as the relevant kinetic and thermodynamic information.

6.2 Experimental

Setup and Chemicals   See 3.2 for more details of time-resolved and steady-state experiments as well as solvents and synthesis of **I-III**.

6.2.1 Low Temperature Measurements

Temperature dependent fluorescence decays and spectra are measured with a home-made cooling apparatus which allows to freeze and control the temperature of four samples in quarz cuvettes by pumping cold nitrogen gas through the cryostat. The temperature precision and stability decreases with cooling and is estimated to be better than 2 K down to 185 K. The reproducibility of the temperature and fluorescence decays was checked in different cooling cycles. The lowest temperature achieved was about 100 K with a stability of better than 5 K. Here, solvents at temperatures in the liquid phase are studied. The temperature dependent relative fluorescence intensities $I_r(T)$ are corrected for the linear increase of the refractive index $n(T)^{25}$ and density $\rho(T)^{25}$ of the solvent relative to room temperature conditions by $I_r(T) = I_r(298K) \times \rho(298K) \times n(298K)^{-1} \times \rho(T)^{-1} \times n(T)$. Due to a narrowing of the absorption band of **I** and **II** with decreasing temperature, the optical density, which could not be measured in the low temperature apparatus, has additionally been corrected as follows. Assuming a temperature independent absorption transition moment, the excitation spectra were normalized to the same area of the first absorption band, such that a factor for the optical density at the band maximum relative to the room temperature value could be obtained. For **III**, this factor remains close to unity, since the band narrowing with cooling is accompanied by a separation of the ^1L_b and ^1CT band at low temperatures (see Ch. 5), while for **I** and **II** it increases up to 1.08 and 1.06 at 168 K. The absolute error of the low temperature fluorescence quantum yields ($\Phi_f(T)$) determined from the integrated intensity area relative to the values at room temperature is estimated to be 10%.

6.2.2 Band-Shape Analysis   The corrected fluorescence spectra $I_f(\lambda_f)$ originally recorded in steps of wavelength λ_f are converted to the wavenumber scale (v_f) by multiplying with λ_f^2 (see 3.2.4) and are afterwards divided by v_f^3 to obtain the reduced intensity $I_f(v_f)/v_f^3$ providing the relevant spectra for the band-shape analysis.[24] The fitting of the spectra was performed with a self-written fit-routine based on the nonlinear least-square method inside the ORIGIN 5.0 program from Microcal. Employing the one-mode approximation the fit function includes ten vibronic bands (j=0-9), even though the results are practically identical, if only seven bands are used. The spectra were fitted with a fixed vibration frequency $\langle v_i \rangle = 1300$ cm^{-1} (active interannular stretching mode) as well as with free parameters. Both procedures yielded similar results and χ^2 tests and the same temperature dependence (except for **III** in diethylether), such that the results of the free fits are presented.

6.2.3 Global Analysis of Emission Decays   Single curve and global analysis of the wavenumber dependent emission decays was carried out with a commercial global analysis program[26] as described in 3.2.6. The iteration procedures including deconvolution of the response function were performed with increasing degrees of freedom. The minimum set of parameter to achieve global χ^2 below 1.2 for all temperatures needs free time shifts, two linked lifetimes and variable amplitudes. However, to further improve the fit of those decays obtained above 180 K in the very blue wing of the spectrum, where emission intensity is low, a short time component of about 100 ps was added. This time component of weak intensity (yield less than 2%) can be due to unfavourable count statistics, stray light or more probably solvation effects (of **CT**). Since it is similarly observed for all three investigated compounds, the kinetic model applied for **III** is not affected and it is therefore ignored in the analysis.

6.3 Results and Discussion

6.3.1 Temperature Dependence of Radiative Back Charge Transfer and Excited State Relaxations for the D-A Biphenyls I-III.

6.3.1.1 Energetics From Fluorescence Band-Shape Analysis

Pursuing the analogy between CT fluorescence and thermal electron transfer processes, Marcus[24] derived eq. 6-1

$$\frac{I_f(v_f, T)}{v_f^3} = \frac{64\pi^4}{3h} n^3 M_f^2 \sum_{j=0}^{\infty} \frac{1}{j!} \left(\frac{\lambda_i'}{hc\langle v_i \rangle} \right)^j e^{-\frac{\lambda_i'}{v_i'}} \sqrt{\frac{1}{4\pi(\lambda_s + \lambda_i')k_BT}}$$
$$\times \exp\left(-\frac{(jhc\langle v_i \rangle + (\lambda_s + \lambda_i') + hcv_f + \Delta G_{-CT})^2}{4(\lambda_s + \lambda_i')k_BT} \right)$$
(6-1)

where v_f and M_f are the fluorescence wavenumber and transition dipole moment, $\langle v_i \rangle$ and λ_i' represent an average vibration wavenumber and the corresponding vibrational reorganization energy within a single mode approximation, $(\lambda_s + \lambda_i')$ is the sum of the low frequency solvent and intramolecular reorganization energy and ΔG_{-CT} is the (negative) free energy change of the fluorescence back charge transfer process. The energetical relations are visualized in scheme 6.3 for a shallow ground state (S_0) potential and a steep ^{1}CT potential. This is a typical situation for electron transfer systems where the force constant for the solute-inner solvation shell interaction is larger in the polar ^{1}CT state $(f_e > f_g)$.[24] Fitting of the temperature dependent fluorescence profiles $I_f(v_f, T)$ of **I**-**III** to eq. 6-1 by varying the four parameters ΔG_{-CT}, $(\lambda_s + \lambda_i')$, λ_i' and $\langle v_i \rangle$ was performed in order to gain information whether the excited state intramolecular relaxation is influenced by cooling.

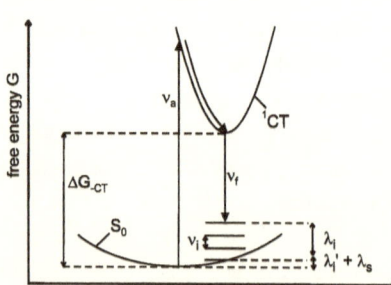

Scheme 6.3 *Illustration of a single radiative back charge transfer process (-CT). See text and eq. 6-1.*

The temperature dependent fluorescence spectra of **I-III** in the nonpolar isopentane/metylcyclohexane 4:1 mixture (IpM) and diethylether (EOE) are shown in Fig.6.1 and 6.2 together with the spectra fitted by eq. 6-1. In Fig. 6.3 and 6.4, the corresponding fit results - ΔG_{-CT}, $\lambda_s + \lambda_{i'}$ and the reduced $HW_f'^2$ are plotted versus the temperature. The important fit data are collected in Tab. 6.1.

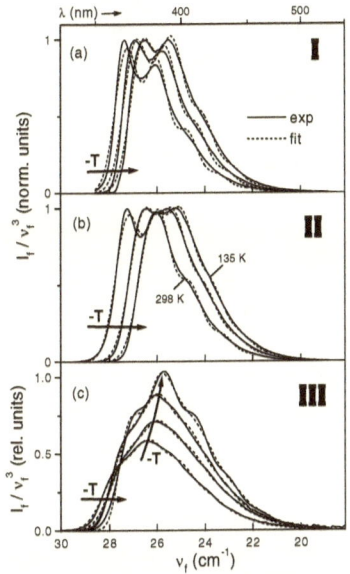
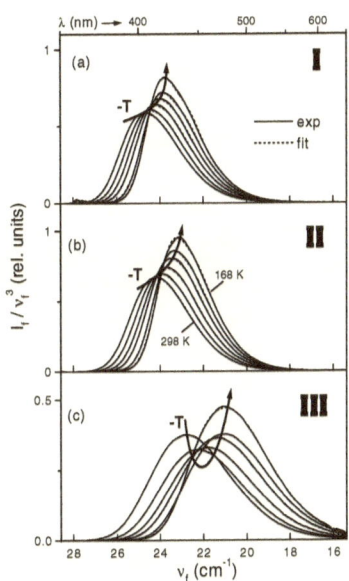

Fig. 6.1 *Fluorescence spectra and fits to eq. 6-1 of I-III in a 4:1 isopentane/methylcycloxane mixture with decreasing temperature (-T) from 298K down to 135 K. Spectra of I and II are normalized while those of III are on a relative intensity scale.*

Fig. 6.2 *Fluorescence spectra and fits to eq. 6-1 of I-III in diethylether with decreasing temperature (-T) from 298 K down to 168 K. All intensities are relative to each other (full correction, see 6.2.1).*

The general influence of three of the four fitted Marcus parameters ΔG_{-CT}, λ_i, $(\lambda_i' + \lambda_s)$ ($<v_i>$ determines the spacings of the vibronic bands and is found polarity independent within a deviation of less than 50 cm^{-1}) can here be followed by a comparison of the obtained band shapes in the nonpolar solvent IpM (Fig. 6.1 and 6.3) with those in the medium polar solvent EOE (Fig. 6.2 and 6.4) for a given molecule/temperature combination (compare, e.g., data at T=298K in Tab. 6.1). Obviously, ΔG_{-CT} only determines the energetical position of the spectrum (v_f^{0-0}) which is somewhat lowered by ($\lambda_i' + \lambda_s$). The width of the vibronic bands and therewith the overall half width HW_f' increases with ($\lambda_i' + \lambda_s$) and λ_i (larger $<v_i>$ broadens HW_f', too; cf. eq. 6-5). The asymmetry of the band shape usually increases with λ_i (compare higher values in Tab.

6.1 for **III** than for **I** or **II**) indicating unfavourable 0-0 and 0-1 vibronic transitions and therewith sizeable changes of the molecular geometry in S_1.

The main changes of the spectra in IpM (Fig. 6.1) with lowering the temperature (-T) are (i) an unexpected loss of structure for **I** and **II** and (ii) a red shift for **I-III**. In case of incomplete structural relaxation at lower temperature, however, an increase of the emission maximum in conjunction with an increase of $-\Delta G_{CT}$ would have been expected (see scheme 6.3) but is not observed (Fig. 6.3a). Moreover, since the planar model compound **I** exhibits the same behaviour, this is not related to a freezing of the rotational relaxation towards planarity in **II** and **III**. The red shift and increasing reorganization energy with decreasing temperature can better be attributed to solute-solvent interactions in the ^{1}CT state as induced by increasing density of the solvent.[27] This effect is absent in BP because the electronic transition occurs between two nonpolar states with negligible solvent influence.

Consequently, only the increase of the Franck-Condon ground state energy given by

$$\Delta E_{FC} = \lambda_s + \lambda_i^{'} + \lambda_i \tag{6-2}$$

produces the fluorescence red-shift with lowering the temperature in IpM (scheme 6.3).

On the other hand, the fit results in diethylether (EOE) shown in Fig. 6.4a and b reveal that the fluorescence red shift with cooling in this solvent (Fig. 6.2) is additionally induced by a decrease of $-\Delta G_{CT}$. On the basis of a continuum dielectric model[28,29] using eq. 6-3

$$-\Delta G_{-CT} = -\Delta G_{-CT}(gas) - \frac{\mu_e^2 - \mu_g^2}{a^3} f(\varepsilon_r) \tag{6-3}$$

with the solvent dielectric constant ε_r, the excited and ground state dipole moments μ_e and μ_g, the Onsager radius a and the Onsager function[28] $f(x)=(x-1)/(2x+1)$, this is understandable with a strong stabilization of the ^{1}CT state by (inner shell) solvation which increases at lower temperatures (see -f(ε_r) vs. T in Fig. 6.4a). Indeed, the temperature dependence of the calculated solvation energy in EOE correlates with that of $-\Delta G_{CT}$ from the band-shape analysis (Tab. 6.1). Moreover, the increase of the dipole moments μ_e from **I** to **III** is well reflected (i) by the increase of the slopes $d(-\Delta G_{CT})/dT$ from **I** to **III** and (ii) by the increase of the reorganization energies ($\lambda_s + \lambda_i'$) from **I** to **III** in the case that a large contribution emanates fom the solvent reorganization energy λ_s according to eq. 6-4

$$\lambda_s = \frac{(\mu_e - \mu_g)^2}{a^3}(f(\varepsilon_r) - f(n^2)) \tag{6-4}$$

From the above considerations, it can be concluded that the energy changes (ΔG_{CT} and λ_s) are only due to the linear increase of solvent polarity with temperature, such that the linear plots of **II** and **I** indicate a temperature independent molecular structure.

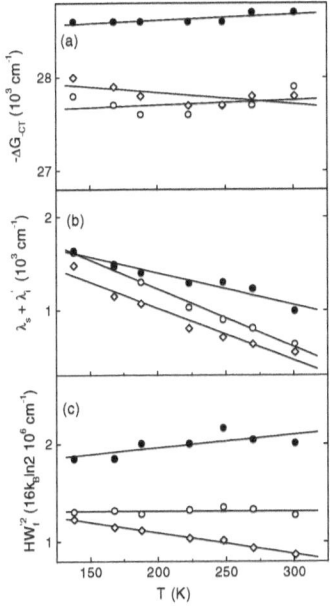
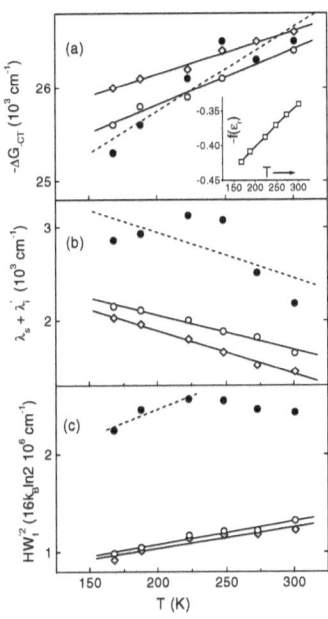

Fig. 6.3 *Temperature dependent results of band-shape analysis for* **I** *(◊),* **II** *(O) and* **III** *(●) in a 4:1 isopentane/methylcycloxane mixture.*

Fig. 6.4 *Temperature dependent results of band-shape analysis for* **I** *(◊),* **II** *(O) and* **III** *(●) in diethylether.*

In preferable cases, a plot of the v_f^3 corrected and squared fluorescence half widths ($HW_f^{'2}$) vs. the temperature (eq. 6-5)

$$\frac{HW_f^{'2}}{16k_B \ln 2} = \frac{\lambda_i' \langle v_i \rangle}{2k_B} + (\lambda_s + \frac{f_g}{f_e}\lambda_i')T \quad (6-5)$$

may be applied to derive information about the reorganization energies.[24] This plot should normally result in a positive slope. In contrast, the unexpected negative slope for **I** in IpM (Fig. 6.3c) shows that this method fails, if a change of the electronic structure is involved.[27] In EOE, a better agreement with the results from eq. 6-1 is obtained. A common feature of the plots shown for $-\Delta G_{CT}$, $(\lambda_s+\lambda_i')$ and $HW_f^{'2}$ versus temperature is that only **III** in EOE exhibits strong deviations from linearity. In addition, only for **III** in EOE a consecutive decrease and increase of the fluorescence intensity with lowering the temperature is observed in Fig. 6.2c. This already (vide infra) points to temperature dependent population of different rotamers in the ^{1}CT state. Since the ground state twist potential of **III** is very flat around φ=90° but very steep towards planarity (see Fig. 3.7), the obtained large values of $(\lambda_i'+\lambda_s)$ and $(f_g/f_e\lambda_i'+\lambda_s)$ below 250 K are consistent with the conclusion in ch.3 that the average rotamer is more planar at low temperatures.

Tab. 6.1 Results (in cm^{-1}) of Band-Shape Analysis for the Radiative Back Charge Transfer (-CT)

	in IpM			in EOE		
	I	II	III	I	II	III[a]
$\langle v_i \rangle$[b]	1260	1275	1315	1285	1275	1300 (1500)
$\lambda_i°$	1140	1335	1995	955	965	1950 (1665)
$(-\Delta G_{-CT})$ at 298 K	27800	27900	28700	26600	26400	26500
$d(-\Delta G_{-CT})/dT$	1.2/K[e]	-0.5/K[e]	-0.6/K[e]	4.6/K	6.0/K	9.0/K (4.8/K)
$-(\mu_e^2-\mu_g^2)/hca^3 \times df(\varepsilon_r)/dT$[d]	2.2/K	2.3/K	3.7/K	6.5/K	7.0/K	11.0/K
$(\lambda_s+\lambda_i')$ at 298 K	565	635	990	1450	1700	2200 (3260)
$d(\lambda_s+\lambda_i')/dT$	-5.6/K	-6.2/K	-3.4/K	-4.6/K	-3.7/K	-4.9/K (-8.3/K)
$(\mu_e-\mu_g)^2/hca^3 \times d(f(\varepsilon_r)-f(n^2))/dT$[d]	-0.2/K	-0.2/K	-0.3/K	-2.6/K	-2.9/K	-5.3/K

[a] free fitting of the spectra for III in EOE leads to a large uncertainty of the results and questionably high values for $\langle v_i \rangle$. Therefore, the presented fit results are obtained by the method usually applied in literature, i.e. using the fixed value $\langle v_i \rangle$=1300 cm^{-1} in accordance with that in IpM. The results derived from free fitting are given in brackets.
[b] constant within a scatter of less than 20 cm^{-1} except III in EOE where a fixed value is used (scatter of free fit procedure is more than 50 cm^{-1}).
[c] in IpM, differences of λ_i between 298K and 168K are 25 cm^{-1}, -120 cm^{-1} and -90 cm^{-1} for I, II and III.; for I and II in EOE, λ_i increases from 870 cm^{-1} at 168 K to 1050 cm^{-1} at 298 K, while λ_i of III varies between 1500 and 1800 cm^{-1} (between 1800 cm^{-1} and 2100 cm^{-1} using $\langle v_i \rangle$=1300 cm^{-1}).
[d] dielectric constants ε_r and refractive indices n are taken from ref. 25 and 30; excited state and ground state dipole moments for I, II and III (μ_e = 21.3 D, 22.0 D, 27.6 D and μ_g = 5.7 D, 5.7 D, 5.0 D) are derived in 3.3.2 using an Onsager radius a = 6 Å. Note, that dielectric continuum models generally fail for nonpolar solvents like IpM.
[e] equals to zero within the error limit of ± 1 cm^{-1}/K

In summary, the temperature dependence of the band-shapes of **I-III** in IpM and EOE support the view that the initial relaxation towards planarity is not hindered with decreasing temperature/increasing viscosity.

6.3.1.2 Analysis of the Fluorescence Lifetimes and Quantum Yields.

Tab. 6.2 Monoexponential or Biexponential Fit Parameters for Typical Fluorescence Decays Obtained by Single-Curve Analysis

solv.	T (K)	v_f (10^{3}cm^{-1})	τ_1 (ns)			τ_2 (ns)	χ^2|DW		
		I-III	I	II	III	III	I	II	III[a]
IpM	188	25.5	1.23	1.38	1.97	-	1.03|2.02	0.96|2.12	1.14|1.95
EOE	188	23.7	1.61	1.68	2.37	5.25	1.20|1.97	0.97|1.89	1.02|1.91
EOE	168	23.8	1.70	1.76	2.02	4.47	0.94|2.05	1.05|1.79	1.09|1.84
BCN	188	18.5	2.08	2.22	(1.90)[a]	12.9	1.12|2.03	1.04|1.94	1.12|1.97

[a] monoexponential fitting for III yields χ^2=4.1 in EOE at 188K, χ^2=1.9 in EOE at 168K (an additional lifetime of 100ps and 1% intensity was added here) and χ^2=1.14 in BCN at 188K (relative amplitude value -0.06 of τ_1 for biexponential fit is negligible).

In Fig. 6.5, the quantum yields of fluorescence (Φ_f) are plotted as a function of inverse temperature for **III** in solvents of different polarity and **II** in EOE. To distinguish between the influence of the radiative rate and nonradiative rate constants, $k_f(T)$ and $k_{nr}(T)$ on $\Phi_f(T)$, the fluorescence time traces $I_f(t)$ were measured as a function of the emission wavenumber v_f and the temperature T for **I-III** in different solvents like EOE, BCN and IpM. Some characteristic fit results are given in Tab. 6.2 (more lifetimes see ch. 10).

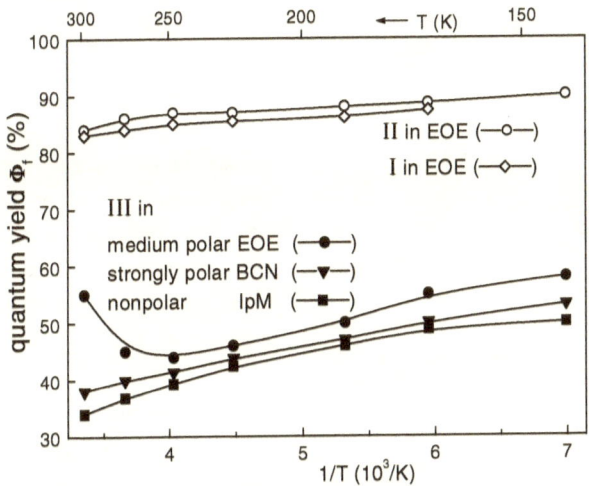

Fig. 6.5 *Fluorescence quantum yields in solvents of different polarity versus reciprocal temperature.*

From the exponential single curve and global analysis fitting procedures, it turned out, that irrespective of solvent, temperature or emission wavenumber, **I** and **II** behave similar and exhibit monoexponential fluorescence decays in the observed nanosecond window of 0.2 ns to 50 ns. Contrary to **I** and **II**, the fluorescence traces of **III** are strongly wavenumber dependent and decay biexponentially in EOE but also in polar solvents (BCN) at lower temperatures (T<188K). This proves that an additional conformational excited state relaxation occurs only for **III** in polar solvents. The spectral effect of this photoreaction from a primary poulated charge transfer species **CT** to a more relaxed species **CTR** is illustrated in Fig. 6.6 by the time-resolved emission spectra of **II** and **III** in EOE at 188 K. They are reconstructed from the wavenumber dependent decay curves $I_f(v_f,t)$ in different time windows Δt_x according to

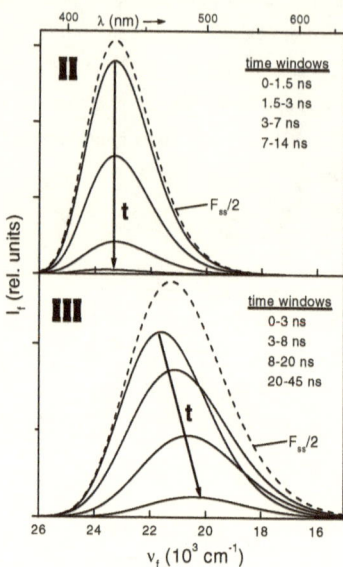

Fig. 6.6 Emission spectra of **II** and **III** in EOE at 188 K as a function of time t. The steady-state (=time integrated) spectrum $F_{ss}(v_f)$ is given in dashed line. The time-resolved spectra of **III** show a successive spectral shift of v_f from 21700 cm^{-1} (Δt=0-3 ns) to 20500 cm^{-1} (Δt=20-45 ns), while those of **II** stay at the position 23400 cm^{-1} equal to the maximum of $F_{ss}(v_f)$.

$$I_f(\Delta t_x, v_f) = F_{ss}(v_f) \frac{\int_{\Delta t_x} I_f(t, v_f) dt}{\int I_f(t, v_f) dt} \quad (6\text{-}5)$$

where $F_{ss}(v_f)$ is the relative steady-state fluorescence intensity and $\int I_f(t,v_f)dt$ represents the sum of all counts collected at v_f. Only **III** exhibits further spectral evolutions even after 20 ns. The fact that the emission spectrum of **II** does not change after 1.5 ns confirms that solvation effects can be excluded on this timescale.

The calculation of the radiative and nonradiative rate constants with eq. 6-6 is straightforward only for **I** and **II** as well as for **III** in nonpolar solvents, where monoexponential fits of the decays yield directly the fluorescence lifetimes $\tau_f(T)$.

$$k_f(T) = \frac{\Phi_f(T)}{\tau_f(T)} \quad k_{nr}(T) = \frac{1 - \Phi_f(T)}{\tau_f(T)} \quad (6\text{-}6\text{ ab})$$

Since both the relative weight and the value for the shorter lifetime τ_1 of **III** in BCN are distinctly smaller than for the long lifetime τ_2,[31] the $k_f(T)$ and $k_{nr}(T)$ values derived with $\tau_f = \tau_2$ are nevertheless representative for the emitting species. However, weights and magnitudes for τ_1 and τ_2 of **III** in EOE are of comparable size (see next section), such that the k_{nr} and k_f data using $\tau_2 = \tau_f$ have only formal character in this case.[32] Let us discuss the $k_f(T)$ and $k_{nr}(T)$ plots in Fig.6.7 and extract the essential informations needed to allow a quantitative characterization of the excited state equilibrium of **CT** and **CTR** in the subsequent section 6.3.2.

First of all, the slight increase of the $\Phi_f(T)$ in Fig. 6.5 with cooling can be traced back to a freezing of the nonradiative channels, while the radiative rates $k_f(T)$ stay constant. Moreover, the quantum yields of **III** (Fig. 6.5) are accidentally similar in all solvents because in nonpolar solvents, high k_f values are paired with high k_{nr} values, whereas in polar solvents low k_f values are paired with low k_{nr} values. The high k_{nr} values in a nonpolar solvent like IpM are explained in ch.5 by a close lying triplet state (1^3B) which becomes relatively higher lying with respect to ^1CT in more polar solvents leading to reduced k_{nr} values. Indeed, the values of k_{nr} for **CT** and

CTR are of similar magnitude ($\approx 10^8$ s^{-1}) in EOE, since the change of k_{nr} in EOE between 270 K and 170 K is small. The intermediate value at 298 K is conceivable with an equilibrium constant (K_{eq}=[CTR]/[CT]) close to unity as derived in ch.3.
The k_f(T) and the related Φ_f(T) data sets are both larger for **I** and **II** because of the planar geometry in S_1 connected with a stronger molecular orbital overlap of the HOMO(D) and LUMO(A) orbitals involved in the ^{1}CT→S_0 transition. The parallel temperature dependence of k_f(T) and k_{nr}(T) for **I** and **II** points to the same full relaxation of **II** to planarity at low temperature as at 298 K. The solvent dependent k_f(T) plots for **III** decrease with solvent polarity, since in nonpolar IpM only **CT**, in medium polar EOE both **CT** and **CTR** and in strongly polar butyronitrile (BCN) mainly the **CTR** (rotamer) species emits (Ch.3).[19,20]

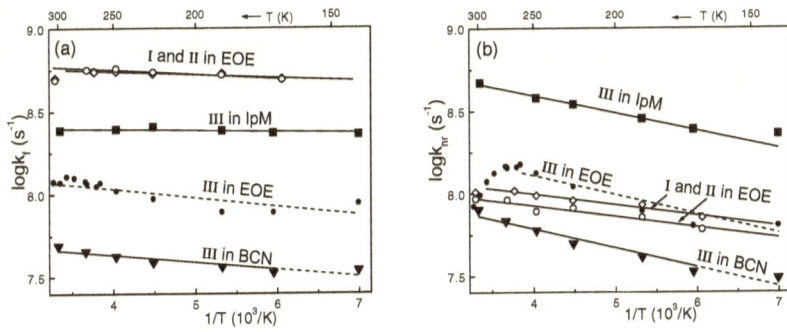

Fig. 6.7 (a) Fluorescence and (b) nonradiative rate constants versus reciprocal temperature. The lines are linear fits to obtain „Arrhenius parameters".

Hence, the photophysical and conformational behaviour at low temperature is equivalent to that at room temperature (ch.3). The small temperature dependence of the k_f and k_{nr} values is essential for the analysis in 6.3.2 of the **CT/CTR** equilibrium of **III** in EOE. It leads to the following boundary conditions:

(1) The small temperature dependence of k_f and k_{nr} for both **CT** and **CTR** corresponds to Arrhenius activation barriers less than RT (= 2.5 kJ/mol). Thus, the temperature influence on the S_1 deactivation rate constants τ_0(CT)$^{-1}$ and τ_0(CT)$^{-1}$ in scheme 6.2 is of minor importance compared to the strong temperature dependence of τ_1 or the derived photoreaction rate constants (Fig. 3.5 or Tab. 6.3 and Fig.6.11). This justifies to make use of the common practice[6] in case of unavailable model compounds to consider τ_0(CT)$^{-1}$ as approximately temperature independent within the above determined accuracy. The same reference value

$\tau_0(CT)$ = 4.3 ns as in Ch.3 is used here, i.e. τ_1 at the glass transition temperature ($T_g \approx$ 160 K), where reaction rates ($k_{CT \to CTR}$) are negligibly small. (The lifetime at 77K used usually as refernce can not be employed here, because the kinetic model changes below T_g due to a freezing of the initial Franck Condon relaxation. see ref. 33 and ch. 10)

(2) The constancy of $k_f(T)$ in lpM gives additional support to the view (see 6.3.1) that the initial relaxation to planarity, which should also take place in EOE prior to solvation, is not hindered by decreasing temperature. The kinetic analysis is therefore based on the precondition that only the **CT** species is populated by the initially excited **FC** species (cf. scheme 6.2). Further, the initial conformational relaxation of **FC** and the subsequent interconversion to **CT** is assumed to be too fast ($k_{FC \to CT}$ << 1/100 ps^{-1}) to be observed by the single-photon counting setup as confirmed by the fast rates $k_{FC \to CT}$ of **I-III** at 298 K (< 1/7 ps^{-1}) found with transient absorption in Ch. 4, as well as supported by the monoexponential fluorescence decay behaviour of **I** and **II** at low temperatures. Since the below derived reaction rate constants $k_{CT \to CTR}$ (1-30 ns in the investigated temperature range from 273 K to 168 K) are more than two magnitudes of order slower than $k_{FC \to CT}$, the kinetic analysis can be simplified by using a 2 species (**CT** and **CTR**) excited state model, where **CT** is instantaneously populated by the excitation pulse.

6.3.2 Temperature Dependence of the Conformational Equilibrium in the Excited State ^{1}CT of the Strongly Pretwisted D-A Biphenyl III in Diethylether

6.3.2.1 The Method to Recover Dual Fluorescence Bands and to Derive the Reaction Rate Constants by Global Analysis of Emission Decays

On the basis of the above defined boundary conditions (two excited species Birks scheme), the kinetic analysis concept of Löfroth[17] is applied to obtain the species associated fluorescence spectra, $SAS_{CT}(\nu_f)$ and $SAS_{CTR}(\nu_f)$, and the rate constants for the forward and back reaction, $k_{CT \to CTR}(T)$ and $k_{CT \leftarrow CTR}(T)$ of the **CT/CTR** excited state equilibrium of **III** in EOE at different temperatures. Therefore, a set of fluorescence decays $I_f(t,\nu_f)$ for a given temperature were recorded at several wavenumbers ν_f and simultaneously fitted to a biexponential function with linked lifetimes τ_1 and τ_2 (eq. 6-7) by the global analysis technique.[26]

$$I_f(t,\nu_f) = \alpha_1(\nu_f)e^{-\frac{t}{\tau_1}} + \alpha_2(\nu_f)e^{-\frac{t}{\tau_2}} \tag{6-7}$$

Typical decay curves and their biexponential fits shown in Fig. 6.8 as well as the temperature dependent fitting results collected in Tab. 6.3 demonstrate that (i) a monoexponential model fails, (ii) the biexponential model with two linked lifetimes is sufficient to achieve good fit quality parameters across the entire fluorescence band, (iii) the wavenumber dependence of the decays is due to a variation of the amplitudes $\alpha_1(v_f)$ and $\alpha_2(v_f)$, (iv) the lifetimes, in particular the short one τ_1, become longer with cooling, (v) τ_1 is observed as risetime in the red side of the spectra.

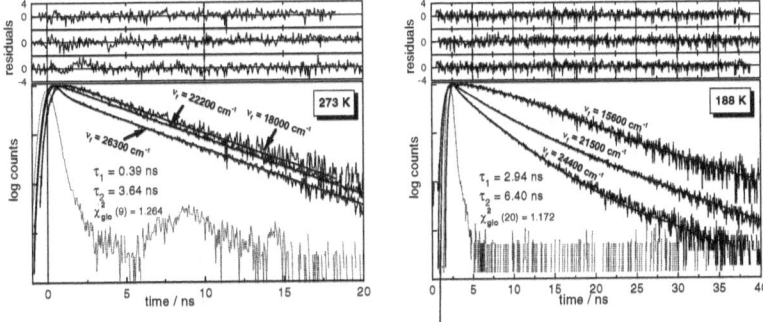

Fig. 6.8 *Wavenumber dependent fluorescence time traces of III in EOE at 273 K and 188 K and their globally fitted curves. The response function is drawn in dotted lines and the residuals (fit quality parameter) are plotted in the upper boxes (at top for the smallest wavenumber v_f).*

Tab 6.3 Fit Results of Temperature Dependent Global Analysis Procedures (see Fig. 6.9).

T [K]	nr.	χ^2_{glo}	τ_1 [ns]	τ_2 [ns]	α_1(blue)[a]	α_1(red)[a]	X [s^{-1}][b]	Y [ns^{-1}][b]
273	9	1.264	0.39	3.64	0.64	-0.42	1.86	1.0
248	13	1.147	0.78	4.23	0.77	-0.35	1.06	0.45
223	7	1.094	1.33	4.97	0.87	-0.44	0.66	0.30
188	20	1.172	2.94	6.40	0.93	-0.38	0.33	0.17
168	13	1.179	3.76	7.42	0.97	-0.29	0.26	0.14

[a] relative amplitudes obtained in the extreme blue and red part of the emission spectrum. It is important to note, that they are not extrapolated values. [b] obtained from eq. 6-12.

The temperature dependent datasets of wavenumber dependent amplitudes and linked lifetimes are then analyzed by the procedure illustrated in Fig. 6.9 for a representative temperature (T=188 K). In Fig. 9.6a, the relative amplitudes corresponding to the short lifetime τ_1 and the longer lifetime τ_2 are plotted versus the wavenumber v_f. A large and continuous variation of the amplitudes is found which is a signature for two excited state species emitting a smooth and structureless spectrum at different wavenumbers. The obervation of negative amplitudes (risetimes) in the low energy region is very important, because it unambiguously proves that a photoreaction from one excited species or distribution to another one takes place. If they were not observed, the two lifetimes could simply be due to two nonreactive species or a single distribution of varying lifetimes. The fact that the observed lifetimes are distinctly different and show a sizeable variation with temperature gives further confidence that the

Ch. 6 Temperature Dependent Fluorescence

hypothesis of nonreactive species can be rejected. Moreover, the absence of excitation wavenumber dependence for the temperature T=183 K was checked by globally analyzing the decays obtained in the absorption range between 29400 cm^{-1} to 33300 cm^{-1} at the emission wavenumbers 25000 cm^{-1}, 21700 cm^{-1}, 18000 cm^{-1}, as well as by the corresponding steady-state measurements. These experiments substantiate that the two lifetimes are not due to stable ground state species and give further support to the view that the initial Franck-Condon relaxation is sufficiently fast to populate exclusively **CT** instead of **CTR**. However, we have to note that the optimum value $\alpha_1(red)=-0.5$ (or $\alpha_2/\alpha_1(red)=-1$),[22] which would evidence the absence of primary **CTR** population,[8,21] is not reached (Tab. 6.3). The fact that the amplitudes or their ratios do not converge in the red edge indicates that there is still spectral overlap of the **CT** and **CTR** spectrum (see Fig. 6.10) which in turn can explain the values of $|\alpha_1(red)|$ somewhat lower than 0.5. It may be pointed out that extrapolation of relative amplitudes $\alpha_1(v_f)$ towards $v_f=0$ to even less negative values have been reported in the literature for photoreactions where direct excitation of the product species seems unlikely, too.[11,34,35] Following the convenient normalization of the amplitude spectra $\alpha_i(v_f)$ with the steady-state spectrum $F_{ss}(v_f)$ suggested elsewhere,[16-18] the decay associated spectra $DAS_i(v_f)$ are simply obtained by:[16]

$$DAS_i(v_f) = \frac{\alpha_i(v_f)}{\sum_i \alpha_i(v_f)\tau_i} F_{ss}(v_f) \tag{6-8}$$

So far, all interpretations are independent of the kinetic model. However, to obtain the fluorescence contributions of the species associated spectra $SAS_{CT}(v_f)$ and $SAS_{CTR}(v_f)$ at wavenumbers v_f

$$SAS_{CT}(v_f) = F_{ss}(v_f) \int_0^\infty \frac{I_f^{CT}}{I_f}(v_f,t)dt \quad SAS_{CTR}(v_f) = F_{ss}(v_f) \int_0^\infty \frac{I_f^{CTR}}{I_f}(v_f,t)dt \tag{6-9}$$

a kinetic scheme is necessary. Assuming ultrafast depopulation of **FC** we can apply scheme 6.2 leading to the following numerical expressions according to Löfroth:[17]

$$SAS_{CT}(v_f) = [DAS_1(v_f) + DAS_2(v_f)]\tau_1\tau_2 Y \tag{6-10}$$

$$SAS_{CTR}(v_f) = [(\tau_1^{-1} - Y)DAS_2(v_f) + (\tau_2^{-1} - Y)DAS_1(v_f)]\tau_1\tau_2 \tag{6-11}$$

In a spectral region where only the precursor species **CT** emits, $SAS_{CT}(v_f)$ in eq. 6-10 can be substituted by the steady state intensity $F_{ss}(v_f)$. Consequently, a plot of the apparent value $Y^{app}(v_f)$ across the emission spectrum

$$Y^{app}(v_f) = \frac{DAS_1(v_f) + DAS_2(v_f)}{F_{ss}(v_f)} \tau_1\tau_2 = \frac{\alpha_1(v_f) + \alpha_2(v_f)}{\alpha_1(v_f)\tau_1 + \alpha_2(v_f)\tau_2} \tau_1\tau_2 \tag{6-12}$$

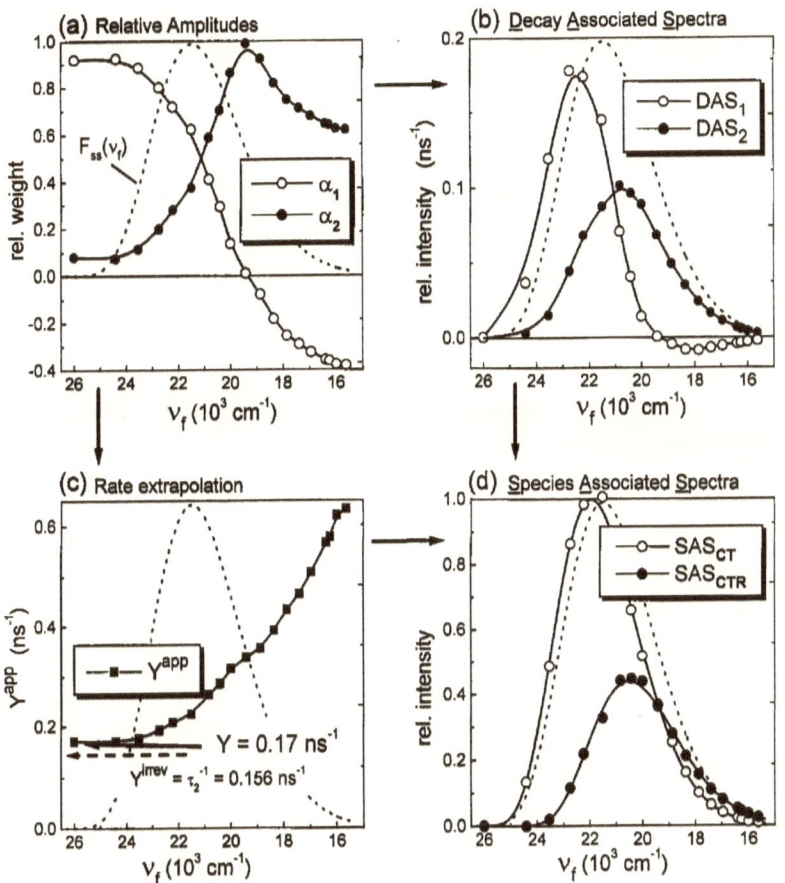

Fig. 6.9 *Diagram of the procedure to derive the photoreaction rate constants and species associated emission spectra from the global analysis results of wavenumber dependent emission decays using the results for* **III** *in EOE at 188 K.*

converges to the true value of Y at high wavenumbers, if the emission contribution of **CTR** approaches zero. The plot of Y^{app} vs. ν_f for **III** in EOE indeed reaches a plateau (Fig. 6.9c) towards the blue side of the emission spectrum for all temperatures. Hence, the remaining unknown values of Y and $X = \tau_1^{-1} + \tau_2^{-2} \cdot Y$ (cf. eq. 3-15) in scheme 6.2 are determined from $Y = Y^{app}(\nu_f \rightarrow \nu_f^{blue})$.

Ch. 6 Temperature Dependent Fluorescence

It may be interesting to note, that this procedure to obtain Y gives the same results as a plot of the frequently used[7-9] ratio DAS_2/DAS_1 (= α_2/α_1) vs. the blue emission edge (eq. 6-13) in the case of strong reversibility or small spectral overlap of the two bands.[22]

$$\frac{DAS_2}{DAS_1}(v_f \to v_f^{blue}) = \frac{\tau_2^{-1} - Y}{Y - \tau_1^{-1}} \qquad (6\text{-}13)$$

However, the plot of Y^{app} is superior to that of the DAS ratio in the (present) case of partial overlap and irreversibility (low DAS_2 or α_2 at v_f^{blue}) because Y^{app} converges already at lower v_f connected with lower derivatives dY^{app}/dv_f than $d(DAS_2/DAS_1)/dv_f$.

In Fig. 6.9c, the value of $Y^{irr} = \tau_2^{-1} = 0.156$ ns^{-1} for the purely irreversible case is indicated. Its small difference to the obtained value Y=0.17 ns^{-1} reveals that the photoreaction **CT→CTR** is in the irreversible regime. Using $Y = Y^{app}(v_f \to v_f^{blue})$ we can calculate the $SAS_{CT}(v_f)$ and $SAS_{CTR}(v_f)$ values (points in Fig. 6.9d or 6.10). The complete SAS spectra (lines in Fig. 6.10) are obtained by nonlinear least-squares fitting of the SAS values to a lognormal function $I_f(v_f)$ of the form given in eq. 6-14[36]

$$I_f(v_f) = I_f^{max} \exp[-\ln(2)\{\frac{\ln(1 + 2\gamma(v_f - v_f^{max})/\Delta}{\gamma}\}^2] \qquad (6\text{-}14)$$

where I_f^{max} and v_f^{max} are intensity and wavenumber of the emission maximum and Δ and γ are width and asymmetry parameters.

Finally, the Birks equations 3-15 can completely be solved, if $\tau_0(CT)$ = 4.3 ns is employed as discussed above in the first boundary condition and as determined in 3.3.3.4.[19] The forward and back reaction rates can then be calculated by

$$k_{CT \to CTR} = \tau_1^{-1} + \tau_2^{-1} - Y - \tau_0(CT)^{-1} \qquad (6\text{-}15)$$

$$k_{CT \leftarrow CTR} = (\tau_1^{-1} - Y)(\tau_2^{-1} - Y)Y^{-1} \qquad (6\text{-}16)$$

The temperature dependent SAS spectra and the Arrhenius plot of the reaction rate constants are shown below in Fig. 6.10 and Fig. 6.11, respectively.

6.3.2.2 Quantitative Characterization of the Excited State Conformational Equilibrium Between Two Charge Transfer Species CT and CTR

Changeover from kinetic to thermodynamic control From integration of the SAS spectra in Fig. 6.10, the fluorescence quantum yields of the **CT** and **CTR** species are calculated and reported in Tab. 6.4. The quantum yield Φ_f^{CTR} of the product species **CTR** and the ratio Φ_f^{CTR}/Φ_f^{CT} increases from 168 K to 248 K, which corroborates the conclusion that the yield of the **CT**→**CTR** photoreaction is enhanced with increasing temperature. At 273 K, however, the ratio Φ_f^{CTR}/Φ_f^{CT} is lower than at 248 K. This indicates the changeover from kinetic to thermodynamic control of the excited state equilibrium above 270 K, which means that the back reaction $k_{CT \leftarrow CTR}$ becomes competitive with the deactivation $\tau_0(\mathbf{CTR})^{-1}$ to the ground state (S_0).[4-6,8] Indeed, the back reaction rate constant $k_{CT \leftarrow CTR}$ (Fig. 6.11) reaches a similar magnitude as the deactivation rate constant to S_0 at 270 K ($10^{8.4}$ s^{-1}) estimated from k_f+k_{nr} in Fig 6.7. The minimum of $\Phi_f(T)$ at ≈ 265 K (Fig 6.6) is therefore a signature for the changeover (T_c in Tab.6.5) from the irreversible reaction mechanism ($T<T_c$) to the reversible reaction mechanism ($T>T_c$) where the back population of the **CT** species leads to an increase of Φ_f. To refine the above conclusions the

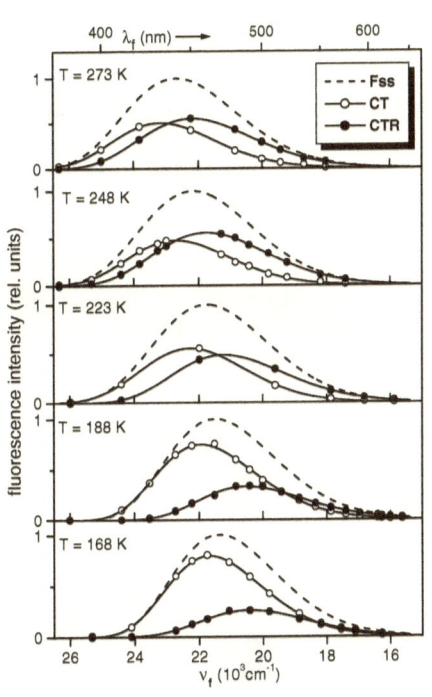

Fig. 6.10 Species associated fluorescence spectra of **CT** and **CTR** and the corresponding steady-state spectra Fss (- -) for **III** in EOE.

Tab. 6.4 Ratios of **CTR** to **CT** Fluorescence Quantum Yields, Radiative Rate Constants and Time-Integrated Excited State Concentrations for **III** in EOE.

T [K]	$f(\varepsilon_r)^a$	η [cP]^a	Φ_f^{CTR}	Φ_f^{CT}	Φ_f^{CTR}/Φ_f^{CT}	k_f^{CTR}/k_f^{CT}	[CTR]/[CT]
273	0.36	0.29	25 %	20 %	1.2	0.7	1.7
248	0.37	0.39	25 %	19 %	1.3	0.7	1.9
223	0.39	0.65	22 %	24 %	0.9	0.6	1.5
188	0.41	1.12	16 %	34 %	0.5	0.8	0.6
168	0.42	1.89	14 %	41 %	0.3	0.9	0.3

^{a} ref. 30

species associated fluorescence quantum yields can be expressed by

$$\Phi_f^{CT} = \frac{k_f^{CT}}{X} \qquad \Phi_f^{CTR} = \frac{k_{CT \to CTR} k_f^{CTR}}{XY} \qquad (6\text{-}17)$$

in the predominantly irreversible kinetic regime, and in the thermodynamic regime by

$$\Phi_f^{CT} = \frac{k_f^{CT} Y}{XY - k_{CT \to CTR} k_{CT \leftarrow CTR}} \qquad \Phi_f^{CTR} = \frac{k_f^{CTR} k_{CT \to CTR}}{XY - k_{CT \to CTR} k_{CT \leftarrow CTR}} \qquad (6\text{-}18)$$

if the conversion efficiencies $\eta_{FC \to CT} \cong 1$ and $\eta_{FC \to CTR} \cong 0$ are assumed (scheme 6.2). No matter which of both regions holds, the ratio of the **CTR** to **CT** quantum yields is given by

$$\frac{\Phi_f^{CTR}}{\Phi_f^{CT}} = \frac{k_f^{CTR}}{k_f^{CT}} \frac{k_{CT \to CTR}}{Y} = \frac{k_f^{CTR}}{k_f^{CT}} \frac{[CTR]}{[CT]} \qquad (6\text{-}19)$$

with $k_{CT \to CTR}$ (Fig. 6.11) and Y (Tab. 6.3) from the kinetic analysis which consequently allows to calculate the ratio of the **CTR** to **CT** radiative rate constants (k_f^{CTR}/k_f^{CT}) and the time-integrated concentration ratios ($[CTR]/[CT]=\int[CTR](t)dt/\int[CT](t)dt$).

It is found (Tab. 6.4) that the ratio of the radiative rate constants k_f^{CTR}/k_f^{CT} (0.7±0.2) is smaller than unity which characterizes the **CT** species as more emissive than the **CTR** species.[37] On the other hand, the concentration ratio [CTR]/[CT] sizeably decreases with cooling after a primary increase of [CTR]/[CT] from higher temperature down to 248 K in the thermodynamically controlled temperature region (T>T$_c$). The peculiar temperature dependence of ΔG_{-CT}, $(\lambda_s + \lambda_i')$ and HW'$_f^2$ from the bandspape analysis (Fig. 6.4) of **III** in EOE above 250 K is therewith explainable with a change of the temperature dependence of the [CTR]/[CT] population ratio.

Reaction activation barrier

The Arrhenius plots of the reaction rate constants in Fig. 6.11 yield activation barriers of about of $E_a(CT \to CTR) = 14$ kJ/mol for the forward and $E_a(CT \to CTR) = 17$ kJ/mol for the back reaction. They are significantly larger than the activation energy for the solvent mobility of EOE ($E_\eta = 7$ kJ/mol) and the Arrhenius barrier of DMABN ($E_a(LE \to CT) = 4$ kJ/mol)[7] demonstrating the existence of a high intrinsic reaction barrier ($E_i^\#$) beside a possible „viscosity barrier" induced by solvent drag.

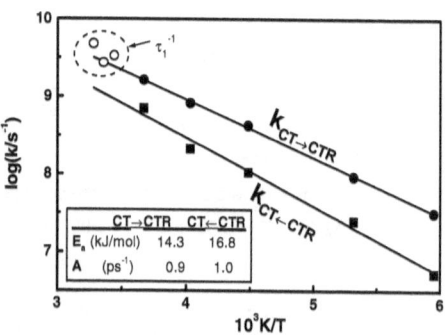

Fig. 6.11 Arrhenius plots of forward and back reaction rate constants. Activation parameters and short lifetimes τ_1 are also given.

The latter conclusion is supported by a comparison of temperature and pressure dependent experiments in ch.7. Besides, the Arrhenius barrier derived in 3.3.3.4 (E_a=15 kJ) assuming purely irreversible kinetics is close to the present value of the detailed analysis because the reaction mechanism is in fact predominantly irreversible.

Thermodynamics

Tab.6.5 Thermodynamic Data for the **CT→CTR** Photoreaction

ΔH	ΔS	ΔG(298K)	T_c^a	E_a	Δv_f^b	$\Delta(\Delta E_{FC})$
-2.5 kJ/mol	-0.7 J/K/M	-2.3 kJ/mol	265 K	14 kJ/mol	850 cm^{-1}	7.5 kJ/mol

[a] changeover temperature from kinetic to thermodynamic control [b] from solvatochromic plot in Fig. 6.12 using ε_r and n at 298 K

$$\Delta H = E_a^{CT \leftarrow CTR} - E_a^{CT \rightarrow CTR} \tag{6-20a}$$

$$\Delta S = R(\ln k_{CT \rightarrow CTR}^{T \rightarrow \infty} - \ln k_{CT \leftarrow CTR}^{T \rightarrow \infty}) \tag{6-20b}$$

$$\Delta G(T) = -RT \ln \frac{k_{CT \rightarrow CTR}(T)}{k_{CT \leftarrow CTR}(T)} \tag{6-20c}$$

$$\Delta(\Delta E_{FC}) = \Delta H - hc\Delta v_f \tag{6-20d}$$

The **CT→CTR** reaction enthalpy ΔH and entropy ΔS, the free reaction enthalpy ΔG at 298 K as well as the difference between the **CTR** and **CT** Franck Condon ground state energy are calculated by eq. 6-20 (see Tab. 6.5 and Scheme 6.4). It is noteworthy that ΔG(298K) is small and close to RT (2.5 kJ/mol) which agrees well with the estimation in 3.3.3.4.[19] Moreover, the change of the enthalpy ΔH (or the change of the potential energy by the reaction **CT→CTR**) is also small and much less than that reported for DMABN in toluene (-10 kJ/mol)[8] or in EOE (-17 kJ/mol).[7] The reason for this large difference in the reaction enthalpy ΔH between **III** and DMABN is due to the fact that for DMABN, a transformation occurs from a locally excited state ^{1}LE

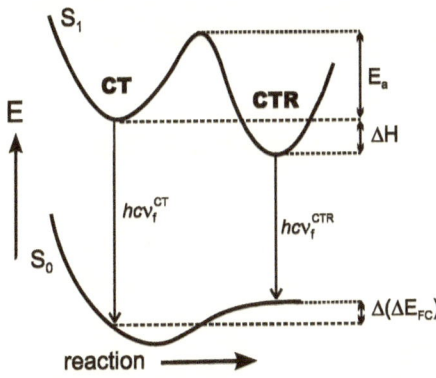

Scheme 6.4 Potential energy diagram of S_0 and S_1 for the photoreaction of **III** in EOE from **CT** to **CTR** in the solvent relaxed $^1CT(S_1)$ state.

(μ_e < 9.9 D)[9] to a ^{1}CT state (μ_e = 17 D)[9] which is strongly stabilized by the solvent, whereas the photoreaction of **III** takes place within a ^{1}CT state which increases its electron transfer character by a conformational change only slightly along the reaction path (see below).

Dipole moments of CT and CTR The dipole moments of the **CT** and **CTR** species are quantified in Fig. 6.12 by solvatochromic plots (Mataga eq. 3-7)[38] of the emission maxima v_f^{max} obtained from the lognormal fits of the SAS$_{CT}$ and SAS$_{CTR}$ values in Fig. 6.10.

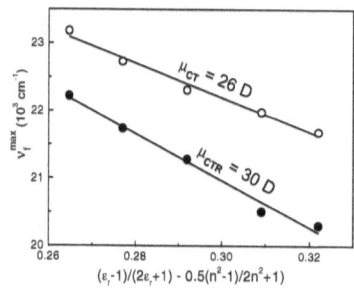

Fig. 6.12 *Lippert-Mataga plot to obtain species associated dipole moments of **CT** and **CTR**.*

The increase of the dipole moment from **CT** (μ_{CT} = 26 D) to **CTR** (μ_{CTR} = 30 D) amounts to 4 Debye revealing the increase of ^{1}ET character accompanied by the photoreaction. Further, these values are consistent with the average dipole moment ($\langle\mu_e\rangle$ = 27.6 D) obtained in 3.3.2 by steady state emission in different solvents.[19] This finding together with the larger radiative rate k_f of **CT** is in agreement with the previous assignment of **CT** to a more planar and **CTR** to a more twisted rotamer as compared to the average conformation in S_0.

Influence of Solvent Polarity, Viscosity and Temperature One might ask about the influence of the solvent properties such as solvent polarity (see $f(\epsilon_r)$ in Tab. 6.4) and viscosity (see η in Tab. 6.4) which both are increased with decreasing temperature. The increase of the polarity preferably stabilizes the more polar **CTR** species and thus could be expected to lead to an acceleration of the **CT**→**CTR** reaction connected with decreasing emission yield. But, the opposite behaviour is observed, since the change of polarity for the chosen conditions is only small as compared to the strong increase of viscosity or decrease of temperature (Tab. 6.4). Due to the increase of the excited state dipole moment along the **CT**→**CTR** path the polarity effect alone reduces the observed barrier, which even supports the existence of a high intrinsic barrier. On the other hand, with regard to the interpretation that an intramolecular rotation is involved in the photoreaction, a slowing down of the rates by the solvent viscosity is very probable and would therefore lead to an apparently higher thermal barrier. As investigated in the next chapter 7, the viscosity is indeed able to slow down the photoreaction.[39] A model including solvent-solute attraction will be proposed (Scheme 7.6) to explain why the initial relaxation towards planarity occuring in all biphenyls is not slowed down by viscosity to such low rates as the intramolecular relaxation from **CT** to **CTR**.

Even though the viscosity effect is not negligible, the temperature is the dominating factor for the photoreaction rates as evidenced by the comparison of temperature with pressure dependent fluorescence experiments (ch.7: cf. Figs. 7.1,7.4,7.7). Consequently, the observed energy barrier and the separated properties of the photoreactive species are in agreement with the calculation results in 3.3.4.3, where a barrier crossing process has been postulated only for **III** in medium polar solvents from a more planar to a more twisted structure as compared to the ground state situation.[19]

6.4 Conclusion

The thermodynamics and the kinetics of the adiabatic photoreaction occuring in the strongly pretwisted donor-acceptor biphenyl **III**, but not in the planar and slightly twisted compounds **I** and **II**, are quantitatively determined by a global fluorescence analysis method. Since the photoreaction is associated with (i) an increase of the dipole moment from 26 D to 30 D and (ii) a decrease of the fluorescence rate constants (factor ca. 0.7), which are both mainly intramolecular properties, the conformational reaction path can mainly be attributed to an intramolecular twisting from a more planar rotamer denoted by the precursor species **CT** to a more twisted rotamer denoted by **CTR**. The „R" of **CT*R*** stands for „*more relaxed*" to identify it as a photoproduct species in the ^{1}CT state which is undoubtedly associated with stronger ^{1}ET character than the precursor **CT** species.

The temperature effects on the equilibrium are such strong, that some deviations from the boundary conditions formulated in 6.3.2 would not affect the interpretation. The main uncertainty of the kinetic model used is the precondition that solely the less relaxed **CT** species is initially populated which requires a fast vibrational relaxation and interconversion **FC→CT** of the Franck-Condon species. Special attention has therefore been given to the influence of temperature on structural flexibility by the analysis of the bandshape and radiative rates for **I**-**III**. Both studies are consistent with an unhindered initial relaxation to planarity for **II** and **III**.

In contrast to DMABN in EOE, separated steady state fluorescence bands are not observed for **III** in EOE, because the difference of the potential energies ΔH (**III**: -2.5 kJ/mol, DMABN: -17kJ/mol)[7] and that of the ground state reorganization energies $\Delta(\Delta E_{FC})$ (**III**: 7.5 kJ/mol, DMABN: 40 kJ/mol)[40] are too similar for the charge transfer species **CT** and **CTR**. Further, a larger energy barrier E_a (**III**: 14.3 kJ/mol, DMABN:[7] 4 kJ/mol) connected with a smaller change of free energy ΔG (at 293K: **III**: -2.3 kJ/mol, DMABN: -6.1 kJ/mol) of **III**, is responsible for two magnitudes of order slower reaction rate constants (at 293K: **III**:$k_{CT \to CTR}=10^{9.4}$ s^{-1}, DMABN: k$^1_{LE} \to ^1_{CT}=10^{10.9}$s^{-1}) as compared to DMABN especially at low temperatures.[7] The reason for these differences between the photoreaction in **III** and DMABN are due to the fact that only the precursor species in DMABN has strongly locally excited character, such that a „steep downhill reaction" connected with a strong change of the electronic character to the more or less pure ^{1}CT state is observed in EOE. For **III**, however, the comparable „steep downhill reaction" ^{1}FC→^{1}CT investigated in Ch. 4 occurs prior to the photoreaction **CT→CTR**, such that small energy changes and a high intrinsic barrier are found, which are characteristic of an equilibrium between two conformer species of similar electronic character.

6.5 References and Notes

(1) (a) Aloisi, G.G.; Elisei, F.; Latterin, L.; Marconi, G.; Mazzucato, U. *J. Photochem. Photobiol.A: Chem.* **1997**, 105, 289. (b) Szymanski, M.; Maciejewski, A.; Kozlowski, J.; Koput, J. *J. Phys. Chem. A* **1998**, 102, 676. (c) Muralidharan, S.; Yates, K. *Chem. Phys. Lett.* **1992**, 192, 571. (d) Kovalenko, S. A.; Ernsting, N. P.; Ruthmann, J. *Chem. Phys. Lett.* **1996**, 258, 445.

(2) Maus, M.; Rettig, W.; Jonusauskas, G.; Lapouyade, R.; Rullière, C. *J. Phys. Chem* **1998**, 102, 7393.

(3) (a) Grabowski, Z.R.; Rotkiewicz, K.; Siemiarczuk, A.; Cowley, D.J.; Baumann, W. *Nouv. J. Chim.* **1979**, 3, 443. (b) Rettig, W. *Angew. Chem. Int. Ed.* **1986**, 25, 971. (c) Dey, J.; Warner, I.M. *J.Phys. Chem. A.* **1997**, 101, 4872 (d) Rotkiewicz, K.; Rechthaler, K.; Puchala, A.; Rasala, D.; Styrcz, S.; Köhler, G. *Chem. Phys. Lett.* **1996**, 98, 15.

(4) Rettig, W.; Chandross, E.A. *J. Am. Chem. Soc.* **1985**, 107, 5617.

(5) Rettig, W. *J. Luminesc.* **1981**, 26, 21.

(6) Rettig, W. *J. Phys. Chem.* **1982**, 86, 1970.

(7) Zachariasse, K.; Grobys, M.; von der Haar, T.; Hebecker, A.; Il'ichev, Yu.V.; Jiang, Y.-B.; Morawski, O.; Kühnle, W. *J. Photochem. Photobiol. A: Chem.* **1996**, 102, 59.

(8) Leinhos, U.; Kühnle, W.; Zachariasse, K.A. *J. Phys. Chem.* **1991**, 95, 2013.

(9) Schuddeboom, W.; Jonker, S.A.; Warman, J.M.; Leinhos, U.; Kühnle, W.; Zachariasse, K.A. *J. Phys. Chem.* **1992**, 96, 10809.

(10) von der Haar, T.; Hebecker, A.; Il'ichev, Yu.V.; Jiang, Y.-B.; Kühnle, W.; Zachariasse, K.A. *Receuil Trav. Chim. Pays-Bas Photochem.* **1995**, 114, 430.

(11) Albinsson, B. *J. Am. Chem. Soc.* **1997**, 119, 6369.

(12) Kasha, M., Discussions *Faraday Soc.* **1950**, 6, 14.

(13) (a) Livesey, A.K.; Licinio, P.; Delaye, M. *J. Chem. Phys.* **1986**, 84, 5102. (b) Livesey, A.K.; Brochon, J.C. *Biophys. J.* **1987**, 52, 693.

(14) (a) Lawton, W.H.; Sylvestre, E.A. *Technometrics* **1971**, 13, 617. (b) Saltiel, J.; Sears, D.F.; Choi, J.-O.; Sun, Y.-P.; Eaker, D.W. *J. Phys. Chem.* **1994**, 98, 35. (c) Spalletti, A.; Bartocci, G.; Masetti, F.; Mazzucato, U.; Cruciani, G. *Chem. Phys.* **1992**, 160, 131.

(15) (a) Ameloot, M.; Boens, N.; Andriessen, R.; Van den Bergh, V.; DeSchryver, F.C. *J. Phys. Chem.* **1991**, 95, 2047. (b) Hermans, E.; DeSchryver, F.C.; Dutt, G.B.; van Stam, J.; de Feyter, S.; Boens, N.; Miller, R.D. New. *J. Chem.* **1996**, 20, 829.

(16) Knutson, J.R.; Walbridge, D.G.; Brand, L. *Biochemistry* **1982**, 21, 4671.

(17) Löfroth, J.-E. *J. Phys. Chem.* **1986**, 90, 1160.

(18) Strehmel, B.; Seifert, H.; Rettig, W. *J. Phys. Chem.* **1997**, 101, 2232.

(19) Maus, M., Rettig, W., Bonafoux, D.; Lapouyade, R. submitted.

(20) (a) Maus, M.; Rettig, W.; Lapouyade, R. *J. Inf. Rec.* **1996**, 22, 451; (b) Rettig, W.; Maus, M. *Ber. Bunsenges. Phys. Chem.* **1996**, 100, 2091.

(21) Birks, J.B. *Photophysics of Aromatic Molecules*, Wiley-Interscience, New York, **1970**.

(22) The ratio of the preexponential factors (or the DAS values, see eq. 6-8) derived from ref. 21 for the biexponential time dependence of the precursor emission is given by

$$\frac{\alpha_2}{\alpha_1}(CT) = \frac{[CT]_{t=0}(\tau_2^{-1} - Y) - [CTR]_{t=0} k_{CT \leftarrow CTR}}{[CT]_{t=0}(Y - \tau_1^{-1}) + [CTR]_{t=0} k_{CT \leftarrow CTR}}$$

and that for the product emission by

$$\frac{\alpha_2}{\alpha_1}(CTR) = \frac{[CTR]_{t=0}(Y - \tau_1^{-1}) - [CT]_{t=0} k_{CT \to CTR}}{[CTR]_{t=0}(\tau_2^{-1} - Y) + [CT]_{t=0} k_{CT \to CTR}}$$

Ch. 6 Temperature Dependent Fluorescence

where α_1 and α_2 are the preexponenential factors for the short and long lifetime τ_1 and τ_2 (cf. eq. 6-7) and $[CT]_{t=0}$ and $[CTR]_{t=0}$ are the initial relative concentrations of the CT and CTR species populated by the FC relaxation. Both equations show that the desired value of Y can only be obtained from the amplitude ratios, if $[CTR]_{t=0}$ equals to zero. Otherwise, experiments with different concentrations of added quenchers which act differently on CT and CTR are necessary (ref. 15). Note also, that α_2/α_1(CTR) for pure CTR emission equals to -1, if $[CT]_{t=0}=0$.

(23) (a) Swiatkowski, G. *phD thesis „Intramolekulare Torsionsrelaxationsprozesse sterisch gehinderter unpolarer Aromaten"* **1982**, Berlin. (b) Swiatkowski, G.; Menzel, R.; Rapp, W. *J. Luminesc.* **1987**, 37, 183.

(24) Marcus, R.A. *J. Phys. Chem.* **1989**, 93, 3078.

(25) Riddick, J.A.; Bunger, W.B.; Sakano, T.K. *Organic Solvents*, John Wiley & Sons **1986**.

(26) Globals Unlimited®, Laboratory of Fluorescence Dynamics at the University of Illinois, **1992**.

(27) The increasing ^{1}ET character (cf. transformation A in scheme 9.2) of I and II in IpM at lower temperature is resonsible for the loss of structure and the absence of a usual band narrowing. Note, that in the low energy region, the fitted spectra exhibit more vibrational structure than the experimental spectra, because the low-energy emitting electronic species contain more ^{1}ET character.

(28) Onsager, L. *J. Am. Chem. Soc.* **1936**, 58, 1486.

(29) Böttcher, C. J. F. *Theory of Electronic Polarisation*, Elsevier, Amsterdam **1973**.

(30) Landolt-Börnstein, *Zahlenwerte und Funktionen aus Physik, Chemie, Astronomie, Geophysik und Technik*, **1969**, Springer Verlag, Berlin. Vol. 2(5), ed. K.H.Schäfer; Landolt-Börnstein, New series IV/6

(31) The wavenumber integrated intensity yield for the exponential terms of the long time component τ_2, which is given by $\int \dfrac{\alpha_2(\nu_f)\tau_2}{\sum_i \alpha_i(\nu_f)\tau_i} d\nu_f$, amounts to 100%, 95% and 70% for III in BCN at 188K (τ_1<50ps, τ_2= 13ns), 168K (τ_1= 3.9ns, τ_2= 15ns) and 143 K (τ_1=4.6ns, 15.4ns), respectively. This clearly demonstrates the dominance of τ_2 with respect to τ_1 for polar conditions also at low temperatures.

(32) In fact, τ_2 in EOE results from two counteracting temperature effects: On the one hand, with cooling the CT→CTR reaction becomes more irreversible introducing enhanced CTR character to τ_2, but on the other hand, the actual concentration of CT increases at lower temperatures giving more weight of CT character to τ_2. In spite of that, τ_2 has to be preferred to other values like the mean value at ν_f^{max} because the contribution of the reaction rates is relatively small over the entire temperature range only in τ_2.

(33) Maus, M.; Rettig, W. *BESSY Annual Report* **1995**, 169.

(34) Yip, R.W.; Wen, Y.-W.; Szabo, A.G. *J. Phys. Chem.* **1993**, 97, 10458.

(35) Heisel, F.; Miehé, J.A. *Chem. Phys.* **1985**, 98, 233.

(36) Fraser, R.D.B.; Suzuki, E. in *Spectral Analysis*; Blackburn, J.A., Ed.; Marcel Dekker: New York. **1970**; p.171.

(37) The reliability of k_f^{CTR}/k_f^{CT} suffers mainly from a possible inaccuracy of the relative amplitudes in the spectral area of high relative fluorescence intensity. In particular, the difficulty to determine the k_f ratio of a system with a relatively small extent of the photoreaction, here at the lowest temperatures, has been stressed and discussed in detail in ref.9.

(38) Mataga, N.; Kaifu, Y; Koizumi, M. *Bull. Chem. Soc. Jpn.* **1956**, 29,465.

(39) Maus, M.; Rettig, W. *J. Inf. Rec.* **1998**, 24(5/6), 461.

(40) For DMABN, $\Delta(\Delta E_{FC})$ = 73kJ/mol was estimated in ref. 7 from $\Delta(\Delta E_{FC}) = hc\nu_a(^1LE) + \Delta H - hc\nu_f(^1CT)$ with the absorption energy $hc\nu_a$ of the ^{1}LE state. This neglects ΔE_{FC} and the stabilization energy of the ^{1}LE state. $\Delta(\Delta E_{FC})$ is therefore recalculated here by eq. 6-20d using ΔH=-17kJ/mol, $\nu_f(^1LE)$=28600 cm^{-1} and $\nu_f(^1CT)$=23850 cm^{-1} from ref. 7 and 10. Note, that the derived value of $\Delta(\Delta E_{FC})$ for III agrees well with the result from the band-shape analysis using $\Delta(\Delta E_{FC})/hc=[\lambda_i(\text{III in EOE})-\lambda_i(\text{III in HEX})+(\lambda_s+\lambda_i')(\text{III in EOE})-(\lambda_s+\lambda_i')(\text{II in EOE})$.

Chapter 7

Pressure and Temperature Dependent Fluorescence of D-A Biphenyls. The Separation of Viscosity and Thermal Control of the Conformational Photoreaction in a Highly Twisted Compound

Abstract:
Steady-state and time-resolved fluorescence experiments as a function of pressure and temperature in triacetine are utilized to separate the influence of a thermal activation barrier from the viscosity control of the slow adiabatic photoreaction occurring for a highly twisted donor-acceptor biphenyl. The kinetic analysis in the irreversible reaction regime reached below 290 K yielded an intrinsic activation barrier of about 9 kJ/mol and a viscosity power law parameter of about 0.2-0.3. These results are consistent with a large amplitude conformational change in the excited state such as a torsional relaxation.

Keywords: fluorescence, pressure, viscosity control, activation energy, biphenyl, conformer

7.1 Introduction

In recent years, high-pressure experiments gained increasing attention to analyze viscosity controlled photoprocesses.[1-8] During the past 25 years, Drickamer et al. demonstrated the usefulness and variability of pressure experiments in analyzing photophysics of several molecules.[1] In addition, the group of Hara investigated various adiabatic photoreactions and stressed the superiority of pressure induced viscosity as compared to the application of solvent mixtures.[2-4] Troe et al. introduced an important variation by measuring isothermic sets of pressure dependent rate constants which enables the extraction of isoviscous activation barriers.[5] It was found that the barrier height in the photoisomerization of trans-stilbene is viscosity dependent through a change of the solvent polarizability. Very recently, Rettig et al. put forward the idea to compare pressure dependent kinetic data sets with the analogous temperature data sets on a common viscosity scale.[6,7] This ratioing approach allows directly to extract the intrinsic thermal barrier and compensates undesirable kinetic effects by viscosity induced changes of solvent properties such as the polarity.

Especially the increase of the density and polarity with increasing viscosity at high pressures is a crucial factor. For example, Zachariasse et al. demonstrated that the photoreaction for dimethylaminobenzonitrile (DMABN) dissolved in diethylether (EOE) is accelerated with increasing pressure/viscosity due to the polarity effect.[8] Thus, the quantitative analysis of such polarity-dependent reactions from a locally excited (^{1}LE) to a ^{1}CT state requires either elaborate corrections or a medium with a much smaller polarity change as compared to the viscosity change. In contrast to EOE, triacetine (TAC) fulfills this requirement with a change of the viscosity over more than 5 orders of magnitudes (η=10 - 3×10^6 mPas) in the pressure range from 1 bar to 5000 bar.[9] Using this solvent, the photoreaction of DMABN is slowed down with increasing pressure, which has been interpreted with a domination of the viscosity instead of the polarity effect.[6,7,10]

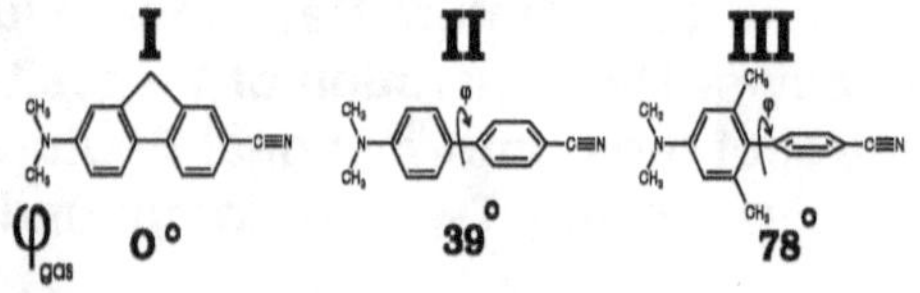

Scheme 7.1 *Molecular Structures of the investigated Donor-Acceptor Biphenyls*

The intramolecular photodynamics of the donor-acceptor biphenyls **I-III** shown in scheme 7.1 have been interpreted with a fast initial relaxation in the primary excited state (^{1}FC; cf. scheme 7.2) towards a more planar structure than in the ground state followed or accompanied by an electron transfer interconversion denoted in scheme 7.2 by the reaction of the kinetic species **FC** into **CT**.[11-13] Only the highly twisted biphenyl **III** in dipolar solvents exhibits an additional relaxation which was attributed to a slow conformational reorganization from a more planar conformation (species **CT**) towards a more twisted structure (**CTR**) as compared to the average ground state conformation (ch.3,6).[12] This implies that a large amplitude motion such as the rotation around the interannular bond is involved in the slow photoreaction, which should be observable by pressure dependent fluorescence experiments.

With reference to the investigations on the excited state equilibrium of **CT** and **CTR** in Ch. 6,[14] this chapter focuses on the question to what extent the adiabatic photoreaction of **III** from species **CT** to **CTR** is controlled, on the one hand by viscosity and, on the other hand, by an intrinsic activation barrier. It starts with a comparative photophysical study of **I-III** in triacetine to obtain evidence for the unhindered initial relaxation towards planarity and to characterize the **CT/CTR** equilibrium by a comparison of pressure and temperature dependent steady-state spectra. Afterwards, the application of the irreversibe reaction scheme 7.2 for the kinetic analysis is discussed followed by a derivation of the theoretical background used to finally analyze the pressure and temperature dependent reaction rate constants $k_{CT \to CTR}$.

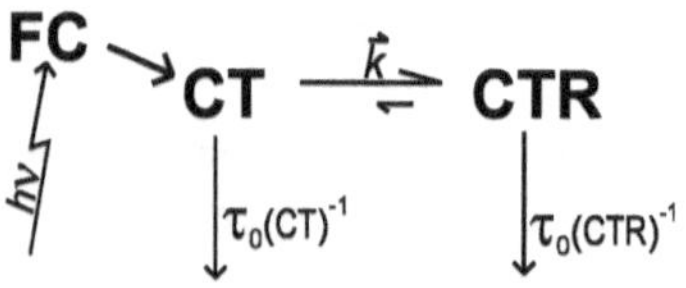

Scheme 7.2 *Kinetic Model for the photoinduced relaxations of* **III** *in the irreversible regime.*

7.2 Experimental

Setup, Materials and Data Analysis   See 3.2 and 6.2[12,15]

7.2.1 High Pressure Equipment   A high-pressure (10 kbar) optical cell made of stainless steel in a shrink construction with four saphire windows was used in a pressure range from 1 bar up to 4.6 kbar and was thermostatized in a copper jacket by water flow from a thermostat. The sample quartz cuvette in the optical path of the cell is cylindric in the upper part such that a teflon piston can be pressed into the cuvette by increasing pressure. The whole cell is connected by high-pressure (10 kbar) pipes to a pressure gauge, a hand-driven high pressure pump and a vessel which supplies the system with spectrograde ethanol as the pressure transmitting medium. The apparatus was limited by the valves to pressures below 5 kbar. More technical details are thoroughly described in ref. 16.

7.2.2 Ti:Sapphire Laser   The third harmonic output of a regenerative mode-locked argon-ion laser-pumped Ti:Sapphire laser in the femtosecond mode at a repetion rate of 81.3 MHz was used to excite the sample at 300 nm and to obtain fluorescence decays with a sub-ps resolution. The time resolution was limited by the microchannel plate yielding a response function of 35 ps. Using a similar single photon counting setup as described in 3.2.6 the signals were sampled in 8192 channels of a multichannel analyzer integrated in a PC (time division 2.7 ps/channel).

7.3 Results and Discussion

7.3.1 Viscosity and Temperature Influence on the Fluorescence and Excited State Relaxations of I-III in triacetine (TAC).

The temperature dependence of the fluorescence spectra, maxima, the globally fitted lifetimes (see 7.3.2) and the derived photophysical properties are shown for **I-III** in Fig. 7.1, 7.2, 7.3 and Tab. 7.1, repectively. A fluorescence band-shape analysis related to Marcus electron transfer theory performed in the same way as in ch. 6, gives no additional important informations for this study. Let us first compare **I** and **II** which are regarded as the model compounds (**I**) where the excited state interannular twist relaxation is absent and (**II**) with the respective fast relaxation towards planarity (ch.3 and 6). Both compounds exhibit the same variations of photophysical behaviour with lowering the temperature, i.e. (i) a slight and continuous increase of the fluorescence intensity or quantum yield (Fig. 7.1ab, Tab. 7.1); (ii) no change of the emission maximum v_f between 300 K and 270 K followed by an increase of v_f at lower temperatures (Fig 1ab, Fig. 7.2) (iii) practically no variation of the fluorescence lifetimes

(Fig. 7.3) and consequently, (iv) similar radiative and nonradiative rate constants (Tab. 7.1). These observations may be taken as evidence that the initial relaxation of **II** and likewise that of **III** towards a more planar conformation in the excited state is not hindered up to extremely high viscosities ($\eta(235\ K) \approx 10^5 mPas$).[9,17,18] The static dielectric constant ε_r, which characterizes the solvent polarity in the relaxed equilibrium of the solvent-solute arrangement through $f(\varepsilon_r)=(\varepsilon_r-1)/(2\varepsilon_r+1)$, increases with cooling from 6.9 ($f(\varepsilon_r)=0.40$) at 300 K up to 8.4 ($f(\varepsilon_r)=0.42$) at 235 K.[19] However, the observed blue-shift of the emission maxima v_f with cooling obviously demonstrates, that full solvent relaxation is not reached during the excited state lifetime of **I-III**, such that $\varepsilon_r(T)$ can not be used as a measure of the solvent polarity in TAC. The nonlinear increase of viscosity η with cooling leads to a sizeable decrease

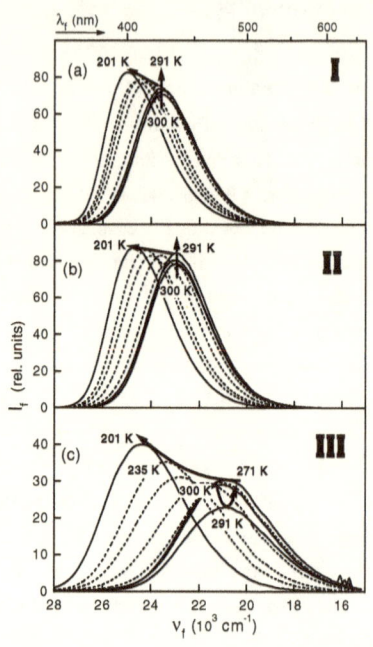

Fig. 7.1 Corrected fluorescence spectra (error $\Delta I_f(v_f) \approx 10\%$) of **I-III** in TAC with decreasing temperature.

of the effective solvent polarity below 270 K where the onset of the blue shift of v_f occurs (Fig. 7.2). Since the average dipole moment increases from **I** to **III**, the slope $-dv_f(T)/dT$ increases in the same direction (Fig. 7.2).

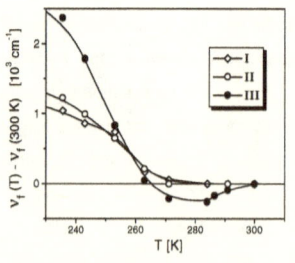

Fig. 7.2 Temperature dependence of the fluorescence maxima.

Only for the highly twisted biphenyl **III**, an initial red shift of v_f is observed from 300 K to 280 K and the resulting minimum in Fig. 7.2 therefore indicates that an additional mechanism is active in **III**. This minimum is in excellent agreement with the minima of the fluorescence quantum yields $\Phi_f(T)$ (see Fig. 7.1c and Tab. 7.1) and of the lifetime $\tau_1(T)$ (Fig. 7.2), which all are observed at (285 ± 5) K. Analogously to the behaviour of $\Phi_f(T)$ and $\tau_1(T)$ in diethylether EOE (ch.6), this temperature (T_c) can be attributed to the changeover from thermodynamic to kinetic control of the photoreaction between the **CT** and **CTR** conformational species. As expected, the higher viscosity of TAC extends the irreversible regime from $T<T_c=265$ K in EOE to $T<T_c=285$ K in TAC, because the back reaction **CT←CTR** is slowed down by the viscosity. The red shift of v_f from 300 K to 280 K is therewith consistent with enhanced population of the **CTR** species emitting at lower frequencies due to its larger dipole moment than that of **CT** ($\mu_{CTR} = 30$ D, $\mu_{CT} = 26$ D). This effect is not observed for **I** and **II**, since they possess only a single and temperature independent conformation.

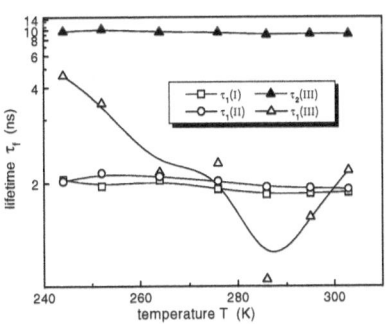

Fig. 7.3 Temperature dependence of the (two) longest lifetime(s) derived from (triexponential) biexponential fitting of the wavenumber dependent fluorescence time traces by the global analysis technique (see 7.3.2).

In order to exclude thermal effects and to solely investigate the viscosity influence on the fluorescence of **III**, which probes the excited state equilibrium of **CT** and **CTR**, the emission spectra are recorded at different pressures p (1 bar to 4 kbar) at a constant temperature of 295 K (Fig. 7.4). The continuous blue-shift of the spectra with pressure guarantees that the increase of the dielectric constant with increasing viscosity or density is not important.

Tab. 7.1 Fluorescence Quantum Yields Φ_f, Radiative $k_f = \Phi_f/\tau_f$ and Nonradiative Rate Constants $k_{nr} = (1-\Phi_f)/\tau_f$ of **I**-**III** in Triacetine.

T (K)	Φ_f (%)			k_f ($10^7 s^{-1}$)			k_{nr} ($10^7 s^{-1}$)		
	I	II	III	I[a]	II[a]	III[b]	I[a]	II[a]	III[b]
300	80	80	59	42	42	9	11	11	6
291	81	80	46	42	42	7	10	10	8
235	88	87	71	44	43	15	7	7	6

[a] using $\tau_f = \tau_1$ (see 7.3.2 and eq. 7-1)
[b] using mean lifetime of τ_1 and τ_2 at emission maximum $<\tau_f>=(\alpha_1\tau_1+\alpha_2\tau_2)/(\alpha_1+\alpha_2)$.

The occurrence of an intensity minimum at intermediate pressure (2kbar) reveals the same mechanism as in the temperature experiment in Fig. 7.1c and therefore proves that the photoreaction between **CT** and **CTR** is viscosity controlled. In particular, the decrease of the fluorescence intensity from the spectrum at lowest viscosity (T=300 K in Fig. 7.1c or p=1 bar in Fig. 7.4) to that with minimum intensity (291 K or 2000 bar) denotes the suppression of the back reaction **CT←CTR** relative to the rate of fluorescence in both experiments. Very importantly, the intensity minimum of the temperature dependent spectra is reached already at a viscosity (η(291 K) = 22 mPas) two orders of magnitude less than for the respective pressure experiment (η(2 kbar,296 K) = 1710 mPas). This vividly demonstrates that viscosity alone can not account for the reduction of the backward photoreaction rate in the temperature experiments in TAC and EOE, but that a strong contribution of a thermal barrier is involved. Of course, the same freezing mechanism is observed for the forward reaction **CT→CTR** by the intensity increase from 291 K to 201 K (Fig. 7.1c) or analogously from 2 kbar to 4 kbar (Fig. 7.4). Also in this case, the lowering of temperature (with the addition of the accompanied increase of viscosity) is more efficient than the increase of viscosity only, since at a common viscosity value of about 10^5 mPas (4kbar in Fig. 7.4 and 235 K in Fig. 7.1), the relative fluorescence intensity is significantly higher in the temperature variation experiment. Note, that the mean nonradiative rate constants <k_{nr}> determined for **III** remain constant with cooling (Tab. 7.1). This points to the absence of strongly temperature dependent radiationless channels and confirms the increase of the time-integrated population of the more emissive **CT** species with cooling (see also the larger k_f values at 235 K than at 291 K)

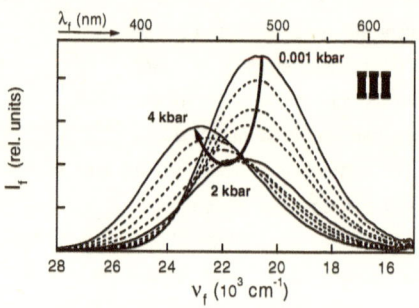

Fig. 7.4 *Corrected fluorescence spectra (error $\Delta I_f(v_f) \approx 15\%$) of* **III** *in TAC with increasing pressure at 295 K.*

To further substantiate and quantify the viscosity control and the thermal activation of the forward photoreaction in **III**, time-resolved emission as a function of temperature and pressure is analyzed in the following.

7.3.2 Analysis of Reaction Rate Constants and its Boundary Conditions

The kinetic analysis rests on scheme 7.2 where the reaction channel from **CT** to **CTR** is active only in **III**, but not in **I** and **II**. Accordingly, the fluorescence decays of **I** and **II** could be well fitted to a biexponential function, while those of **III** needed one additional exponential term to achieve fits of comparable quality ($\chi^2 \leq 1.2$) as demonstrated in Fig. 7.5.

$$I_f(t,v_f) = \alpha_s(v_f)e^{-\frac{t}{\tau_s}} + \alpha_1(v_f)e^{-\frac{t}{\tau_1}} + \alpha_2(v_f)e^{-\frac{t}{\tau_2}} \quad (7\text{-}1)$$

The decay curves in Fig. 7.5 were recorded using a Ti:sapphire laser single photon counting setup which allowed to determine the short time component τ_s with a better precision than with the BESSY setup (see 3.2) used for the lifetime measurements below. Although slightly larger, the magnitudes of the shortest lifetimes τ_s decrease in the same order as the interconversion lifetime $\tau_{FC \to CT}$ obtained by the transient absorption measurements (ch.4), namely from **I** to **III**. This indicates that the $^1FC \to {}^1CT$ interconversion is followed by solvation processes which are dominantly observed in the present fluorescence experiment. Both experiments of **I-III** in TAC also agree that the additional relaxation process in **III** ($\tau_2 =$ 8 ns), which is considerably slower than the solvent relaxation time of TAC (Tab. 4.3: $\tau_l \approx 150$ ps; $\tau_s \approx 690$ ps),[10] is absent in **I** and **II**.

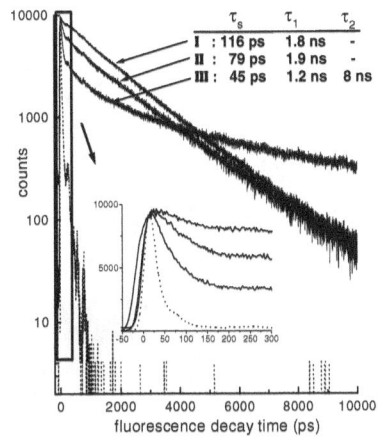

Fig. 7.5 *Fluorescence decay curves of* **I-III** *in triacetine at 298 K ($\lambda_{ex}/\lambda_{em}$ = 300 nm / 395 nm). The apparatus response function (fwhm = 35 ps) of the single photon counting setup using a fs Ti:sapphire laser is shown as a dotted line.*

The complex decay behaviour of **III** is shown in Fig. 7.6 by two typical sets of emission decays recorded across the entire fluorescence spectrum. Both sets are obtained for isoviscous conditions, one at lowered temperature (a) and the other at elevated pressure. Note, that the time traces at high pressure (b) always decay faster than those at low temperature (a) for comparable wavelengths. This agrees with the conclusion above that the photoreaction is more effectively frozen by a temperature reduction due to the additional influence of a thermal barrier. Moreover, the appearance of τ_2 as a risetime in the long wavelength region ensures that the adiabatic photoreaction is indeed active.

Unfortunately, the kinetic analysis of the photoreaction rate constants $k_{CT \to CTR}$ can not be done in the same way as in the previous chapter 6, because the high viscosity of triacetine leads to (i) the observation of the additional short lifetime (τ_s) which is attributed to solvation processes slower than the **FC→CT** interconversion (ch. 4 and 5) and (ii) a much stronger wavenumber dependence of the relative amplitudes α_i and the associated lifetimes which is the well-known consequence of non-exponential relaxation components in solvents with high friction. For molecules like DMABN showing dual fluorescence bands due to a $^1LE \to {}^1CT$ adiabatic photoreaction, it was suggested to determine the average survival time $<\tau>$ of LE

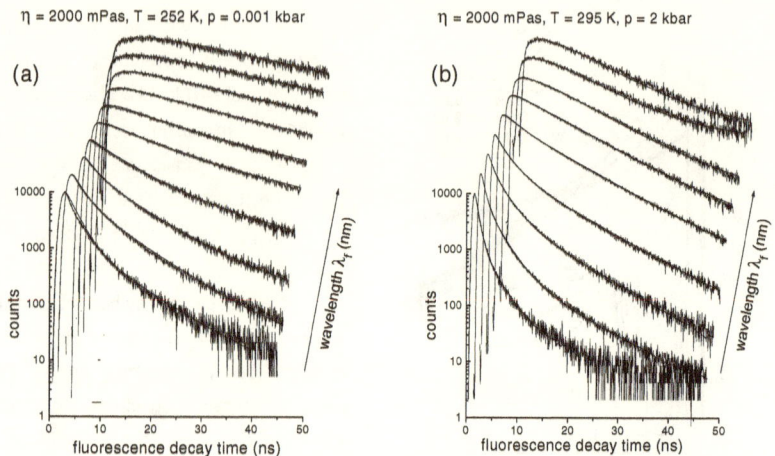

Fig. 7.6 *Wavelength dependent (λ_f = 380-610 nm) fluorescence time traces and global analysis fits of* **III** *in triacetine for isoviscous conditions under (a) reduced temperature and (b) increased pressure.*

$$\langle \tau \rangle = \sum_i \alpha_i \tau_i \qquad (7\text{-}2)$$

from a multiexponential fit for the decay of the ^{1}LE band and to calculate the ^{1}LE→^{1}CT reaction rate constant by $k^1_{LE \to \, ^1CT} = \langle \tau \rangle^{-1} - \tau_{ref}^{-1}$.[6,7,3]

However, this method is not applicable here because in contrast to DMABN, there is spectral overlap of the **CT** and **CTR** bands (ch.6) such that $\langle \tau \rangle$ would obtain „wrong" contributions of the long lifetime belonging to **CTR** even in the irreversible regime and (ii) the solvation process of the **CT** species responsible for the shortest lifetime τ_s is not involved in the **CT→CTR** photoreaction of **III**, but follows the fast (Ch.4: $\tau_{FC}(\mathbf{III})$=10 ps at 298 K)[11] **FC→CT** interconversion and may be regarded, in a simplified picture, as completed prior to a significant onset of the **CT→CTR** reaction. Hence, a method is needed which excludes the wrong contributions of both solvation and **CTR** decay. Here, the wavenumber dependent fluorescence decays (Fig. 7.6) for a given system (molecule/pressure/temperature) are simultaneously fitted by the global analysis procedure (cf. Fig. 6.9) using variable relative amplitudes corresponding to three (or two for **I** and **II**) linked lifetimes τ_s, τ_1, τ_2 (τ_s, τ_1). Since they are obtained from averaging over the entire spectrum and not only for a single wavenumber as eq.7-2, each lifetime is a representative average for one of the three lifetime distributions resulting from physically meaningful three-excited species kinetics (cf. scheme 7.2). Next, the first boundary condition for a possible solution of the kinetics is the application of the irreversible reaction scheme **FC→CT→CTR** which is based on the following experimental facts:

Concerning the back reaction **FC←CT**:

This neglection is justified, since the lifetime of the **FC** species (τ_{FC} = 10 ps) and the solvent reorientation times are distinctly shorter than the relaxation time of the **CT→CTR** photoreaction ($\tau_s^{-1} > 10 \times \tau_1^{-1}$),

Concerning the back reaction **CT←CTR**:

(i) The temperature and pressure dependent fluorescence quantum yields determined from the spectra shown in Fig. 7.2 and 7.5 allow to mark out the viscosity range for the irreversible regime. The calculation of the rate constants $k_{CT \rightarrow CTR}$ is therefore restricted to the viscosity range η larger than 200 mPas for the pressure and larger than 50 mPas for the temperature dependent experiments. The success of this simplified irreversible scheme (ch.3) has been demonstrated for **III** even in the less viscous (η < 2 mPas) solvent diethylether where practically the same activation barrier is obtained with a full analysis[14] taking into account the back reaction (ch.6).

(ii) Since TAC (ε_r=7) is more polar than EOE (ε_r=4.2) and the product species **CTR** has a higher dipole moment than **CT** (μ_{CT}=26D; μ_{CTR}=30D), the change of the solvent stabilization energy ΔE_{solv}(CT→CTR) at 298 K can be estimated to change from -22 kJ/mol in EOE to -26 kJ/mol in TAC on the basis of a dielectric continuum model (eq.7-3).[20]

$$\Delta E_{solv}(CT \rightarrow CTR) = -\frac{\mu_{CTR}^2 - \mu_{CT}^2}{a^3} f(\varepsilon_r) \qquad (7\text{-}3)$$

As a result, the polarity-induced higher energy barrier for the back reaction **CT←CTR** and lower barrier for the forward reaction **CT→CTR** should shift the irreversible regime to lower viscosity and higher temperatures as compared to EOE. Anticipating viscosity dependent reaction rates, the higher viscosity should even have a stronger effect by a considerable decrease of the backward rate relative to the viscosity independent rate constant τ_0(CTR)$^{-1}$ to S_0. This exactly corresponds to the different changeover temperatures (T_c(EOE)=265 K; T_c(TAC)= 291 K) derived above (7.3.1) and in 6.3.2.2.

(iii) Whereas τ_1 changes strongly with the large viscosity variations, τ_2 is found relatively constant ($\log(\tau_2^{-1} / s^{-1})$ = 8.5 ± 2%) indicating only a minor contribution of the rate constant $k_{CT \leftarrow CTR}$ to τ_2^{-1}. Therefore, τ_2 is essentially the lifetime of the **CTR** species τ_0(CTR).

Consequently, the desired rate constant for the **CT→CTR** photoreaction is given by:

$$k_{CT \rightarrow CTR} = \tau_1^{-1} - \tau_0(CT)^{-1} \qquad (7\text{-}4)$$

with the intrinsic lifetime τ_0(**CT**) = 4.5 ns as estimated from the extrapolation of τ_2(T) vs. T which reaches a plateau at the glass transition temperature (T_g≈228 K)[10] where $k_{CT \rightarrow CTR}$ is suppressed (second boundary condition). The value obtained for $\tau_0(\mathbf{CT})$ is similar to that used for EOE in ch.6 (4.3 ns).

All rate constants plotted in Fig. 7.7 and 7.8 are derived from global analysis. A time-saving way is used for the construction of Fig. 7.9. The lifetimes τ_1 are obtained from single curve fits in the blue spectral region (λ_f=380 nm) using a fixed value of τ_2 from fits in the red spectral part (λ_f = 550 nm) where the amplitudes α_2 are large. To elucidate the compatibility of both methods, results from global and single curve analysis are compared in Tab. 7.2, which correspond to Fig. 7.7 and 7.8.

7.3.3 The Method to Distinguish between Viscosity and Thermal Control of a Photoreaction

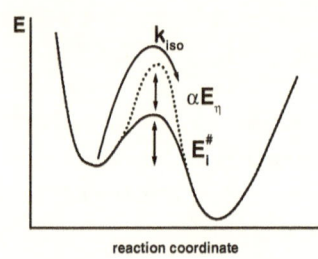

Scheme 7.3 Adiabatic photoreaction across a thermal ($E_i^{\#}$) and diffusive (αE_η) energy barrier.

Scheme 7.3 illustrates that the kinetics of adiabatic photoreactions or photoisomerizations can be controlled by an intrinsic thermal barrier $E_i^{\#}$ or the solvent viscosity inducing a diffusion barrier αE_η or by a mixture of both. The temperature and viscosity dependence of the photoreaction rate $k_{iso}(\eta,T)$ is well descibed by the frequently utilized eq. 7-5.

$$k_{iso}(\eta,T) = D\eta^{-\alpha} e^{-\frac{E_i^{\#}}{RT}} \text{ with } 0 < \alpha < 1 \quad (7\text{-}5)$$

where D is a viscosity independent intramolecular diffusion constant and α is a measure for the strength of viscosity control.[21] After replacing the viscosity by the well-known Andrade eq. 7-6

$$\log\eta(T) = \log\eta_0 + \frac{E_\eta}{R\ln 10} \cdot \frac{1}{T} \quad (7\text{-}6)$$

we arrive at a description of the rate constant in dependence of the temperature only:

$$\log k(T) = [-\alpha \log\eta_0 + \log D] - \frac{\alpha E_\eta + E_i^{\#}}{R\ln 10} \frac{1}{T} \quad (7\text{-}7)$$

Eq. 7-7 has the form of the conventional Arrhenius equation. One can easily see that the slope of the usual plot of logk vs T^{-1} results in an apparent activation barrier E_a which contains a relative contribution α of the activation energy for the viscous flow of the solvent (solvent mobility E_η) beside the intrinsic activation barrier.

$$E_a = \alpha E_\eta + E_i^{\#} \quad (7\text{-}8)$$

Hence, to obtain $E_i^\#$ the fraction α of E_η has to be separated. According to eq. 7-5 the parameter α can be determined, if the viscosity is varied at a constant temperature. This can experimentally be achieved by varying the pressure p, such that α is directly obtained from the slope α_p of a plot logk(p) vs. logη(p) at a fixed temperature:

$$\log k(p) = [\log D - \frac{E_i^\#}{R\ln 10 T_{const}}] - \alpha_p \log \eta(p) \tag{7-9}$$

On the other hand, an apparent value α'_T can be obtained for α, if the temperature dependent rate constants k(T) are plotted against logη(T) by

$$\log k(T) = [\frac{E_i^\#}{E_\eta}\log\eta_0 + \log D] - \alpha'_T \log\eta(T) \tag{7-10}$$

where $\alpha'_T = \alpha + E_i^\#/E_\eta$ contains a thermal contribution. Consequently, a comparison of both plots (eq.7-9 and 7-10) reveals the involvement of an intrinsic energy barrier which can then simply be calculated by

$$E_i^\# = (\alpha'_T - \alpha_p) E_\eta \tag{7-11}$$

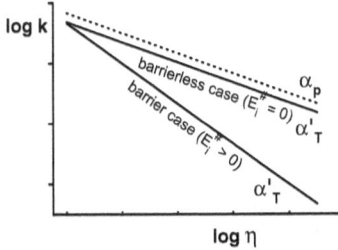

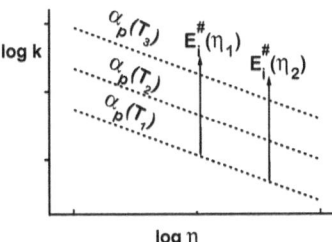

Scheme 7.4 *Comparison of rate constants for a pressure (-----) and two temperature (——) dependent data sets exemplifying the ideal behaviour for a barrierless and thermally activated reaction.*

Scheme 7.5 *Plots of isothermic rate constants to derive temperature dependent values of α and viscosity dependent $E_i^\#$.*

Scheme 7.4 illustrates that in case of a barrierless reaction ($E_i^*=0$) both plots are parallel ($\alpha'_T = \alpha_p$) while in the barrier case the slope of the temperature dependent data set should be larger than for the pressure dependent one. The reliability of the experimental results may be checked by inspection of the difference between the intercepts of both plots which should be close to zero as follows from eq. 7-12:[22]

$$\Delta \log k(\eta = 1) = -E_i^\# \left(\frac{1}{\ln 10 R T_{const}} + \frac{\log \eta_0}{E_\eta}\right) \cong 0 \quad \text{for } E_\eta >> |\log\eta_0| \tag{7-12}$$

Ch. 7 Pressure and Temperature Dependent Fluorescence

An alternative way is to determine $E_i^{\#}$ from an Arrhenius plot at isoviscous conditions adjusted by the appropiate pressure (cf. eq. 7-5: $E_i^{\#}$ = -dlnk/dRT).[5] The measurement of isotherms as indicated in scheme 7.5 further allows to evaluate the influence of the solvent viscosity or better of the density on the barrier height.

The results of both methods employed to the photoreaction of **III** are discussed in the following.

7.3.4 Separated Thermal Activation Energy and Viscosity Dependence of the Conformational Photoreaction in III.

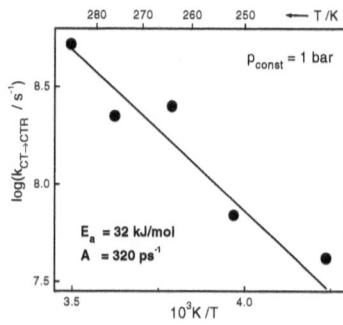
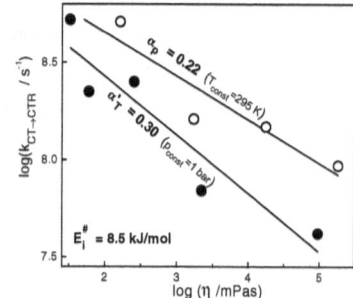

Fig. 7.7 Arrhenius plot of forward reaction rate constants in triacetine and apparant activation parameters analogously to the plot in diethylether (Fig. 6.11).

Fig. 7.8 Rate constants of the photoreaction of **III** in triacetine obtained by variation of pressure (O) and temperature (●) as a function of a common logarithmic viscosity scale

The conventional Arrhenius fit for the **CT→CTR** reaction rate constants in TAC shown in Fig. 7.7 yields values for the activation barrier E_a and the preexponential factor A which are both considerably larger than those in EOE as solvent (cf. Fig. 6.11). Since EOE and TAC are of similar polarity but of different viscosity, this already indicates a strong viscosity influence (cf. eq. 7-7). Following eq. 7-9 and 7-10, the rate constants $k_{CT→CTR}$ are plotted in Fig. 7.8 as a

Tab. 7.2 Results of logk vs. logη(T) and logk vs. logη(p) Plots and Activation Parameters

method	Δlogk(η=1)	$α_p$	$α'_T$	Δα	$\langle E_a \rangle$ / kJmol^{-1}	$\langle E_i^{\#} \rangle^a$ / kJmol^{-1}	$\langle \Delta V^{\#} \rangle$ / Å^3
global	0.07	0.22	0.30	0.08	32 ± 3	8.5 ± 3	21.4
$λ_f$ = 420 nm	-0.05	0.24	0.32	0.08	33 ± 3	8.5 ± 3	23.6
$λ_f$ = 380 nm	0.008	0.29	0.34	0.05	37 ± 3	5.5 ± 4	27.8

[a] derived from eq. 7-11 by using the average value of the solvent mobility $\langle E_η \rangle$=105 kJ/mol which results from the range of $E_η$=50-160 kJ/mol in the considered temperature region.

function of a common logarithmic viscosity (η) scale for the pressure (p) and temperature (T) dependent data using η(T) and η(p) derived on the basis of ref. 9,17,18. As expected, the rate constants sizeably decrease with increasing pressure which is conceivable with a large amplitude motion associated with the **CT→CTR** reaction. The slope of the temperature variation data set (a'$_T$ = 0.30) is significantly larger than that of pressure (a$_p$ = 0.22) demonstrating that with lowering the temperature, the rates are slowed down by a thermal barrier in addition to the pure viscosity effect. Because the solvent mobility E_η varies between 60-150 kJ/mol in the considered range of η, the intrinsic thermal barrier $E_i^{\#}$ can be estimated from eq. 7-11 between 5-11 kJ/mol as indicated in Tab.7.2 by an average value $\langle E_i^{\#} \rangle$ obtained for $\langle E_\eta \rangle$=105 kJ/mol. In addition to the results from a global analysis of the emission decays, the parameter obtained from single curve analysis at the observation wavelengths (λ_f) 420 nm and 380 nm are reported. The pressure and temperature plots of each of the three methods coincide at η=1 (eq. 7-12) giving confidence to the derived parameters. Obviously, the global and the λ_f = 420 nm single curve analysis delivers practically identical results, at least as far as the parameters (Δlogk(η=1), $\Delta\alpha$, $\langle E_i^{\#} \rangle$) obtained from ratios between temperature and pressure data sets are considered. This supports the proposal elsewhere,[6] that the ratio procedure minimizes undesirable effects such as viscosity dependent polarity changes and that those parameters derived from a single data set (α_p, α'_T, $\langle E_a \rangle$, $\langle \Delta V^{\#} \rangle$) should therefore be viewed with caution. Nevertheless, the values for λ_f=380 nm differ from the global and the λ_f=420 nm method, especially because of a stronger pressure/viscosity dependence of the rate constants. It seems then, that the high energy (conformer) reactants are exposed to a larger diffusional energy barrier connected with a larger activation volume $\Delta V^{\#}$ (=-RTdlnk(p)/dp) than the low energy species. This observation might be understandable with the detection of conformers being in a non-equilibrium nuclear (solvent and intramolecular) arrangement in a steep part of the reactant well far away from the transition state. However, in view of the limited data, a more detailed discussion would be too speculative. In any case, all methods agree within the error limit of about 3 kJ/mol that a substantial thermal barrier $E_i^{\#}$ is involved in the photoreaction of **III**. The fact that no thermal activation was found for DMABN in TAC with the corresponding experiments[6,7,10] strengthens the above conclusion in accordance with the steady-state pressure and temperature dependent experiments (Fig. 7.4).

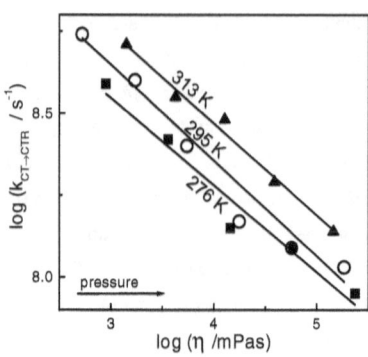

Fig. 7.9 *Isotherms of the viscosity dependence of the photoreaction rate constants.*

Next, the influence of viscosity on the activation barrier for isoviscous conditions (scheme 7.5) is analyzed by three isotherms of logk(p) versus η presented in Fig. 7.9 with the results given in Tab. 7.3. Within the precision of the lifetime measurements the slopes of the linear fits are equal around $\langle \alpha_p \rangle$ = 0.28 ± 0.01. As a consequence, the intrinsic activation barrier is rather independent of the solvent density in the investigated high viscosity region. It can therefore be concluded that in the investigated range of solvent properties, the photodynamics are dominated by viscosity and temperature changes without a detectable influence of polarity or polarizability changes on the barrier.

Tab. 7.3 Results of Isothermic Plots

method	α_p(276 K)	α_p(295 K)	α_p(313 K)	$\langle E_i^{\#} \rangle$ / kJmol^{-1}	$dE_i^{\#}/dp$ / Jbar^{-1}
λ_f=380 nm	0.27	0.29	0.28	9 ± 4	-0.5 ± 1

The possible existence of an intrinsic energy barrier $E_i^{\#}$ for the **CT→CTR** photoreaction of **III** in such medium polar solvents could be reconstructed in ch.3 (Fig. 3.10) by a counteraction of electronic with solvation and steric energy contributions. However, on the basis of the kinetic scheme 7.2, how can we understand that the initial torsional relaxation prior to the population of **CT** is nearly unaffected by increasing viscosity while that accompanying the **CT→CTR** can considerably be slowed down by viscosity? Scheme 7.6 may help to give a qualitative explanation. The initial intramolecular rotation around the interannular single bond is supposed to take place within a few picoseconds within the weakly polar and primary populated state (^{1}FC) as supported by transient absorption experiments (ch. 4). Due to the small dipole moment of ^{1}FC the solute-solvent interactions are weak and thus the rotation can proceed effectively unhindered within the holes of the solvent structure as it has been observed for the parent unsubstituted biphenyl BP.[23] By contrast, the dipole moment of the ^{1}CT state formed after the electron transfer interconversion (see ch.3) is large (ch. 6: μ_{CT} = 26D) leading to stronger coulombic interactions with the solvent dipoles and a contraction of the solvent cage. It can therefore be expected that a large amplitude motion such as the „*back rotation*" to a more twisted geometry in ^{1}CT is distinctly more affected by the microviscosity than the primary „*forward rotation*" to a more planar conformation in ¹FC. Although qualitative, this model is consistent with the experimental results of **I-III**.

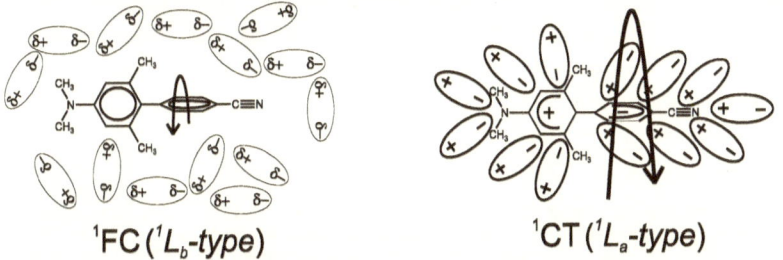

¹FC (¹L_b-type)   ¹CT (¹L_a-type)

Scheme 7.6 *Possible solvent-solute interaction model for photoinduced torsional relaxations of III to explain unhindered intramolecular rotation towards planarity in the primary excited ¹FC state and viscosity controlled „back rotation" to the twisted structure in the polar ¹CT state. (see text)*

7.4 Concluding Remarks

From the comparison of pressure with temperature dependent fluorescence kinetics and spectra in the highly viscous triacetine, it can unambiguously be concluded that the adiabatic photoreaction from the more emissive (more planar) **CT** to the less emissive (more twisted) **CTR** species is viscosity controlled and connected with a substantial thermal barrier in a medium polar environment. The steady-steady state spectra made it possible to define the irreversible regime for the **CT→CTR** reaction, such that a quantitative kinetic analysis could be performed. Thereby, effects of polarity and viscosity induced barrier shifts could be excluded in the temperature and pressure range investigated. In contrast to the corresponding experiments on DMABN in triactine, the existence of a thermal barrier is found as large as about 9 kJ/mol. The viscosity dependence expressed by the power law parameter α = 0.2-0.3 determined for the adiabatic photoreaction in **III** resembles the outcome of several pressure studies on excited state bond twisting reactions where values of α between 0.2 and 0.6 have been found.[3-6] Interestingly, the group of K.Hara, who is one of the pioneers regarding such pressure experiments, worked out that the values of α predominantly found for ¹LE→¹CT reactions in less viscous solvents can contain a sizeable contribution of solvent relaxation and that only the limiting values typically around α=0.2-0.3 in the high viscosity regime should be considered as due to the *intrinsic* viscosity dependence of the barrier crossing process.[3] Since the present investigations are conducted at extremely high viscosity and the change of solvation in the **CT→CTR** reaction is only minor, this underlines the involvement of an conformational change which can reasonably be assigned to an interannular torsional relaxation towards a more twisted structure. The apparant discrepancy to the view that the relaxation towards planarity is essentially unhindered by increasing (micro-)viscosity in the liquid solvent phase as proposed in the case of BP,[23] **II** in all solvents, **III** in nonpolar solvents and generally in the primary populated nonpolar ^{1}FC state, may be attributed to stronger solvent interactions of the solute in the highly polar ^{1}CT state. Within the suggested model, the solvent influence might be not only due to (outer shell) solvent drag but additionally to an enlargement of the effective rotor sizes by inner shell (polar) solvent-solute attraction in the ^{1}CT state. Of course, this friction hypothesis requires further experiments probing volume changes, such as polarity dependent time-resolved anisotropy and photoacoustics.

7.5 References

(1) (a) Drickamer, H.G.; Frank, C.W. *Electronic Transitions and the High Pressure Chemistry and Physics of Solids; Chapman and Hall*: London **1973**. (b) Rollinson, A.M.; Drickamer, H.G. *J. Chem. Phys.* **1980**, 73, 5981. (c) Dreger, Z.A.; Yang, G.; White, J.O.; Li, Y.; Drickamer, H.G. *J. Phys. Chem. A* **1997**, 101, 9511.

(2) (a) Hara, K.; Arase, T.; Osugi, J. *J. Am. Chem. Soc.* **1984**, 106, 1968. (b) Hara, K.; Obara, K. *Chem. Phys. Lett.* **1985**, 117, 96.

(3) (a) Hara, K.; Bulgarevich, D.S.; Kajimoto, O. *J. Chem. Phys.* **1996**, 104, 9431. (b) Hara, K.; Kometani, N.; Kajimoto, O. *J. Phys. Chem.* **1996**, 100, 1488. (c) Kometani, N.; Kajimoto, O.; Hara, K. *J. Phys. Chem. A* **1997**, 101, 4916.

(4) (a) Hara, K.; Kiyotani, H.; Bulgarewich, D.S. *Chem. Phys. Lett.* **1995**, 242, 455. (b) Hara, K.; Kiyotani, H.; Kajimoto, O. *J. Chem. Phys.* **1995**, 103, 5548. (c) Hara, K.; Ito, N.; Kajimoto, O. *J. Phys. Chem. A* **1997**, 101, 2240.

(5) (a) Schroeder, J.; Troe, J.; Vöhringer, P. *Chem. Phys. Lett.* **1993**, 203, 255. (b) Schroeder, J.; Schwarzer, D.; Troe, J.; Vöhringer, P. *Chem. Phys. Lett.* **1994**, 218, 43.

(6) Rettig, W.; Fritz, R. and Braun, D.: *J. Phys. Chem. A* **1997**, 101, 6830.

(7) Braun, D.; Rettig, W. *Chem. Phys.* **1994**, 180, 231.

(8) Zachariasse, K.A.; Growbys, M.; v.d.Haar, T.; Hebecker, A.; Il'ichev; Y.V.; Morawski;O.; Rücker,I.; Kühnle,W. *J. Photochem. Photobiol. A Chem.* **1997**, 105, 373.

(9) Groubert, E.; Charles, E. *C.R. Acad. Sci.* **1969**, 269, 1454.

(10) Braun, D.; *PhD Thesis*, HU Berlin, **1995**; Verlag Dr. Köster, Berlin, (ISBN 3-89574-119-1).

(11) Maus, M.; Rettig, W.; Jonusauskas, G.; Lapouyade, R. and Rullière, C. *J. Phys. Chem.* **1998**, 102, 7393.

(12) Maus, M., Rettig, W., Bonafoux, D.; Lapouyade, R. *submitted*.

(13) (a) Maus, M.; Rettig, W.; Lapouyade, R. *J. Inf. Rec.* **1996**, 22, 451; (b) Rettig, W.; Maus, M. *Ber. Bunsenges. Phys. Chem.* **1996**, 100, 2091. (c) Maus, M.; Rettig, W.; *J. Inf. Rec.* **1998**, 24(5/6), 461.

(14) Maus, M.; Rettig, W. *to be published*.

(15) Globals Unlimited®, Laboratory of Fluorescence Dynamics at the University of Illinois, **1992**.

(16) Fritz, R. *PhD thesis „Temperatur- und druckabhängige Untersuchungen zur Bestimmung des freien Volumens in niedermolekularen organischen Gläsern und Polymeren mittels adiabatischer Photoreaktionen fluoreszierender Sondenmoleküle"* TU Berlin; Köster Verlag, Berlin, **1994**

(17) *Landolt-Börnstein, Zahlenwerte und Funktionen aus Physik. Chemie, Astronomie, Geophysik und Technik*; ed. K.H.Schäfer, **1969**, Springer Verlag, Berlin. Vol. 2(5).

(18) Gegiou, D.; Muszkat, K.A.; Fischer, E. *J. Am. Chem. Soc.* **1968**, 90, 12.

(19) Ras, A.M.; Bordewijk, P. *Recl. Trav. Chim. Pay-Bas* **1971**, 90, 1055.

(20) Böttcher, C. J. F. *Theory of Electronic Polarisation*, Elsevier, Amsterdam **1973**.

(21) Bagchi, B.; Oxtoby, D.W. *J. Chem. Phys.* **1983**, 78, 2735.

(22) For the present conditions, the theoretical value using literature data for TAC (ref. 9,17,18) is about 10^{-3}.

(23) (a) Swiatkowski, G. *PhD thesis „Intramolekulare Torsionsrelaxationsprozesse sterisch gehinderter unpolarer Aromaten"* **1982**, Berlin. (b) Swiatkowski, G.; Menzel, R.; Rapp, W. *J. Luminesc.* **1987**, 37, 183.

Chapter 8

The Influence of Conformation and Energy Gaps on Optical Transition Moments in D-A Biphenyls.

Abstract:
Transition moments of absorption (M_a) and fluorescence (M_f) of three donor-acceptor biphenyls (**I-III**) are compared with those obtained by quantum chemical calculations to verify the previous conclusions of photoinduced twisting relaxations. The configurational analysis in terms of an exciplex description (analysis of electronic structure using reference states) reveals that existing experimental formalisms are insufficient to quantitatively derive the remarkably strong coupling of the electron transfer state (^1ET) with ^1L_a and S_0. Using AM1/CI optimized geometries, excited state structural relaxations other than twisting (e.g. bond-length shortening) are shown to lead to a better π-conjugation connected with increased M_f values. On the other hand, energy gap dependent ZINDO/S-CI calculations show that solvent polarity induced S_1 decoupling from ^1L_a reduces M especially for highly twisted rotamers. Nevertheless, in the experimental range of energy gaps between ^1L_a and ^1ET of around 10000 cm^{-1}, the polarity induced decrease of M_f plays a minor role. The experimental M_f values in polar solvents are indeed consistent with a photoinduced relaxation of **II** towards planarity and of **III** towards perpendicularity.

8.1 Introduction

Exciplex theory concerned with the electronic structure of molecular excited π-complexes has been well established since the pioneering works of Mulliken[1] and Murrel[2] appeared. The excited state of an electron donor-acceptor (D-A) complex is generally described as a linear combination of zeroth-order locally excited (^1LE) states in the D and A part and of zeroth-order electron transfer states where an electron is transferred either from the D unit to the A unit ($^1ET = D^+A^-$) or in the reverse direction from the A to the D unit ($^1RET = D^-A^+$)[2,3]:

$$^1\Psi^*_{DA\ exciplex} = (C_0\ ^1\Psi_{DA}) + C_1\ ^1\Psi_{D+A-} + C_2\ ^1\Psi_{D-A+} + C_3\ ^1\Psi_{LE(D)} + C_4\ ^1\Psi_{LE(A)} \qquad (8\text{-}1)$$

where the contribution (C_0^2) of the neutral ground state ($S_0 = DA$) is usually small in the case of a large energy gap between S_0 and the excited states.

The extent of electron transfer character (C_1^2) is of practical importance to derive the coupling elements with the excited states ($V^1_{LE^{-1}ET}$) and the ground state ($V^1_{ET\text{-}S0}$) which in turn are related to electron transfer probabilities („golden rule expression")[4] and the spatial arrangement between D and A.[5-9] Experimentally accessible quantities to analyze electronic coupling are the radiative (or nonradiative) rate constant k or the related electronic transition moment M. In the simplified case of an exciplex without intermolecular orbital overlap ($V^1_{ET\text{-}S0} = 0$) and coupling only with the A unit ($V^1_{ET^{-1}LE(D)} = 0$), the experimental transition moment M_{obs} is exclusively given by that of the $S_0 \rightarrow\ ^1LE(A)$ transition:

$$M_{obs} = C_4 M_{LE(A)} \qquad (8\text{-}2)$$

Ch. 8 Influence of Conformation and Energy Gaps on Transition Moments 131

Applying eq. 8-3 derived by Beens and Weller[3]

$$C_1^2 = 1 - C_4^2 = \left\{1 + \left(\frac{V_{^1LE(A)-^1ET}}{v_{LE(A)} - v_{obs}}\right)^2\right\}^{-1} \quad (8\text{-}3)$$

the coupling matrix element $V^1_{^1LE-^1ET}$ between ^1LE and ^1ET can then be calculated from the observed transition energy of the exciplex v_{obs} and local A unit $v_{LE(A)}$. Eqs. 8-1 to 8-3 demonstrate the dependence of the transition moment M on the energy gap $\Delta E^1_{^1LE-^1ET} = v_{LE(A)} - v_{obs}$ between ^1LE and ^1ET states, which is varied by the polarity of the solvent.

These solvent effects on M (or k) have very recently been exploited to work out new formalisms for more complicated exciplex-type borderline cases. Gould, Young, Mueller, Albrecht and Farid[5] proposed an experimental method to derive both $V^1_{^1LE-^1ET}$ and $V^1_{^1ET-^1S0}$ from plots of k_f/v_f vs. $(v_f/(v_{LE}-v_f))^2$ where k_f and v_f are the fluorescence rate constant and average fluorescence energy and v_{LE} is the $S_0 \rightarrow ^1LE$ transition energy. The only assumption is that both coupling elements do not vary with the solvent polarity, which is, of couse, not fulfilled, if the conformation is polarity dependent. Simultaneously, a similar method has been put forward by Bixon, Jortner and Verhoeven[6] which is based on a pertubative treatment of the $^1LE-^1ET$ coupling and therefore requires an ^1LE contribution of less than 10%. Both methods have been succesfully applied to exciplexes[5] and σ-bridged D-A molecules.[6,7]

However, it is a new challenge to evaluate ^1LE and ^1ET character or coupling matrix elements for directly linked π-systems where additional difficulties arise due to strong molecular orbital overlap between D and A and polarity dependent conformational energy minima. First attempts to experimentally determine such coupling elements, wavefunction contributions and most probable conformations in linked D-A biaryl systems have been published during the past year by Herbich and Kapturkiewicz (H/K formalism).[8,9]

Here, the coupling strength between nonpolar ^1L_a and ^1ET states in the D-A biphenyls I-III is analyzed in terms of an „exciplex description" (eq. 8-1) by a comparison of quantum chemical calculations on a semiempirical level with experimental transition moments. This should provide a possibility to test the universal applicability of the H/K method. As a main goal, the influence of solvent polarity and structural changes on the transition moments of I-III will be evaluated in order to confirm the previous conclusions of the polarity dependent rotational excited state relaxations (ch.3).[10-12]

8.2 Experimental and Semiempirical Calculations

Steady-state spectra, extinction coefficients and fluorescence rate constants are reported in chapter 2[13] and 3.[11]

CNDO/S-SCI calculations including 49 singly excited configurations have been performed with the QCPE program #333 modified to use the original CNDO/S parametrization[14] and to calculate the excited state dipole moments. To evaluate the effect of the changing energy gaps with solvent polarity, ZINDO/S-SCI computations[15] including a self consistent reaction field (SCRF) calculation (gas phase and SCRF methods A_1 or C with different Onsager parameters) prior to CI of 122 singly excited configurations have additionally been applied. Because the gas phase transition moments calculated with ZINDO/S (M_{Zindo}) are $\sqrt{2}$ times larger than those obtained with CNDO/S (M_{ondo}), the M_{Zindo} values are divided by $\sqrt{2}$. Using the AM1 Hamiltonian contained in the AMPAC program,[16] all input geometries were fully optimized in the ground state by the Newton algorithm and in the excited state by the Broyden-Fletcher-Goldfarb-Shanno (BFGS) algorithm including a CI expansion of 100 (see 6.2).[12] All quantum chemical computations were executed on a HP 735 and a SGI workstation.

8.3 Results and Discussion

8.3.1 Experimental Transition Dipole Moments

To derive information about the relative change of the average twist angle $<\varphi>$ between ground and excited state S_0 and S_1, an observable with an unambiguous correlation to $<\varphi>$ is necessary which is similar in S_0 and S_1. The transition energy, e.g., is an inappropriate observable in this respect since it does not only change with the twist angle φ but even more with the counteracting solvation energy (see ch. 3 and Fig. 3.10).[11] An useful observable should be the electronic transition moment because it exhibits a strong and monotonically increasing angular variation for the $2^1A(^1CT)$ state (cf. Figs. 8.1, 8.2 and 8.4).[10] Therefore, using the transition moment for absorption M_a (probing $<\varphi>$ in S_0) and fluorescence M_f (probing $<\varphi>$ in S_1) the desired comparison should be possible. According to eq. 8-4[17]

$$M_a = \sqrt{\frac{hc10^3 \ln 10}{8\pi^3 N_L n} \int \varepsilon(\nu_a) d\ln \nu_a} \quad (8\text{-}4a)$$

$$M_f = \sqrt{\frac{3h}{64\pi^4} \frac{k_f}{n^3 \nu_f^3}} \quad (8\text{-}4b)$$

the transition moments are obtained from the experimental data in ch.3[11] and collected in Tab. 8.1. Two trends can be followed in the absorption transition moments. In the horizontal comparison for a given solvent, the transition moments decrease from **I** to **III** undoubtedly

Ch. 8 Influence of Conformation and Energy Gaps on Transition Moments   133

ascribable to the enhanced ground state twist angle. In the vertical (solvent polarity) comparison of M_a, a very small increase within the experimental error limits of 0.1-0.2 D is observed for **I-III**. The same holds for M_f of **I** from EOE to ACN and M_f of **II** from HEX to ACN. Whether this slight but methodical increase of the transition moments M_a and M_f for **I** and **II** with solvent polarity is due to electronical (decoupling from locally excited states) or conformational effects is discussed below. In any case, it seems not to be related to phenyl twisting since the effect is also observed for **I**. The distinctly stronger M_f for **I** in EOE as compared to HEX points to a significant ^1L_b-^{1}CT mixing in HEX when ^{1}CT is not sufficiently stabilized by the solvent. The strong decrease of M_f for **III** with solvent polarity has been correlated with a dependence of the excited state mean twist angle $<\varphi>$ on polarity in ch.3.[10,11]

Tab. 8.1 Experimental Transition Moments (in D) of Absorption (M_a) and Fluorescence (M_f)

solvents	$F_2(\varepsilon_r)$	M_a			M_f		
		I	II	III	I	II	III
HEX	0.09	6.0	5.4	2.8	5.1[a]	6.2	4.3
EOE	0.25	6.1	5.6	2.9	6.3	6.4	3.4
ACN	0.39	6.2	5.7	3.0	6.6	6.6	2.3
EtOH	0.38				6.7	6.6	2.4

[a] solvent abbreviations represent n-hexane (HEX), diethylether (EOE), acetonitrile (ACN) and ethanol (EtOH).
[b] M_f(I) is perturbed due to mixing of the less allowed ^1L_b into ^{1}CT.
[c] Solvent polarity function $F_2(\varepsilon_r,n)=(\varepsilon_r-1)/(2\varepsilon_r+1)-(n^2-1)/(4n^2+2)$ using ε_r and n data from ref. 18

The comparison between M_a and M_f for each compound reveals a sizeable change only for **II** and **III** which are able to twist around the central single bond. For the fluorene derivative **I**, restricted to planarity, M_a and M_f are similar (except in HEX because a part of the fluorescence is "lost" due to strong ^1L_b-^{1}CT mixing, see below) from which it can be concluded that the differences between M_a and M_f observed for **II** and **III** can mainly be attributed to changes of the twist angle φ.

In spite of this striking argument for a dominating M vs. φ correlation, a non-negligible influence of the solvent dependent energy gaps between the zeroth-order ^{1}ET and locally excited (^{1}LE) states on M might be important which is investigated in the following.

8.3.2 Electronic Coupling Between ^{1}ET and ^1L_a in Dependence of the Twist-Angle

A comparison of the absorption spectra and CNDO/S-CI results (state energies and extinction coefficients) of **I-III** in Fig. 8.1 directly illustrates the electronic coupling of ^{1}CT with the higher lying ^1L_a state(s). Experiment and calculation agree that with decreasing twist angle φ ($=\varphi_{D-A}$), the intensity (ε_{max}) of the $S_0 \rightarrow {}^1$CT transition increases while the $S_0 \rightarrow {}^1L_a$ transition looses nearly all intensity with enhanced planarity. This means, that with decreasing twist angle φ, the ^{1}CT state mixes with the higher lying ^1L_a states „stealing their intensity".[20] The potential energy curves of ^{1}CT and ^1L_a further demonstrate their strong electronic interaction. The repulsion between both curves increases with planarity leading to a distinctly stronger stabilization of planar conformations in ^1CT than in the ground state (S_0). Note, that the electronic repulsion between S_0 and ^1CT is negligible due to the large energy difference $\Delta E_{S_0\text{-}^1CT}$ and that the repulsion between ^1L_b and ^1CT is symmetry forbidden.

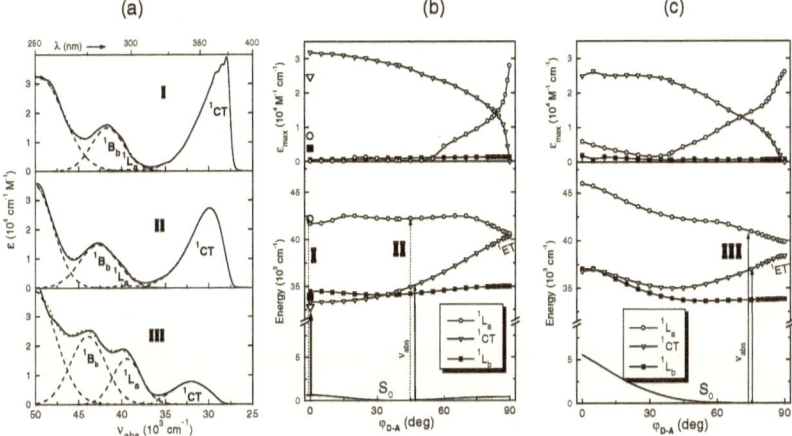

Fig. 8.1 (a) Absorption spectra (——) and decomposed bands (- - -) of the ^1CT ^1L_a and ^1B_b states and fitted curves (····) using Gaussian functions. (b,c) Molar extinction coefficients (upper panel) and state energies of S_0, the lowest lying ^1L_b and ^1L_a states of **II** and **III** obtained from AM1 and CNDO/S-CI as in ch. 2 and 3.[11,13] All corresponding results for **I** (enlarged symbols at φ=0°) are also shown. Due to strong mixing between the local S_4 and S_5 states $^1L_a(A)$ and $^1L_a(D)$, as well as the S_1 and S_2 states $^1L_b(D)$ and $^1L_b(A)$ at nonperpendicular twist angles, the data only of the relevant lowest state with non-vanishing transition intensity are plotted (see text).[19]

At 90°, the delocalized ^1L_b-type, ^1L_a-type and ^1CT states converge to the reference states $^1L_b(D)$, $^1L_b(A)$, $^1L_a(D)$, $^1L_a(A)$ of local character[19] and to the electron transfer states $^1ET(D^+A^-)$ and $^1RET(D^-A^+)$ transferring an electron from the dimethylaniline donor (D) to the benzonitrile acceptor (A) unit and in the reverse direction, respectively (for more details, see Fig. 2.1, ch. 2[13] and Appendix A). Following the formalism in the Appendix A (eq. A-6), the wavefunction of the spectroscopic ^1CT state can then be expressed by

$$^1\Psi_{CT}(DA) = \sum_{j=D,A} C_a^j {}^1\Psi_{L_a}(j) + \sum_{j=D,A} C_b^j {}^1\Psi_{L_b}(j) + C_e {}^1\Psi_{ET}(D^+A^-) + C_r {}^1\Psi_{RET}(D^-A^+) + \sum_k C_k {}^1\Psi_k \quad (8\text{-}5)$$

which resembles the wavefunction of an exciplex in eq. 8-1 with the exception that the ground state contributions are inherently included in the single wavefunctions of these *reference states*. For the sake of clearness, let us neglect the small contributions of the states S_k and consider only the linear combinations of the localized ^1L_b and ^1L_a states, such that eq. 8-5 formally simplifies to

$$^1\Psi_{CT} = C_a {}^1\Psi_{L_a} + C_b {}^1\Psi_{L_b} + C_e {}^1\Psi_{ET} + C_r {}^1\Psi_{RET} \quad (8\text{-}6)$$

Ch. 8 Influence of Conformation and Energy Gaps on Transition Moments   135

with the reference contributions C_a^2, C_b^2, C_e^2, C_r^2 as calculated with CNDO/S by eq. A-6 in the Appendix A and as presented in Tab. 8.2.

Tab. 8.2 Contributions of the Reference States to the ^{1}CT State of I-III

character	$\varphi_{D-A} = 0°$		$\varphi_{D-A} = 40°$		$\varphi_{D-A} = 80°$		$\varphi_{D-A} = 90°$	
	II	I	II	III	II	III	II	III
^1L_a	45.04%	39.7%	41.3%	38.3%	28.3%	18.4%	0.00%	0.00%
^1L_b	0.00%	5.5%	0.00%	0.8%	0.00%	0.00%	0.00%	0.00%
^{1}ET(D$^+$A$^-$)	40.4%	35.5%	46.8%	49.2%	63.9%	73.0%	79.3%[a]	84.3%[a]
^{1}RET(D$^-$A$^+$)	12.8%	11.0%	10.1%	7.1%	2.3%	1.2%	0.32%	0.04%
^{1}ET(D→A)$_{net}$	27.5%	24.5%	36.8%	42.1%	61.6%	71.8%	79.0%	84.3%
sum	98.2%	91.8%	98.3%	95.4%	94.5%	92.6%	79.6%	84.3%

[a] since only pure HOMO(D)→LUMO(A) contributions are considered, the values do not approach 100%. The rest of ≈ 20% belongs to higher ^{1}ET(D$^+$A$^-$) states.

The high contributions of the ^1L_a states to the spectroscopic ^{1}CT state at nonperpendicular angles φ_{D-A} even at 80° quantitatively reveals the strong coupling with the lowest lying electron transfer states of ^{1}A symmetry. Needless to say, that the different ^{1}B symmetry of the ^1L_b states prevents an admixture, except for I and for angles in the crossing region of ^1L_b and ^{1}CT. The strong mixing of the reference ^1L_a states with the lower lying ^{1}ET is well understandable with the low energy gap $\Delta E^1_{La^{-1}ET}$ between them (Fig. 8.1 b and c). Since $\Delta E^1_{La^{-1}ET}$ is less for II (=0.029eV) than for III (=0.185 eV), the contribution of the ^1L_a state to ^{1}CT is larger in II at a given φ_{D-A}.[21] On the other hand, the relatively small contribution (C_r^2 < 13%) of the ^{1}RET state representing the electron transfer from the HOMO(A) to the 2UMO(D) orbital (Fig. 2.1)[13] is simply due to its high energy (6.03 eV for II and 5.96 eV for III) connected with a large energy difference to the ^{1}ET state (Tab. 8.3).

The fact that the sum of the considered contributions is more than 90% at nonperpendicular angles (at φ_{D-A}=90° a symmetry change and orbital decoupling occurs connected with enhanced mixing of higher ^{1}ET states) indicates that the ^1L_a, ^{1}ET and ^{1}RET zeroth-order states provide a good basis to describe the properties of the spectroscopic ^{1}CT state. In the next section, this is demonstrated for the $S_0 \to {}^1$CT transition moment M^1_{CT}.

8.3.3 Description of Transition Moments with Reference States

Within the above proposed concept of a state analysis with ground state coupled reference states, the $S_0 \to {}^1$CT transiton moment M^1_{CT} should be reproducible by

$$M_{CT}(\varphi, r_b, \sum_i \Delta E_i) = $$
$$C_a(\varphi, r_b, \Delta E_{1_{L_a}-^1ET})M_{^1L_a} + C_b(\varphi, r_b, \Delta E_{^1L_b-^1ET})M_{^1L_b} + C_e(\varphi, r_b, \sum_{i=^1L_a,^1L_b,^1RET}\Delta E_{^1ET-i})M_{^1ET}(r_b, \varphi) + C_r(\varphi, r_b, \Delta E_{^1RET-^1ET})M_{^1RET}(r_b, \varphi)$$

(8-7)

which accounts for the twist angle (φ), interannular bond length (r_b) and energy gap ($\sum \Delta E_i$) dependent coupling of ^{1}ET with the excited states ^1L_a, ^1L_b, ^{1}RET and the conformation dependent transition moment to the ground state S_0. Due to preferential mixing of ^{1}ET with the ^1L_a states $\sum \Delta E_i$ is mainly determined by $\Delta E^1_{La^{-1}ET}$.

The reference transition moments are determined as follows. Because the interaction between the localized $^1L_b(D)$ and $^1L_b(A)$ as well as between the $^1L_a(D)$ and $^1L_a(A)$ states leads to a cancelling of the transition moments for the upper states, the values of M'_{Lb} (1.1 D for **II** and 0.6 D for **III**) and M'_{La} (5.2 D for **II** and 5.05 D for **III**) are taken from the respective lower lying states at $\varphi_{D-A}=90°$.

In order to describe the spectroscopic transition moment M'_{CT} separately by M'_{ET} (and M'_{RET}) resulting solely from ^1ET-S_0 (^1RET-S_0) coupling and, on the other hand, by the contribution of M'_{La} deriving from ^1L_a-^1ET coupling, a twist angle φ and bond length r_b dependent expression is needed for $M'_{ET}(\varphi,r_b)$ and $M'_{RET}(\varphi,r_b)$. According to Dogonadze et al.[22] the electronic coupling integral V_{AB} for an electron transfer between two molecular complexes A and B through the directly connected atoms a and b can be evaluated by

$$V_{AB} = S_{ab}\beta_{ab}c_a c_b \tag{8-8}$$

where S_{ab} is the overlap integral, β_{ab} is the resonance integral and c_a and c_b are the atomic orbital coefficients, each parameter corresponding to the connecting atoms a and b, respectively. Since the electronic transition moment M is generally related to V by[1,4]

$$M = V \frac{\Delta\mu}{\Delta E} \tag{8-9}$$

and the normalized overlap integral $S(\varphi,r_b)$ for the interannular bond in biphenyls can be expressed by[23]

$$S(\varphi, r_b) = \cos(\varphi) k(r_b) \tag{8-10a}$$

with $k(r_b) = 0.771 + 1.446(1.54 - r_b)$ (8-10b)

we obtain after substitution

$$M'_{ET}(\varphi,r_b) = \cos(\varphi) k(r_b) \beta_0 c_{h(d)} c_{l(a)} \frac{\mu_{ET^0} - \mu_{S_0}}{E_{ET^0} - E_{S_0}} \tag{8-11a}$$

$$M'_{RET}(\varphi,r_b) = \cos(\varphi) k(r_b) \beta_0 c_{h(a)} c_{l+1(d)} \frac{\mu_{RET^0} - \mu_{S_0}}{E_{RET^0} - E_{S_0}} \tag{8-11b}$$

using the standard resonance integral $\beta_0 = -2.39$ eV of two neighbouring benzene carbon atoms,[23,24] the pπ atomic orbital coefficients $c_{h(d)}$ and $c_{l(a)}$ ($c_{h(a)}$ and $c_{l+1(d)}$) of the D-A connecting atoms d and a in the HOMO(D) and LUMO(A) orbitals (HOMO(A) and 2UMO(D) orbitals), the dipole moment difference $\Delta\mu_{ET}^0$ ($\Delta\mu_{RET}^0$) and the energy difference ν_{ET}^0 (ν_{RET}^0) between the zeroth-order states S_0 and $^1ET^0$ ($^1RET^0$), i.e. the reference states ^1ET

Tab. 8.3 Reference State Energies and Dipole Moments in the Gas Phase at $\varphi_{D-A}=90°$[a]

state	E(eV)		μ(D)	
	II	III	II	III
S_0	0.054[b]	0.005[b]	5.2[b]	5.0[b]
^1L_a	5.03	4.94	6.3	5.8
^1L_b	4.55	4.47	6.9	5.3
^1ET (D$^+$A$^-$)	5.00	4.76	30.7	29.5
^1RET (D$^-$A$^+$)	6.03	5.96	-17.6	-15.0

[a] obtained by CNDO/S [b] obtained by AM1

Tab. 8.4 Resonance Parameters[a]

	$c_{h(d)}$	$c_{l(a)}$	$c_{h(a)}$	$c_{l+1(d)}$	$k(r_b)$[b]
II	0.4835	0.5335	0.4838	0.5677	0.8881
III	0.5322	0.5360	0.4814	0.5469	0.8780

[a] obtained by CNDO/S [b] obtained by AM1

and ^{1}RET at 90° where the coupling matrix elements with S_0 are zero. The latter four parameters are taken from CNDO/S for the decoupled geometries at φ=90° and collected in Tab. 8.3 and 8.4.

The obtained reference transition moments resulting from coupling with the ground state only are plotted in Fig. 8.2 and compared with the CNDO/S calculated $S_0 \rightarrow {}^1CT$ spectroscopic transition moments M^1_{CT} as a function of φ_{D-A}, respectively.

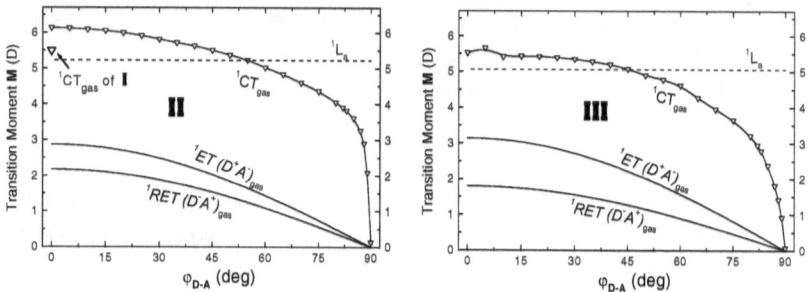

Fig. 8.2 Transition moments (to S_0) of the reference states ^1L_a, ^1ET and ^1RET coupled only with S_0 and the spectroscopic ^1CT state of **II** and **III** in the gas phase.

It shows that the major intensity of the $S_0 \rightarrow {}^1CT$ transition is gained through $M^1_{L_a}$, while $M^1_{ET}(\varphi)$ is mainly responsible for the angular dependence of M^1_{CT}. Note, that even for more twisted geometries M^1_{CT} remains distinctly larger than expectable for a pure (C_e^2=100%) ^1ET transition stressing the eminent ^1L_a character of ^1CT over the entire angular range. Comparing **II** with **III**, the smaller contribution of $M^1_{L_a}$ in M^1_{CT} of **III**[21] leads, on the one hand, to smaller M^1_{CT} values and, on the other hand, to a more smooth dependence of M^1_{CT} on φ_{D-A} near perpendicularity (cf. Fig. 8.1).

To verify the applicability of eq. 8-7 and 8-11, some derived M_{CT} values at different φ_{D-A} are compared in Tab. 8.5 with the relevant values obtained from CNDO/S and ZINDO/S using a CI calculation. The satisfying agreement between the respective M^1_{CT} data supports the validity of the proposed concept. Of course, in the region of more perpendicular angles, the deviations are larger due to the steep dependence of M^1_{CT} on φ_{D-A} (Fig. 8.2) and less contributions of the considered reference states (Tab. 8.2).

Tab. 8.5 Verification of Eq. 8-7 and Compatibility of CNDO/S with ZINDO/S.

method	φ_{D-A} = 90°		φ_{D-A} = 80°		φ_{D-A} = 40°		φ_{D-A} = 0°	
	II	III	II	III	II	III	II	I
M^1_{CT}(eq.8-7)	0.00	0.00	3.24	2.67	5.41	5.23	6.09	a
M^1_{CT}(cndo)	0.13	0.07	4.05	3.21	5.66	5.20	6.14	5.49
M^1_{CT}(zindo)	0.006	0.04	3.56	2.80	5.66	5.19	6.19	5.50

[a] due to specific ^1L_b-^1L_a mixing the zeroth-order value of $M^1_{L_a}$ of **I** must be less than for **II** and **III**, but is not directly available. A value of $M^1_{L_a}$=4.6 D is needed to achieve M^1_{CT}(eq. 8-7) = 5.5 D using $M^1_{ET}(0)$ and $M^1_{RET}(0)$ of **III**.

8.3.4 Influence of Structural Relaxations other than φ_{D-A} Twisting on M_{CT}

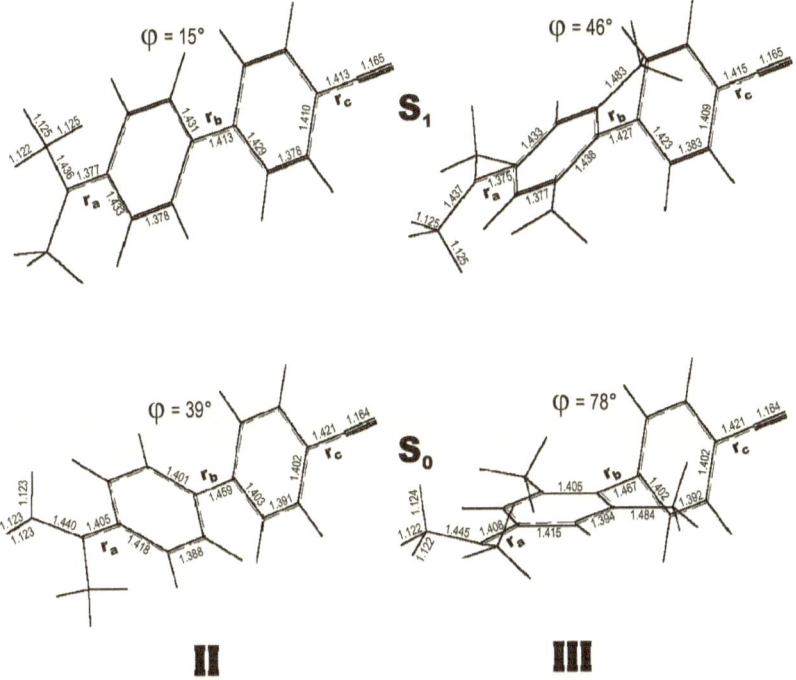

Fig. 8.3 With AM1 optimized structures (C_2 symmetry) of **II** and **III** in S_0 and $S_1(^1CT)$.

The comparison of the optimized structures in S_0 and $S_1(^1CT)$ reveals that the biphenyls tend to achieve a better π-electron conjugation between the chromophoric molecular subgroups in the ^1CT state as compared to S_0 (absence of solvent stabilization !). Therefore, the connecting bonds r_a, r_b and r_c are shortened in S_1 and the symmetric dihedral angle (α_N) between the methyl groups of the aminonitrogen and the plane of the linked benzene unit as well as the interannular angle φ_{D-A} are more planar. As a consequence of the better overlap between the atomic π-orbitals involved in the radiative back electron transfer from the A to the D unit, the calculated M^1_{CT} at a given φ_{D-A} are increased by a factor of 1.1-1.2 for the fluorescence transition (Tab. 8.6). In particular, the shorter bond length r_b gives the major contribution to the increase of M^1_{CT} as revealed by the ratios of the resonance parameter $k(r_b)$ being close to the ratios of M^1_{CT}. In addition, the photoinduced relaxation of all intramolecular coordinates except φ_{D-A} is calculated to result in a Stokes-shift of 3300 cm^{-1} - 3700 cm^{-1}, which is rather independent of the compound and depends only slightly on the twist angle. Note, however, that this effect is exaggerated by the calculations since the experimental Stokes-shift between the 0-0 transition of the absorption and fluorescence bands of **I** in HEX is only 460 cm^{-1} (Tab. 3.1).[11] Nevertheless, the shortening of the bond lengths in the excited state can be assumed to partially contribute to an increase of the experimental fluorescence transition moment with respect to that of absorption (Tab. 8.1).

The most important result of Tab. 8.6 for the present analysis is that the ratio M_f/M_a, scarcely depends on φ_{D-A}, such that the shape of the M^1_{CT} vs. φ_{D-A} curves in Fig. 8.2 is practically not changed by these additional structural coordinates.

Tab. 8.6 Comparison of CNDO/S Calculated Transition Parameters for AM1 Optimized Geometries in S_0 (absorption) and S_1 (fluorescence)[a] at Fixed Twist Angles φ.

	I/φ = 0°	II/φ = 0°	II/φ = 39°	III/φ = 40°	III/φ = 50°
$M_a(S_0 \rightarrow {}^1CT)$	5.5 D	6.1 D	5.7 D	5.2 D	4.9 D
$M_f(S_0 \leftarrow {}^1CT)$	6.7 D	7.1 D	6.2 D	5.8 D	5.4 D
M_f/M_a	1.24	1.17	1.11	1.11	1.11
$k_{CT}(r_b)/k_{S0}(r_b)$[b]	1.07	1.08	1.07	1.07	1.05
$v_{flu} - v_{abs}$	-3700 cm^{-1}	-3600 cm^{-1}	-3300 cm^{-1}	-3300 cm^{-1}	-3300 cm^{-1}

[a] S_1 optimization yields ^{1}CT as the lowest excited singlet state for φ_{D-A}<60°. [b] ratio of resonance parameter (eq. 8-10b)

8.3.5 Influence of the ^{1}ET-^{1}L$_a$ Energy Gap on M$_{CT}$

According to eq. 8.7 the energy gap $\Delta E^1_{La^{-1}ET}$ is the third main parameter to control M^1_{CT}. To simulate the effect of a varying $\Delta E^1_{La^{-1}ET}$ on the spectroscopic transition moment M^1_{CT}, ZINDO/S-CI calculations of $\Delta E^1_{La^{-1}ET}$ (calculated for the D-A decoupled geometry at φ_{D-A}=90°, i.e. no coupling of ^{1}ET with S_0 and ^1L_a) and M^1_{CT} (calculated for the D-A coupled geometry at a given φ_{D-A}, i.e. coupling of ^{1}ET with S_0 and ^1L_a) were performed including various Onsager reaction field parameter. Since the reaction field strongly stabilizes the ^{1}ET state, while the stabilization of the ^1L_a state is small, a wide range of 1-10×10^3 cm^{-1} for $\Delta E^1_{La^{-1}ET}$ could be covered. The compatibility of the twist angle dependence of ZINDO/S gas phase transition moments with that obtained by CNDO/S is confirmed in Tab. 8.5.

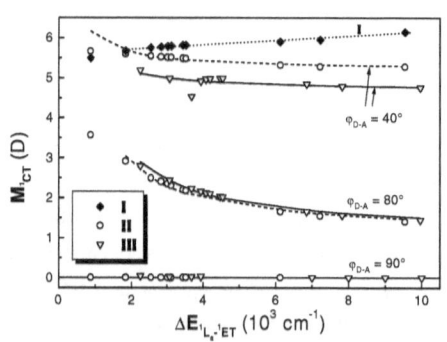

Fig. 8.3 $S_0 \rightarrow {}^1$CT transition moments M^1_{CT} of **I** (♦), **II** (o) and **III** (▽) as a function of the energy gap $\Delta E^1_{La^{-1}ET}$ between the lowest lying reference states ^1L_a and ^{1}ET calculated by ZINDO/S for different Onsager field parameters. Using the M value of the lowest ^1L_a at 90° for M^1_{La}, the data are fitted to eq. 8-12. Such obtained "M^1_{ET}" and "$V^1_{La^{-1}ET}$" values turn out to be of no physical meaning (see text).

The results in Fig. 8.3 show that for planar and medium twisted geometries, the energy gap $\Delta E^1_{L_a}$-$^1_{ET}$ has practically no influence on M^1_{CT}. In this angular region, the reference state ^1ET is an intense transition itself (eq. 8-11 and Fig. 8.2), such that the $\Delta E^1_{L_a}$-$^1_{ET}$ dependent interaction of ^1ET and ^1L_a does not lead to a sizeable decrease of M^1_{CT}. The interaction with the weakly allowed ^1L_b states is even more crucial. E.g., the sudden drop of M^1_{CT} for **III** at $\Delta E^1_{L_a}$-$^1_{ET}$ =3700 cm^{-1} is due to a crossing with the upper ^1L_b state. Especially M^1_{CT} of the D-A fluorene **I** essentially depends on ^1ET-^1L_b mixing as discussed previously (Fig. 8.1).[11,13] The increase of M^1_{CT} for **I** with $\Delta E^1_{L_a}$-$^1_{ET}$ in Fig. 8.3 is therefore due to the enhanced energy gap $\Delta E^1_{L_b}$-$^1_{ET}$ (lower C_b in eq. 8-7) and furthermore confirms the experimental observation of the strong increase of the fluorescence transition moment M_f of **I** from HEX to ACN (Tab 8.1). In **II** and **III**, however, this ^1ET-^1L_b decoupling with increasing solvent polarity only plays a minor role but may partially contribute to an increase of the experimental M_a and M_f (Tab. 8.1) with solvent polarity.

An influence of the energy gap $\Delta E^1_{L_a}$-$^1_{ET}$ on M^1_{CT} only appears for more twisted geometries. Recently, solvent polarity dependent M^1_{CT} of donor-bridge-acceptor molecules have been analyzed[6,7] by a formalism of perturbative coupling of the ^1ET state with a close lying, highly allowed locally excited state, which is equivalent to using eq. 8-12 for **II** of **III** at fixed twist angles φ:

$$M_{^1CT}(\varphi) = M_{^1ET}(\varphi) + M_{^1L_a} V_{^1L_a-^1ET}(\varphi) \frac{1}{\Delta E_{^1L_a-^1ET}} \tag{8-12}$$

The lines in Fig. 8.3 are derived from free fitting of the calculated M^1_{CT} vs. $(\Delta E^1_{L_a}$-$^1_{ET})^{-1}$ data with eq. 8-12. However, the fact that the intercepts obtained for infinite $\Delta E^1_{L_a}$-$^1_{ET}$ (**II**: 1.1 D at 80° and 5.2 D at 40°) are more than twice as large as the expected $M^1_{ET}(\varphi)$ values from eq. 8-11a (**II**: 0.5 D at 80° and 2.2 D at 40°) demonstrates that the ^1L_a-^1ET coupling for **I-III** is too strong to be treated in a perturbative way. This is not surprising since Bixon, Jortner and Verhoeven worked out[6] that the contribution of the allowed LE state should not exceed 2%, which is hardly fulfilled here even for φ_{D-A}=80° and $\Delta E^1_{L_a}$-$^1_{ET}$ = 10000 cm^{-1} (see Tab. 8.2: %^1L_a = 20-30% at φ_{D-A} = 80°). Moreover, the nonvanishing value of C_a in eq. 8-7 prevents a direct application of the purely experimental method proposed recently by Herbich and Kapturkiewicz (H/K),[8] where the coupling elements $V^1_{L_a}$-$^1_{ET}$ and V_{S_0}-$^1_{ET}$ can only be evaluated on the assumption that M^1_{CT} of the fluorescence approaches M^1_{ET} in a polar solvent. Let us apply, nevertheless the H/K formalism for the absorption case:

It was suggested to calculate the contribution of the locally excited ^1L_a state (C_a^2 in eq. 8-7) to ^1CT by using eq. 8-13[8]

$$1 - C_a^2 = \frac{\varepsilon_{max}(D-A)}{\varepsilon_{max}(ref)} \tag{8-13}$$

where ε_{max}(D-A) and ε_{max}(ref) represent the molar extinction coefficients at the absorption maxima of the delocalized $S_0 \rightarrow {}^1L_a$ transition of the composite D-A molecule and a reference value for the localized ^1L_a transition without intensity loss to the $S_0 \rightarrow {}^1CT$ transition. The ε_{max}(ref) value can be obtained from the calculated ε_{max}(D-A) values at φ_{D-A}=90° in Fig. 8.1 (ε_{max}(ref) = 27950 M^{-1}cm^{-1} for **I** and **II**; 26055 M^{-1}cm^{-1} for **III**). The value of ε_{max}(ref) for **II** agrees with the sum of the experimental ε_{max} data of dimethylaniline ($\varepsilon_{max}(^1L_a)$ = 16200 M^{-1}cm^{-1})[25] and benzonitrile ($\varepsilon_{max}(^1L_a)$ = 12000 M^{-1}cm^{-1})[25] giving further confidence in the calculation results. Using the ε_{max}(D-A) data from the deconvoluted ^1L_a bands in Fig. 8.1 (480 M^{-1}cm^{-1}, 770 M^{-1}cm^{-1} and 15750 M^{-1}cm^{-1} for **I**, **II** and **III**) the ^1L_a contribution C_a^2 to the ^1CT bands of **I-III** obtained from eq. 8-13 are 98%, 97% and 40%. These values are by far too large since Tab 8.2 reveals that the ^1L_a character of **I** and **II** can be at most 40% and that of **III** at most 20% (Note, that in solvents, especially in the polar ones, the ^1L_a character should even decrease.). The reason for the failure of eq. 8-13 in this case is the remarkably strong mixing of ^1L_a and ^1ET in **I-III** which leads to a complete loss of the $S_0 \rightarrow {}^1L_a$ absorption intensity at intermediate twist angles (Fig. 8.1). For the acridine compounds investigated by H/K, the contribution of ^1L_a to the ^1CT band according to eq 8-13 is only 25-40% which seems sufficiently low to apply the H/K formalism. The present results therefore indicate that the contribution of the locally excited state should not exceed 40% to justify the H/K formalism.

In contrast to the influence of the variation of the interannular bond length r_b on M^1_{CT}, the magnitude of the change of ΔE^1_{La}-$^1_{ET}$ or solvation on M^1_{CT} depends on the twist angle φ and the compound (Fig. 8.3). Thus, a calibration curve M^1_{CT} vs φ is expected to change its shape in dependence on ΔE^1_{La}-$^1_{ET}$ (see Fig. 8.4). Nevertheless, even in the twist region of large φ_{D-A}, the change of M^1_{CT} with the twist angle is much stronger than with ΔE^1_{La}-$^1_{ET}$, such that all sizeable variations of the experimental transition moments of **II** and **III** can, as proposed in ch. 3,[10-12] indeed be traced back to changes of the twist angle φ.

8.3.6 Evaluation of S_0 and S_1 Twist Angles in Solvents

In order to simulate the influence of increased energy gaps in HEX, EOE and ACN on the M^1_{CT} vs. φ_{D-A} curves, M^1_{CT} is calculated by ZINDO/S for various twist angles φ using three different Onsager field parameters. Latter are empirically adjusted in such a way that the calculated transition energies of the model compound **I** restricted to planarity correspond to the 0-0 fluorescence energies of **I** in HEX, EOE and ACN. The resulting ΔE^1_{La}-$^1_{ET}$ values of **I-III** are situated between 6-10×10^3 cm^{-1} (cf. Fig. 8.3). The calculated M^1_{CT} values of the „non-twistable" **I** in these solvents are between 5.9 D and 6.2 D being in sufficient agreement with the experimental values. Thus the $M^1_{CT}(\varphi_{D-A})$ curves are used as obtained without any correction.

The comparison of the $M^1_{CT}(\varphi_{D-A})$ curves obtained for HEX, EOE and ACN in Fig. 8.4 with those for the mixed ^1CT in the gas phase and the pure $^1ET(D^+A^-)$ in ACN shows that the „sudden decoupling" of ^1L_a and ^1ET near $\varphi_{D-A}=90°$ does not take place in the solution phase, such that the $M^1_{CT}(\varphi_{D-A})$ curve adapts the shape of a cosine function. The change of the wavefunction coefficients C_a and C_e (eq. 8-7) is smoother than calculated by CNDO/S and the resulting less steep dependence of the permanent dipole moments on φ should therefore lead to broader energy minima near perpendicularity as discussed in ch. 3.3.3 (see Fig. 3.10).[11] One might be astonished about the high M^1_{CT} values around $\varphi_{D-A}=0°$, even though the weight of ^1ET to ^1CT increases with polarity. This behaviour is understandable, if the increase of M^1_{ET} with decreasing energy E^1_{ET} (eq. 8-11) is considered. For example, M^1_{ET} of **II** (**III**) at $\varphi=0°$ increases from 2.9 D (3.1 D) in the gas phase to 3.9 D (4.1 D) in ACN (compare $M^1_{ET}(\varphi_{D-A})$ in Fig. 8.4 and 8.2). Decoupling of ^1ET from the less allowed ^1L_b but less efficient decoupling of ^1ET from ^1L_a of the more planar geometries may additionally enhance M^1_{CT} at $\varphi_{D-A}=0°$ in polar solvents.

The M^1_{CT} curves may be used as rough calibration curves to derive spectroscopic twist angles for the ground ($<\varphi_M(S_0)>$) and excited state ($<\varphi_M(S_1)>$) from the experimental transition

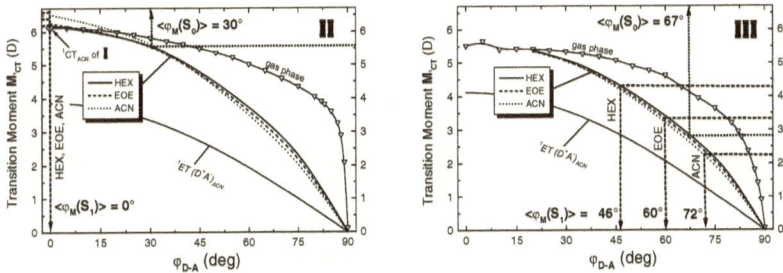

Fig. 8.4 With ZINDO/S calculated calibration curves M^1_{CT} vs φ_{D-A} to derive spectroscopic twist angles $<\varphi>$ from the experimental transition moments of **I-III** in HEX, EOE and ACN (Tab. 8.1). The arrows denote the correlations of the calculated M^1_{CT} with M_a ($\cdots$) and M_f (- - -)
For comparison, the calculated M^1_{CT} curves of ^1CT in the gas phase and in the case of the pure ^1ET (HOMO(D)→LUMO(A)) state in ACN[26] are additionally shown.

moments M_a and M_f in Tab. 8.1. It is justified to use the calculated M^1_{CT} curves for HEX to correlate the M_a as well as the M_f values in HEX because of only a small Stokes-shift which does not significantly change $\Delta E^1_{La-}{}^1_{ET}$. Hence, values of about 0°, 30° and 67° are obtained for the ground state twist angles $<\varphi_M(S_0)>$ of **I**, **II** and **III**, respectively (Fig. 8.4). These angles are very reasonable and comparable to literature data. For example, the twist angle obtained for **III** in S_0 is in full agreement with that given for 2,2'-dimethylbiphenyl.[27] For biphenyl itself, the reported twist angles φ ranging between 15-35° in solution[27,28] are in line with the value for **II** in S_0. Furthermore, the values $\varphi_M(S_0)$ are consistent with the mean twist angles (weighted with $\eta(\varphi)$) obtained from AM1 (Tab. 3.5: $\langle\varphi_\eta(S_0)\rangle$ = 34.5° for **II** and 72° for **III**). Excellent agreement is found between the experimental M_f values and the calculated M^1_{CT} values of **II** at φ_{D-A}=0° (note, that the increase of M_f with polarity in Tab. 8.1 is correctly reproduced by the calculated M^1_{CT}). In accordance with the preliminary interpretation in ch. 3,4,6,7, this points to the excited state relaxation of **II** in the solvent phase from a twisted geometry ($\approx$ 30°) into a geometry with an equilibrium angle φ around 0°.

For **III**, the interpretation of a polarity-dependent twist relaxation in S_1 given in ch. 3[10-11] can be confirmed: In nonpolar solvents, the conformational relaxation pathway in S_1 is also towards planarity ($\langle\varphi_M(S_1)\rangle\approx$46°) due to $\langle\varphi_M(S_1)\rangle < \langle\varphi_M(S_0)\rangle$, while in strongly dipolar solvents, it leads to a conformation ($\varphi\approx$72°) more twisted than in S_0. The obtained intermediate twist angle in medium polar EOE ($\langle\varphi_M(S_1)\rangle$=60°) is consistent with the proposed equilibrium of two rotamer distributions. The only difference to the previous results (Tab. 3.5) are the derived absolute values of $\langle\varphi_M(S_0)\rangle$ and $\langle\varphi_M(S_1)\rangle$ for **III**, which are all approximately 15° smaller in the present evaluation including the solvent influence (Fig. 8.4).[29]

8.4 Concluding Remarks

The quantum chemical configuration analysis in terms of an modified „exciplex description" (contribution of reference states with inherent ground state coupling) serves as a helpful tool to describe the transition moments of **I-III**. The comparison with absorption and fluorescence results leads to the conclusion that the coupling of the zeroth-order states ^{1}ET and ^1L_a in **I-III** is too strong to analyze the experimental transition moments with perturbation theory[6] or with a formalism recently proposed by Herbich and Kapturkiewicz.[8]

Taking into account the influence of structural relaxations other than twisting around the interannular bond φ and of solvent polarity induced decoupling from the intensity borrowing ^1L_a state on the fluorescence transition moment M_f, the previous conclusions in 3.4[10-12] regarding photoinduced twist relaxations can be confirmed.

8.5 References and Notes

(1) Mulliken, R.S. *J. Am. Chem. Soc.* **1952**, 74, 811.

(2) Murrel, J.N. *J. Am. Chem. Soc.* **1959**, 81, 5037.

(3) Beens, H.; Weller, A. In *Organic Molecular Photophysics*; Birks, J.B., Ed.; Wiley: New York, 1975; Vol.2, Chapter 4.

(4) for reviews see: (a) Newton, M.D. *Chem. Rev.* **1991**, 91, 767. (b) Barbara, P.F.; Meyer, T.J.; Ratner, M.A. *J. Phys. Chem.* **1996**, 100, 13148. (c) Hush, N.S. *Coord. Chem. Rev.* **1985**, 64, 135.

(5) Gould, I.R.; Young, R.H.; Mueller, L.J.; Albrecht, A.C.; Farid, S. *J. Am. Chem. Soc.* **1994**, 116, 8188.

(6) Bixon, M.; Jortner, J.; Verhoeven, J.W. *J. Am. Chem. Soc.* **1994**, 116, 7349.

(7) Verhoeven, J.W.; Scherer, T.; Wegewijs, B.; Hermant, R.M.; Jortner, J.; Bixon, M.; Depaemelare, S.; De Schryver, F.C. *Receuil des Travaux Chimique des Pays-Bas* **1995**, 114, 443.

(8) (a) Herbich, J.; Kapturkiewicz, A. *J. Am. Chem. Soc.* **1998**, 120, 1014. (b) Herbich, J.; Kapturkiewicz, A. *Chem. Phys. Lett.* **1997**, 273, 8.

(9) Kapturkiewicz, A.; Herbich, J.; Karpiuk, J.; Nowacki, J. *J. Phys. Chem. A* **1997**, 101, 2332.

(10) (a) Maus, M.; Rettig, W.; Lapouyade, R. *J. Inf. Rec.* **1996**, 22, 451. (b) Rettig, W.; Maus, M.; Lapouyade, R. *Ber. Bunsenges. Phys. Chem.* **1996**, 100, 2091.

(11) Maus, M., Rettig, W., Bonafoux, D.; Lapouyade, R. *submitted for publication*.

(12) Maus, M.; Rettig, W.; Jonusauskas, G.; Lapouyade, R.; Rullière, C. *J. Phys. Chem.* **1998**, 102, 7393.

(13) Maus, M.; Rettig, W. *Chem. Phys.* **1997**, 218, 151.

(14) Del Bene, J.; Jaffe, H.H. *J. Chem. Phys.* **1968**, 48, 1807; **1968**, 48, 4050; **1968**, 49, 1221; **1969**, 50, 1126.

(15) ZINDO/S SCF/CI program package available from M.C. Zerner; University of Florida, Gainesville, FL.; see for ref.: (a) Zerner, M.C.; Loew, G.H.; Kirchner, R.F.; Mueller-Westerhoff, U.T. *J. Am. Chem. Soc.* **1980**, 102, 589. (b) Zerner, M.C.; Karelson, M.M. *J. Phys. Chem.* **1992**, 96, 6949.

(16) AMPAC 5.0, Semichem, 7128 Summit, Shawnee, KS 66216 (1994); Dewar, M.J.S.; Zoebisch, E.G.; Healy, E.F.; Stewart, J.P. *J. Am. Soc.* **1985**, 107, 3902.

(17) Birks, J.B. *Photophysics of Aromatic Molecules*, Wiley-Interscience, New York, **1970**.

(18) Riddick, J.A.; Bunger, W.B.; Sakano, T.K. *Organic Solvents*, John Wiley & Sons **1986**.

(19) The first two excited singlet states are of ^1L_b-type, the higher one being mainly localized on benzonitrile (A) and the lower one on dimethylaniline (D). Both states are of B symmetry within the C_2 point group and therefore cannot cross each other. Their low extinction coefficients across the whole angular range demostrates the ^1L_b typical weakly allowed excitation nature. Moreover, they mix only very weakly with each other as evidenced by a small exchange of their electron configurations (see ch. 2). A completely different mixing behaviour is observed for the zeroth-order $^1L_a(D)$ and $^1L_a(A)$ states. They mix such strongly that even at φ=90° it is not possible to discriminate between acceptor and donor ^1L_a states. The twist angle φ dependent energies of these two close lying ^1L_a states for **II** are: at φ=90° $v(S_4)$ = 40200 cm^{-1}, $v(S_5)$ = 42200 cm^{-1}; at φ=70°:$v(S_4)$ = 42100 cm^{-1}, $v(S_5)$ = 42400 cm^{-1}; at φ=0°:$v(S_4)$ = 41000 cm^{-1}, $v(S_5)$ = 42400 cm^{-1}.

(20) Note, that this is not an effect of vibronic coupling in the Herzberg-Teller sense (Herzberg, G.; Teller, E. *Z Phys. Chem.* **1933**, B21, 410.) because the equal symmetry ^1A of the ^1L_a and ^1CT states allows them to mix electronically without vibrational symmetry breaking (except at exact φ=90° where the molecular point group changes from C_2 to C_{2v} with the irreproducible representation A_2 for ^1CT and A_1 for the ^1L_a states). Otherwise, the coupling effect could not be that pronounced.

(21) The smaller ionization potential (IP) of the donor group in **III** (from AM1: IP(3,5-dimethyldimethylaniline)=8.2 eV, IP(dimethylaniline)=8.3 eV) is responsible for the lower lying zeroth-order ^1ET state and more ^1ET character to ^1CT of **III**, respectively.

(22) Dogonadze, R.R.; Kutznetsov, A.M.; Marsagishvili, T.A. *Electrochim. Acta* **1980**, 25,1.

(23) Suzuki, H. *Electronic Absorption Spectra and Geometry of Organic Molecules*, Academic Press, New York **1967**.

(24) Parr, R.G. *J. Chem. Phys.* **1952**, 20, 1499.

(25) DMS UV-Atlas, VCH-Butterworth, Weinheim-London **1967**.

(26) The $M^1{}_{ET}(\varphi)$ curves are calculated by eq. 8-11 and the data in Tab. 8.4 and 8.4 using a solvent-corrected energy of ^1ET according to $E^1{}_{ET}(ACN) = E^1{}_{ET}(Gas) - E_{solv}(ACN)$ with $E_{solv}(ACN) = 1.3$ eV for **II** and 1.15 eV for **III** as obtained from ZINDO/S.

(27) Suzuki, H. *Bull. Chem. Soc. Jpn.* **1959**, 32, 1340 and 1350.

(28) (a) Kurland, R.; Wise, W.B. *J. Am. Chem. Soc.* **1964**, 86, 1877. (b) Eaton, V.J.; Steele, D. *J. Chem. Soc. Faraday Trans.* II **1973**, 69, 1601. (c) Le Gall, Suzuki, H.; *Chem. Phys. Lett.* **1977**, 46, 467. (d) Uchimura, H.; Tajiri, A.; Hatano, M. Bull. *Chem. Soc. Jpn.* **1981**, 54, 3279 and *Chem. Phys. Lett.* **1975**, 34, 34. (e) d'Amibale, A.; Lunazzi, L.; Boicelli, A.C.; Macciantell, D. *J. Chem. Soc. Perkins Trans. II* **1973**, 69, 1396. (f) Lim, E.; Li, Y.H. *J. Chem. Phys.* **1970**, 52, 6416. (g) Barrett, R.M.; Steele, D. *J. Mol. Struct.* **1972**, 11, 105.

(29) It is noteworthy that the experimental transition moments M_a and M_f are mean values of a rotamer distribution ($<M_{exp}> = \int M(\varphi) \eta(\varphi) d\varphi$) similar to the dipole moment $<\mu>$ (eq. 3.9). As a consequence, at higher twist angles (polar solvents) the spectroscopically determined twist angles $<\varphi_{exp}>$ are possibly underestimated in regard to the real equilibrium angle φ_{eq} or the mean twist angle $<\varphi_\eta> = \int \varphi \eta(\varphi) d\varphi$ because the symmetry around $\varphi = 90°$ is not taken into account. Thus the observed transition moment $<M_{exp}>$ gives more weight to the less perpendicular angles φ (e.g. a broad distribution around $<\varphi_\eta> = \varphi_{eq} = 90°$ does not yield $<M_{exp}> = 0$). The underestimation of $<\varphi>$ due to the convolution integral of $<M_{exp}>$ is small apart from the perpendicular case because there the slope $dM/d\varphi$ is rather flat (Fig. 8.4).

Chapter 9

Conformation and Energy Gap Dependent Electron Transfer Interactions in Flexible D-A Biaryls: The Case of Two Twisted 9-(dimethylanilino) phenanthrenes

Abstract:
Absorption and solvent dependent fluorescence transition moments (M_a and M_f) and energies of two differently twisted 9-(dimethylanilino)phenanthrenes are determined and compared with the relevant phenanthrene properties and with quantum chemical calculations (AM1 and CNDO/S-CI) to derive the electronic and molecular structure of the first excited singlet state (S_1). The increase of M_f by changing the solvent polarity from nonpolar to medium polar can be attributed to a solvent-induced change of the S_1 nature from the phenanthrene ^1L_b-type to ^1L_a-type. The observed relation $M_f > M_a$ points to enhanced coupling of the zeroth-order electron transfer state (^1ET) with the ground state (S_0) due to a photoinduced relaxation towards planarity in accordance with the calculation results. Further increase of the solvent polarity leads to a strong decrease of M_f below M_a for the more twisted compound explainable with enhanced electron transfer interactions at the expense of ^1L_a character associated with a narrower perpendicular rotamer distribution in S_1.

Keywords: donor-acceptor biaryls, state mixing, conformational relaxation, absorption, fluorescence

9.1 Introduction

Numerous studies on donor-acceptor biaromatic compounds linked by a single bond have revealed the importance of photoinduced intramolecular electron transfer (ET) interactions on the photophysics.[1-9] However, the consequences on the electronic and molecular structure are frequently under controversial discussion. The fluorescence transition probability which is directly related to the fluorescence transition moment (M_f), has been applied to characterize the nature of the emitting state.[3,4,10,11] Large M_f values of twisted donor-acceptor biaryl compounds reveal that the radiative back ET can be strongly allowed by substantial π-electron interactions through the interannular single bond responsible for large coupling matrix elements between the zeroth-order ^1ET and the ground state (S_0).

Electronic coupling with strongly allowed local transitions is frequently discussed as a source to further increase the radiative back ET probability.[3,4,10,11] However, strong electronic coupling of the ET transition with a local ππ* transition less allowed than ET has not yet been reported for D-A compounds containing a polycyclic aromatic unit. For anthryl derivatives, such as p-(9-anthryl)-N,N-dimethylaniline (ADMA),[4] it is not surprising that electronic coupling occurs with the ^1L_a state localized in the anthryl unit, since the ^1L_a state of anthracene is the lowest lying electronically excited state (S_1) situated 350 cm^{-1} below the less allowed ^1L_b state as recently found by polarized two-photon excitation in solid argon (Ar).[12] On the other hand, in the polycyclic aryls pyrene[13] and carbazole,[14] the weakly allowed ^1L_b state is lower lying by 3700 cm^{-1} and 4300 cm^{-1} than the second excited state ^1L_a. Nevertheless, an influence of the local ^1L_b state has not been reported for the radiative back ET of biaromatic D-A carbazole[3] or pyrene[4,5] derivatives.

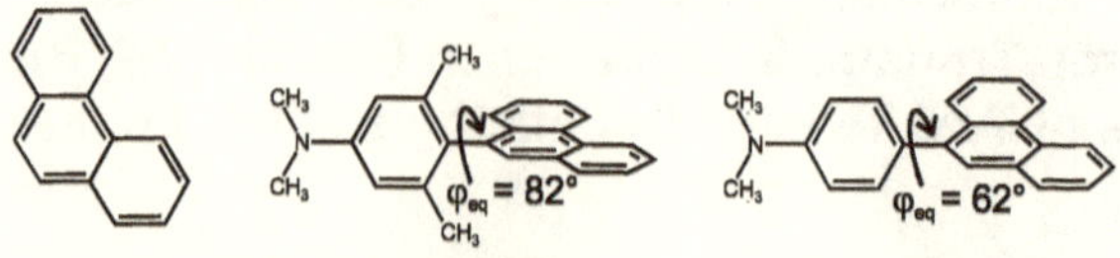

Scheme 9.1 *Molecular structures of the investigated compounds phenanthrene (P), DM9DP and 9DP.*

We extend the series of D-A biaryls with two differently twisted dimethylaniline-phenanthrene D-A derivatives (9DP and DM9DP in scheme 9.1). Phenanthrene (P) has the largest energy gap (5500 cm^{-1} in Ar[14]) between the lowest lying ^1L_b and the higher lying ^1L_a state. Thus, a dominating influence of the zeroth-order ^1L_b state of P should be observable in the fluorescence of the composite molecules 9DP and DM9DP. Photophysics, transient absorption and nonradiative processes have already been studied previously.[1,2] The goal of this contribution on 9DP and DM9DP is to analyze the change of the S$_1$ electronic structure by solvent-induced inreasing ET interactions and the consequences on the excited state conformation with regard to the interannular twist angle φ. Therefore, quantum chemical calculations are compared with experimental transition dipole moments and energies.

9.2. Experimental and Calculations

Most of the steady-state spectra, fluorescence quantum yields and lifetimes on 9DP and DM9DP have been published previously.[1,2] Additional quantum yields of DM9DP are determined relative to the reported value in diethylether and the relevant lifetimes are measured using synchrotron radiation (see 3.2).[7,15]

CNDO/S-SCI (49 singly excited configurations) calculations of transition dipole moments and energies have been performed with the QCPE program #333 modified to use the original CNDO/S parametrization[16] and to calculate the excited state dipole moments. All input geometries were fully optimized in the ground state by the Newton algorithm with the AM1 Hamiltonian within the AMPAC program.[17]

Similarly to **I-III** (Fig. 2.1), the title compounds 9DP and DM9DP are regarded as composite molecules consisting of a donor D (dimethylaniline in 9DP and 3,5-dimethyl-dimethylaniline in DM9DP) and an acceptor A (phenanthrene) moiety (see Fig.9.4).[6] The same configurational analysis treatment as for **I-III** in ch. 2 and 8 has been performed using the

9.3 Results

9.3.1 Spectroscopic Transition Moments and Energies

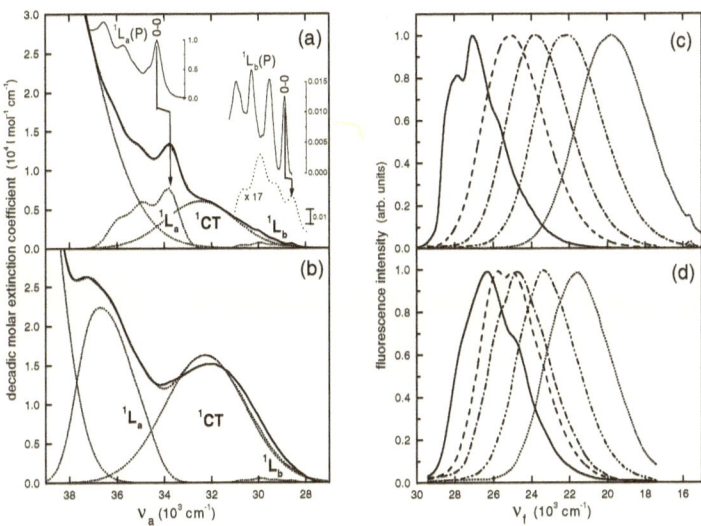

Fig. 9.1 Absorption spectra of (a) DM9DP and (b) 9DP in iso-octane (———). The decomposed ^1L_a, ^1CT and ^1L_b bands together with the fit curves are plotted as dotted lines. For comparison, the $^1L_b(P)$ and $^1L_a(P)$ parts of the pure phenanthrene absorption spectrum are shown in (a) as an inset and the correlation of the 0-0 transitions are denoted.
Fluorescence spectra of (c) DM9DP and (d) 9DP in iso-octane (———), dibutylether (- - -), diethylether (- · -) ethylacetate (tetrahydrofuran for 9DP) (-··-) and acetonitrile (·····).

The steady state absorption and fluorescence spectra of DM9DP and 9DP are shown in Fig. 9.1. The absorption spectrum of DM9DP exhibits obvious features of the phenanthrene (P) spectrum of which the $^1L_a(P)$ and $^1L_b(P)$ (four of the six prominent vibronic bands) bands are also plotted in Fig 9.1a. It is therefore straightforward to assign the first three visible bands in DM9DP to $^1L_b(A)$, $^1L_a(A)$ and to the delocalized charge transfer band ^1CT arising from the interaction between the dimethylaniline (D) and phenanthrene (A) moiety. For 9DP (Fig. 9.1b), the same assignment to ^1CT can be made for the intense low-energy band at 32500 cm^{-1}, since it shows typical features of a charge transfer transition, such as a broad and structureless shape and a solvent polarity induced red shift (v_a^{max}(ACN) = 31350 cm^{-1}).[1] Consequently, the band at 35000 cm^{-1} can be attributed to a more delocalized ^1L_a transition, while the ^1L_b transition is hidden underneath the ^1CT band. The spectra are decomposed using j gaussian functions with j=3 for ^1L_a, j=1 for ^1CT and j=4 for the ^1L_b bands. Gaussian functions (eq.2-1) were

Ch. 9 The Case of 9-(Dimethylanilino) phenanthrenes

preferred to others (eq.6-1 or 6-14) because they only need a minimal set of parameters. Moreover, the maxima of the L_a and L_b vibronic bands (Tab. 9.1) could accurately be determined by the minima of the second derivative spectra $\varepsilon''(v)$ as described in 2.3.2,[6] such that they could be fixed for the non-linear least squares spectral fits. The same procedure has been applied to 9DP, except that the parameter for the hidden L_b vibronic bands are kept constant with the values from the fit for DM9DP. The absorption transition moments M_a of these decomposed bands have been computed according to eq. 9-1[18]

$$M_a^2 = \sum_j \frac{3hc2303}{8\pi^3 Nn} \int \frac{\varepsilon(v_a)dv_a}{v_a} \qquad (9\text{-}1)$$

with the Planck constant h, velocity of light in a vacuum c, Avogradro number N, refractive index n of the solvent, wavenumber of absorption v_a and the decadic molar extinction coefficients $\varepsilon(v_a)$. Likewise, the fluorescence transition moments M_f are calculated by eq. 9-2[18] using the experimentally determined fluorescence quantum yields Φ_f, band maxima v_f and lifetimes τ_f. Some characteristic values for the derived transition moments are collected in Tab. 9.1.

$$M_f^2 = \frac{3h}{64\pi^4 n^3} \frac{\Phi_f}{\tau_f v_f^3} \qquad (9\text{-}2)$$

Tab. 9.1
(a) Experimental Transition Dipole Moments in Debye Units and (b) Transition Energies in 10^3 cm^{-1}.

(a)	Absorption			Fluorescence[a]			
	$M_a(^1L_a)$	$M_a(^1L_b)$	$M_a(^1CT)$	$M_f(OCT)$	$M_f(BOB)$	$M_f(EOE)$	$M_f(ACN)$
P[b]	2.8	0.38	-	0.33	n.d.	n.d.	n.d.
DM9DP	2.0	0.53	2.0	0.6	2.9	2.3	1.4
9DP	3.4	n.d.	3.8	1.6	3.7	4.4	4.0

(b)	$v_{L_a}^{0-0\,c}$	$v_{L_b}^{0-0\,d}$	v_{CT}^{max}	v_f^{0-0} (OCT)[e]	v_f^{max} (BOB)	v_f^{max} (EOE)	v_f^{max} (ACN)
P	34.2	28.9	-	28.9	n.d.	n.d.	n.d.
DM9DP	33.8	28.6	32.3	28.2[f]	25.1[f]	23.8[f]	19.7[f]
9DP	35.0	n.d.	32.5	≈ 26.6[f]	25.3[f]	24.6[f]	21.3[f]

[a] solvent abbreviations: iso-octane (OCT), dibutylether (BOB), diethylether (EOE) and acetonitrile (ACN)
[b] values determined with eq. 9-1 and 9-2 using literature data (ref. 20).
[c] two further vibronic bands are located at: 36.1, 37.2 for 9DP; 35.0 and 36.0 for DM9DP; 35.65 and 36.43 for P.
[d] five further vibronic bands are located at: 29.24, 29.94, 30.68, 31.45 and 32.05 for DM9DP; 29.54, 30.96, 31.75 and 32.42 for P.
[e] further vibronic bands are located at: 25.94 for 9DP; 27.0 and 25.7 (average high frequency vibration v_i obtained from band-shape fitting with eq. 6-1 is 1250 cm^{-1}) for DM9DP; 27.54, 26.14, 24.78 (a_1 progression) and 28.08, 26.73 (b_2 progression) for P.[21]
[f] For consistency with the previous paper (ref. 1,2), the quantum corrected fluorescence maxima are determined from the maxima on the wavelength scale.

Tab. 9.2 Results of the Solvatochromic Plots Using an Onsager Radius of 5.7 Å[a]

	DM9DP		9DP	
	$v_f(F_2)$	$v_f(L)$	$v_f(F_2)$	$v_f(L)$
$\langle\Delta\mu_{eg}\rangle$ (D)	22.1	14.3	15.9	10.2
$\langle\mu_e\rangle$ (D)	22.7	14.9	16.7	11.0
μ_g (D)[b]	1.2	1.2	1.5	1.5
intercept (cm^{-1})	30.49	28.69	27.67	26.68
slope (cm^{-1})	27.36	11.46	14.25	5.85
correl[a]	-0.996	-0.988	-0.930	-0.931

[a] using data for all solvents in Fig. 9.2 [b] obtained by AM1

The fluorescence spectra of DM9DP and 9DP (Fig. 9.1cd) exhibit a substantial red shift accompanied by a broadening of the band half widths (DM9DP: fwhm = 3200-4400 cm^{-1}; 9DP: 3500-3900 cm^{-1}) with increasing solvent polarity, which indicates the strong ^{1}ET character in the emitting state. The solvatochromic plots for DM9DP and 9DP which are equivalent to those of **I-III** in Fig. 3.3 c and d are shown in Fig. 9.2. Using the Onsager radius a = 5.7Å as elsewhere[1] together with calculated (AM1) ground state dipole moments μ_g 1.2 D for DM9DP and 1.5 D for 9DP, the linear fits according to eq. 3-7 and 3-8, respectively, yields the results in Tab. 9.2. The average excited state dipole moment $\langle\mu_e\rangle$ of DM9DP is larger than for 9DP revealing stronger electron transfer interactions in the more twisted biaryl DM9DP.[22]

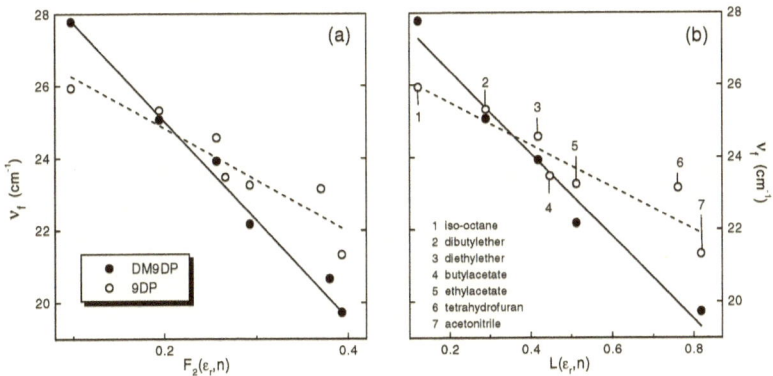

Fig. 9.2 Solvatochromic Plots of DM9DP and 9DP and linear fits according to eq. 3-7 and 3-8.

9.3.2 Quantum Chemical Calculations

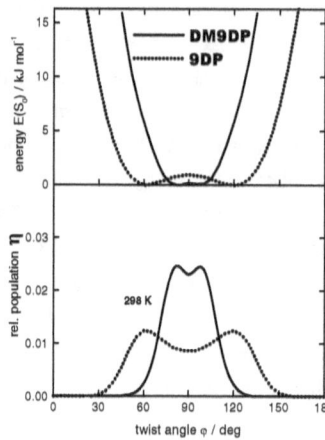

Fig. 9.3 Ground state twist potentials derived from AM1 optimizations (upper panel) and the corresponding rotational distributions η(φ) according to the Boltzmann eq. 3-18 (lower panel) for DM9DP(——) and 9DP(· · ·).

Tab. 9.3 Data from S0 Twist Potentials[a]

	φ_{eq}	$E_i^{\#}(\varphi=0°)$	$E_i^{\#}(\varphi=90°)$	$\langle\varphi_\eta\rangle$
DM9DP	82°	107 kJ	0.2 kJ	79°
9DP	62°	36 kJ	1.0 kJ	64°

[a] obtained by AM1/gas phase/298K; for details see 3.3.4.1

Fig. 9.3 shows the calculated ground state (S$_0$) twist potential and the corresponding rotamer distribution η(φ) determined from Boltzmann's law (eq. 3-18) at 298 K for DM9DP and 9DP. The equilibrium twist angle φ_{eq} for DM9DP in S$_0$ is 82° (62° for 9DP). Taking into account the rotamer convolution integral (eq. 3-19), a mean twist angle $\langle\varphi_\eta\rangle$=79° for DM9DP and $\langle\varphi_\eta\rangle$=64° for 9DP is obtained ($\langle\varphi_\eta\rangle$ of 9DP agrees with the angle (64.2°) from the x-ray structure[1] of 9DP). All data in Tab. 9.3 describing both potentials demonstrate that the rotamer distribution of DM9DP is clearly more confined to φ=90° than that of 9DP.

Fig. 9.4 depicts the relevant molecular orbitals (MOs) of 9DP involved in the low energy transitions of both D-A phenanthrene compounds at φ=90° and at the equilibrium twist angle between the electron donor unit dimethylanilino (D) and the acceptor unit phenanthryl (A). It can be seen that for 90° the orbitals are localized on the subchromophores and consequently the transitions on one subunit can be assigned to local transitions of phenanthrene (A) and dimethylaniline (D). The local singlet transition of lowest energy is the ^1L_b(A) consisting of the one-electron transitions 1→-2 and 3→-1. The second local singlet transition is the ^1L_a(A) connected with 1→-1 and 3→-2 configurations. Transitions between the subunits are connected with a full intramolecular electron transfer (^{1}ET) at 90° and the one of the lowest energy consists of the 2→-1 or HOMO(D)→LUMO(A) configuration. For deviations from 90° the transitions are linked with the same configurations but the orbitals become delocalized over the subunits, especially the occupied MOs 1, 2 and 3 (Fig. 9.4). As a result all the transitions get delocalized character but nevertheless can be regarded as keeping their substantial (spectroscopic) properties over the whole angular range (e.g. transition moments and their directions), such that they can be correlated with the pure ^1L_a, ^1L_b, ^1B_b, etc. states of D and A at φ=90°. The corresponding properties of the first lowest excited singlet states are presented in Fig. 9.4.

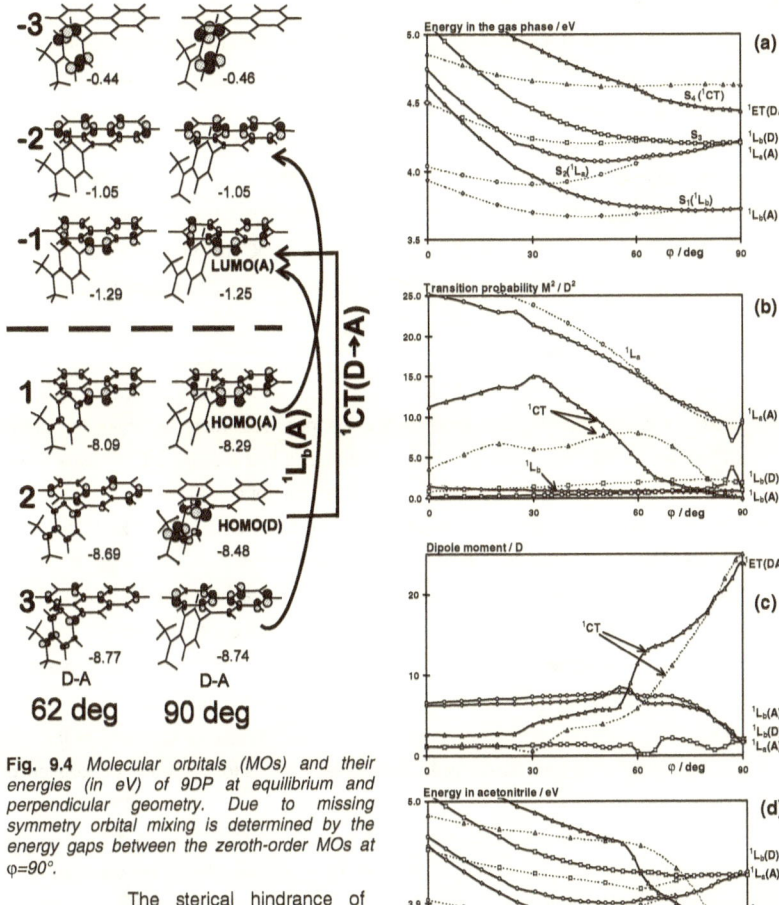

Fig. 9.4 Molecular orbitals (MOs) and their energies (in eV) of 9DP at equilibrium and perpendicular geometry. Due to missing symmetry orbital mixing is determined by the energy gaps between the zeroth-order MOs at $\varphi=90°$.

Fig. 9.5 a) Energies, b) squared transition moments and c) dipole moments of the S1-S4 states obtained from CNDO/S and AM1 results as a function of the twist angle φ for DM9DP (———) and 9DP (· · ·). d) Calculated energies in acetonitrile using eqs. 3-3 and 3-4.

The sterical hindrance of more planar geometries is less effective in the excited states (Fig. 9.5a). Particularly the ^1L_a-type S_2^{calc} states, which contain most of the $^1L_a(A)$ character, exhibit a strong tendency to more planar angles φ ($\varphi_{eq} = 25°$ for 9DP and $\varphi_{eq} = 50°$ for DM9DP).

At $\varphi=90°$, dimethylaniline (donor D) and phenanthrene (acceptor A) moieties are decoupled which results in rather pure locally excited

S_1^{calc}-S_3^{calc} states $^1L_b(A)$, $^1L_a(A)$, $^1L_b(D)$ and the zeroth-order $S4^{calc}$ state $^1ET(D^+A^-)$, where one electron is transferred from D to A. Due to the low symmetry of both molecules, strong electronic coupling between all states is possible at non-perpendicular twist angles, via CI and MO mixing within the SCF/CI procedure. Geometrical deformations at $\varphi < 90°$, which are mainly (i) bending of the whole molecule with respect to the linking bond and (ii) loss of planarity of the phenanthrene unit to a boatlike structure (for more details see ref. 1), even amplify the electronic coupling effects. Thus the ^1ET (here pure HOMO(D)→LUMO(A)) character of the $S4^{calc}$ state ^1CT is reduced from 82% at $\varphi=90°$ to 39% for DM9DP (10% for 9DP) at $\varphi=60°$. On the other hand, the ^1ET character of S_2^{calc} increases through MO mixing from 2% at $\varphi=90°$ to 23% (26% for 9DP) at 60°. The first four excited singlet states at $\varphi=60°$ absorb 67% (48% for 9DP) of the pure HOMO(D)→LUMO(A) configuration indicating that the ^1ET character is distributed over all excited states in the gas phase. The smaller ionization potential (IP) of the donor group in DM9DP (from AM1: IP(DM9D)=8.2 eV, IP(9D)=8.3 eV) is responsible for a lower lying zeroth-order ^1ET state (Fig 3a), which vice versa introduces more ^1ET character to the S_1^{calc}-S_4^{calc} states of DM9DP. Likewise, it can be anticipated that a stabilized ^1ET state in the solution phase leads to a further increase of the ^1ET character for these low lying states. Hence beside the twist angle, the solvation coordinate is able to influence the mixing behaviour through a change of the energy gaps between the states.

In order to get an idea of solvation effects, the solvation energies of the excited states (ΔE_{solv}) were calculated by the Onsager[23] eqs. 3-3 and 3-4 using the calculated dipole moments $\mu(\varphi)$ in Fig. 9.5c, an exaggerated Onsager cavity radius $a=4.7$ Å and the dielectric constant $\varepsilon_r=37.5$ for acetonitrile. The resulting excited state energies (eq. 3-4) in acetonitrile are plotted in Fig. 9.5d. In accordance with the minimum overlap rule,[24] the dipole moments of the ^1CT state are maximized for perpendicular twist angles leading to strong solvent stabilization of highly twisted conformers.

As an attempt to evaluate coupling effects in solution, the CNDO/S parametrization was adjusted (in particular, the Coulomb integral of the nitrogen was varied from the standard value -12 eV down to -3 eV), such that the HOMO(D) became energetically situated below HOMO(A) associated with a stabilized ^1CT state. This simulates a solvent stabilization of the ^1ET state. The resulting mixed MOs at $\varphi=62°$ are then qualitatively different from those in Fig 9.4, namely orbital 3 loses its donor character while the MOs 1 and 2 both gain a large weight of HOMO(D). Consequently, the 1→-1 transition, which contributes to the $^1L_a(A)$ state, adopts more ^1ET character leading to a $^1L_a/^1CT$ state that is energetically close to $S_1(^1L_b(A))$ and connected with a larger dipole and transition moment than in the $^1L_b(A)$-type state. Such a change of coupling (in the gas phase ^1CT mainly interacts with ^1L_b, see ref.1) may occur in solvents and could lead to a level crossing of S_1 ($^1L_b(A)$) and $S_2(^1L_a/^1CT)$ in medium polar solvents (see 9.4.2).

9.4 Discussion

9.4.1 Influence of Electron Transfer Interactions on Absorption and Emission Bands.

The $^1L_b(A)$ and $^1L_a(A)$ absorption bands of DM9DP (Fig. 9.1a) have (i) similar shapes, (ii) consist of the same progressions with similar intensity distribution (e.g. the six vibronic bands with 700 cm^{-1} spacings in the $^1L_b(P)$ band) and (iii) have comparable extinction coefficients as those of the pure P spectrum (not decomposed). The slight red shift of the $^1L_b(A)$ and $^1L_a(A)$ 0-0 vibronic bands of 300 cm^{-1} and 400 cm^{-1} is in quantitative agreement with 9-phenylphenanthrene.[1] This shows the dominating phenanthrene character of the transitions involved in DM9DP, although the additional ^{1}CT band „steals" part from the intensity of the $^1L_a(A)$ band as revealed by the larger transition moment $M_a(^1L_a)$ in P (Tab. 9.1). On the other hand, the features of the P spectrum are blurred in the spectrum of 9DP (Fig. 9.1a) demonstrating much stronger electronic interactions of the phenanthryl (A) with the dimethylanilino (D) unit than in DM9DP.

The two intense ^1L_a and ^{1}CT absorption bands of DM9DP and 9DP have to be correlated with the calculated S_2^{calc} and S_4^{calc} states (Fig. 9.5). This means, that the S_2^{calc} and S_4^{calc} of the gas phase calculations exchange their character in solution comprehensible with a solvent stabilization of the zeroth-order ^{1}ET state. It is interesting to note, that the calculations still reproduce the experimentally observed stronger electronic interaction of the ^1L_a and ^{1}CT states in the less twisted 9DP which leads to a larger ^1L_a-^{1}CT energy splitting ($\Delta\nu^{calc}(\varphi=62°)=4360$ cm^{-1}, $\Delta\nu^{exp}=2500$ cm^{-1}) than in DM9DP ($\Delta\nu^{calc}(\varphi=82°)=2180$ cm^{-1}, $\Delta\nu^{exp}=1500$ cm^{-1}).

The calculations and absorption spectra attribute rather pure ^1L_b character to the S_1 state of DM9DP and 9DP. However, the fluorescence in nonpolar iso-octane does not display the mirror-image relationship as in P.[21] A comparative vibrational analysis reveals that the progression polarized in the short-axis of P (totally symmetric a_1 vibration (C_{2v} point group) of about 1400 cm^{-1}) is supressed in DM9DP and 9DP, while the long-axis polarized b_2 vibration (830 cm^{-1}), which is of low intensity in the fluorescence spectrum of P, is the main accepting mode in the ground state of DM9DP and 9DP. This is consistent with a symmetry change of S_1 from $^1A_1$ (pure $^1L_b(P)$-type) to more $^1B_2$ symmetry and thus indicates increased state mixing after solute and solvent relaxation in the excited state of DM9DP and 9DP.

9.4.2 Squared Transition Moments as Indicators for Molecular and Electronic Structure

In order to analyze the change of the electronic and/or moleculare structure in dependence of the solvent polarity, the squared transition moments M_f^2 are plotted in Fig. 9.6 versus the polarity parameter $F_2(\varepsilon,n)$ of eq.3-7b.

Generally, an increase of the squared transition dipole moment M_f^2 or the energy corrected radiative rate $k_f/\nu_f^3 n^3$, which is proportional to M_f^2, reflects a change of the S_1

wavefunction (Ψ_{S1}) which can be induced by (i) electronic and (ii) conformational sources, which are in the present case

(i) enhanced contribution of the allowed states ^1L_a and delocalized ^1CT (zeroth-order ^1L_a-^1ET coupling)

(ii) more planar twist angle φ between D and A moieties (zeroth-order S_0-^1ET coupling).

Of course, both effects depend on each other, but the comparison of M^2 for different conditions (calculation vs. absorption vs. fluorescence, 9DP vs. DM9DP vs. P, different solvent polarity) may allow us to derive the most probable electronic and molecular structure.

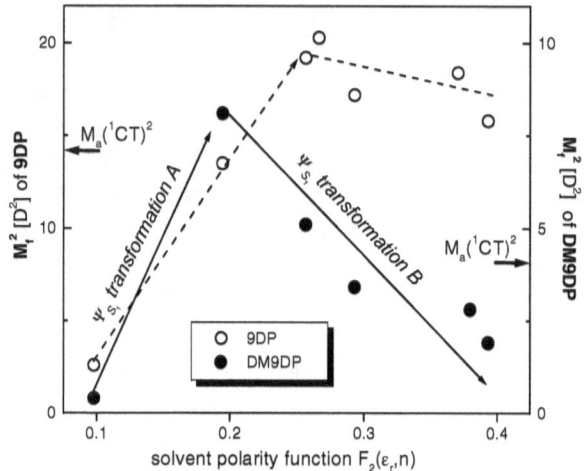

Fig. 9.6 *Plots of squared fluorescence transition moments M_f^2 vs. the solvent polarity parameter F_2 (eq.2) for DM9DP (•) and 9DP (o). The values of the squared absorption transition moments $M_a(CT)^2$ of the CT band in iso-octane are indicated by arrows for DM9DP at the right and for 9DP at the left axis. Transformation A indicates the change of the wavefunction of S_1 (Ψ_{S1}) from ^1L_b to $^1L_a/^1ET$ character and transformation B indicates the change from ^1L_a to ^1ET character (see text and scheme 2).*

First of all, the squared transition dipole moments M_f^2 and M_a^2 of 9DP are about twice as large as those of DM9DP over the whole solvent polarity range (Fig. 9.6) being simply due to a less twisted conformer distribution in all states (Fig. 9.3 and 9.5a). However, the ratio increases at higher solvent polarity ($F_2>0.25$) indicating that the electronic and/or molecular structure change differently for both compounds. The consecutive increase (Ψ_{S1} transformation A) and decrease (Ψ_{S1} transformation B) of the transition dipole moments with increasing solvent polarity shown in Fig 9.6, distinguishes between a nonpolar ($F_2<0.1$), medium polar ($F_2=0.2$-0.25) and highly polar ($F_2>0.35$) region.

Nonpolar region ($F_2 < 0.1$)

The observed fluorescence transition moments of 9DP and DM9DP are slightly higher than for the pure $^1L_b(P)$ state (Tab. 9.1). Agreement is found between M_f of DM9DP (0.6 D) with that of 9-phenylphenanthrene (0.6 D),[1] with $M_a(^1L_b)$ (0.53 D) of the decomposed $^1L_b(A)$ absorption band (Fig.9.1) and roughly with the calculated M (0.9 D) of ^1L_b, which consists of a small amount of ^1ET and ^1L_a electron configurations. By contrast, for both compounds DM9DP and 9DP, the derived transition moments of the ^1CT bands (Tab. 9.1) are distinctly larger. These comparisons imply that the main character of the emitting state is of the $^1L_b(A)$-type with an admixture of the more allowed states ^1L_a and ^1CT. Which of both states delivers the main intensity to S_1 is hard to decide. A simple state symmetry argument speaks in favour of ^1CT mixing, since the angle between the transition moment directions of the $^1L_b(A)$ with the ^1CT state is about 30°, in contrast to the near perpendicular orientation with the ^1L_a state.

In any case, S_1 has dominant ^1L_b character, which, according to the similar twist potentials of the calculated S_0 and S_1 in Fig. 9.3, should result in only minor excited state angular relaxations ($\varphi_{eq}(S_1^{calc})$ = 80° for DM9DP and 50° for 9DP). The more planar distribution of 9DP as compared to DM9DP (Fig. 9.3) is also in accordance with the 6 times higher M_f^2, if some interaction with the allowed ^1L_a and ^1CT transitions is considered (Fig. 9.5b).

Medium polar region (F_2=0.2-0.25)

From nonpolar to medium polar solvents, M_f^2 is strongly increased by a factor of about 8 for 9DP in diethylether ($F_2 \approx 0.25$) and even 20 for DM9DP in dibutylether ($F_2 \approx 0.2$) indicating a vanishing contribution of the ^1L_b state to the emitting wavefunction. This can be explained by an enhanced ^1CT character of S_1, because due to the high dipole moments of the HOMO(D)→LUMO(A) configuration (^1ET), ^1CT is strongly lowered by increasing solvent polarity. However, the delocalized ^1CT alone can not account for these high M_f^2 since the calculated M^2 of ^1CT (Fig. 9.5b) in the accessible angular region as well as $M_a(^1CT)^2$ from absorption (Fig. 9.6) are distinctly lower. The increase of M_f^2 is therefore caused by an enhanced ^1L_a character which is „transferred" to S_1 through solvent-induced ^1ET interactions. This corresponds to the mixing behaviour which is theoretically derived (9.3.2) by simulating a lower lying ^1ET. Further, the fact that the ^1ET contributions are intrinsically higher for the methylsubstituted compound DM9DP explains why the full decoupling of S_1 from ^1L_b is reached already at lower polarity than for 9DP (see maximum of M_f^2 at different F_2 in Fig. 9.6).

Concerning the conformational structure, the increased ^1L_a character of S_1 should enforce both molecules to relax towards a more planar structure similar to the calculated S_2 state (Fig. 3a: $\varphi_{eq}(^1L_a)$= 50° for DM9DP and 25° for 9DP). This is also consistent with both ratios M_f^2/M_a^2 larger than unity (Fig.9.6).

Highly polar region ($F_2>0.35$)

In highly polar solvents such as acetonitrile, M_f^2 of DM9DP is reduced again to almost 20% as compared to dibutylether (Ψ_{S_1} transformation B in Fig. 9.6), while M_f^2 of 9DP decreases only insignificantly. Increased ^{1}CT character of Ψ_{S_1} can account for the small M_f^2 of DM9DP, since the calculated M^2 of ^{1}CT (Fig. 9.5b) are smaller than those of ^1L_a, especially for the more twisted conformers which are preferentially stabilized by the high dipole moments (Fig. 9.5c and 9.5d).

The involved excited state relaxation can either be traced back to (i) solvation alone or (ii) solvent-assisted narrowing of a perpendicular rotamer distribution. Thereby, the main source of reduced transition moments for S_1 emission is (i) an increased energy gap to the higher lying, allowed states (less zeroth-order ^1L_a-^{1}ET coupling) or (ii) less D-A molecular orbital overlap (less zeroth-order ^{1}ET-S_0 coupling). In view of the sizeable decrease of M_f^2 for DM9DP, it is reasonable to conclude that the excited state rotamer distribution of DM9DP in highly polar solvents ($F_2>0.35$) is distinctly more twisted than in less polar solvents. However, since the decrease of M_f^2 with regard to $M_a(^1CT)^2$ is only moderate (Fig.9.6), it is not possible to ascertain that it is also more twisted than in the ground state. For 9DP in highly polar solvents, the M_f^2 values remain even larger than $M_a(^1CT)^2$ of the CT absorption band, although a solvent-induced decoupling from ^1L_a can be anticipated. This indicates that 9DP still exists in a more planar conformation than in the ground state.

It seems then, that a sufficiently large amount of ^{1}ET character of S_1 (in DM9DP mainly induced by a more twisted conformation than in 9DP) is a prerequisite to observe the polarity induced decrease of M_f and to prevent the photoinduced relaxation to a more planar structure. The same behaviour is found for the donor-acceptor biphenyls pair **II** and **III** (ch. 8). For **III**, the ^{1}ET character in S_1 is even larger than for DM9DP (e.g. compare Fig. 3.3/Tab.3.2 with Fig. 9.2/Tab.9.2: $\mu_e(\mathbf{III})$ = 28 D > μ_e(DM9DP) = 23 D; $\nu_f(\mathbf{III})$ < ν_f(DM9DP) although locally excited states are higher lying in **III**, see Fig.2.1b and Fig. 9.5a.) which consequently explains why a rotamer distribution more twisted in S_1 than in S_0 is clearly observed for **III** in polar solvents but not for DM9DP (note also that a relaxation in the subnanosecond range is found only for **III**).

In summary, the solvent-induced bimodal changes of the fluorescence transition moments (or M_f^2) can be understood by solvent-induced changes of the weights of the locally excited (LE) phenanthrene ^1L_b(A) and ^1L_a(A) states and the zeroth-order ^{1}ET state to the wavefunction $^1\psi$ of the emitting state as shown in scheme 9.2.

$$\Psi_{S1}(ET/L_a/L_b/S_0) \xrightarrow{A} \Psi_{S1}(ET/L_a/S_0) \xrightarrow{B} \Psi_{S1}(ET/S_0)$$

$$^1LE \qquad\qquad\qquad ^1CT \qquad\qquad\qquad ^1ET$$

→ solvation or solvent polarity

Scheme 9.2 *Schematic representation of the solvent-induced transformation of S_1 nature in terms of zeroth-order electronic state mixing.*
Transformation A: decoupling from ^1L_b assisted by rotation to planarity
Transformation B: decoupling from ^1L_a assisted by rotation to perpendicularity
The S_1 relaxation to a more twisted structure than in S_0 seems to occur only in the case of large ^1ET interactions.

9.5 Conclusions

For 9DP and DM9DP in nonpolar solvents, weak electronic interaction of the ^1L_b(A)-type S_1 state with $^1L_a/^1CT$ occurs within an angular distribution similar to the ground state. In medium polar solvents, the ^1L_a state gains more ^1CT character by electronic mixing with the energetically lowered ^1ET, such that its solvent stabilization „transfers" enhanced ^1L_a character to S_1 (Ψ_{S1} transformation A). This results in higher transition moments (M_f) and a more planar angular distribution which both are characteristics of the ^1L_a state (Fig.9.5). As the solvent polarity is further increased, the ^1ET character is strongly enhanced leading again to smaller M_f values for DM9DP (Ψ_{S1} transformation B). Due to a less twisted structure of 9DP, the coupling of the zeroth-order ^1ET with the ground state is stronger with the consequence of a larger contribution to the total M_f and correspondingly smaller relative changes of M_f. This can explain why the transformation B (scheme 9.2) is not necessarily reflected in a decrease of M_f for 9DP (Fig. 9.6).

In polar solvents, the main photoinduced relaxation coordinate of D-A biaryls containing a large subunit, such as phenanthrene,[1,2] carbazole,[3] anthracene,[4] or pyrene,[5] seems to be solvation rather than twisting towards solvent stabilized perpendicular rotamers. In contrast to small size D-A biaryl compounds,[7-9,24,25] the driving force in the excited state towards planarity even prevails because of a better π-electron delocalization and smaller solvent stabilization due to a less effective dipole field (e.g. by a comparably larger dipole moment and Onsager radius in eq. 3-3 and 3-4). Hence, due to less ^1ET interactions in DM9DP (φ_{eq}=82°) as compared to the similarly twisted **III** (φeq=78°) an excited state relaxation to a structure more twisted than in S_0 is experimentally observed on a nanosecond-timescale for **III** (ch.3,6) but not for DM9DP.

Generally, solvent dependent transition moments do not need to result from multiple minima on the S_1 potential surface, but can simply be due to a single minimum associated with solvent-induced changes of mixing strength between S_1 and higher lying states. As a consequence of this solvent-induced coupling, the excited state potential energy minimum can shift to planarity or perpendicularity depending on the states involved which in turn additionally influences the transition moment. Analogously to vibronic coupling,[26] the mechanism of solvent-induced electronic coupling can alter the radiative strength of the first excited state. This kind of mixing is considerably stronger and moreover controllable through the polarity of the medium.

9.6 References and Notes

(1) A. Onkelinx, F.C. De Schryver, L. Viane, M. Van der Auweraer, K. Iwai, M. Yamamoto, M. Ichikawa, H. Masuhara, M. Maus, W. Rettig *J. Am. Chem. Soc.* **1996**, 118, 2892.

(2) A. Onkelinx, F. Schweitzer, F.C. De Schryver, H. Miyasaka, M. Van der Auweraer, T. Asahi, H. Masuhara, H. Fukumura, A. Yashima, K. Iwai, *J. Phys. Chem. A* **1997**, 101, 5054.

(3) A. Kapturkiewicz, J. Herbich, J. Karpiuk, J. Nowacki, *J. Phys. Chem. A* **1997**, 101, 2332.

(4) J. Herbich, A. Kapturkiewicz, *J. Am. Chem. Soc.* **1998**, 120, 1014 and references therein.

(5) J. Dobkowski, J. Waluk, W. Yang, C. Rullière, W. Rettig, *New. J. Chem.* **1997**, 21, 429.

(6) M. Maus, W. Rettig, *Chem. Phys.* **1997**, 218, 151.

(7) M. Maus, W. Rettig, D. Bonafoux, R. Lapouyade, *submitted for publication*.

(8) M. Maus, W. Rettig, R. Lapouyade, *J. Inf. Rec.* **1996**, 22, 451.

(9) W. Rettig, M. Maus, R. Lapouyade, *Ber. Bunsenges. Phys. Chem.* **1996**, 100, 2091.

(10) I.R. Gould, R.H. Young, L.J. Mueller, A.C. Albrecht, S. Farid, *J. Am. Chem. Soc.* **1994**, 116, 8188.

(11) M. Bixon, J. Jortner, J.W. Verhoeven, *J. Am. Chem. Soc.* **1994**, 116, 7349.

(12) J. Wolf, G. Hohlneicher, *Chem. Phys.* **1994**, 181, 185.

(13) M. S. Gudipati, J. Daverkausen, G. Hohlneicher, *Chem. Phys.* **1993**, 174, 143.

(14) M. S. Gudipati, J. Daverkausen, M. Maus and G. Hohlneicher, *Chem. Phys.* **1994**, 186, 289 and references therein.

(15) M. Vogel, W. Rettig, *Ber. Bunsenges. Phys. Chem.* **1987**, 91, 1241.

(16) J. Del Bene and H.H. Jaffé, *J. Chem. Phys.* **1968**, 48, 1807; **1968**, 48, 4050; **1968**, 49, 1221; **1969**, 50, 1126.

(17) AMPAC 5.0, **1994** Semichem, 7128 Summit, Shawnee, KS 66216.

(18) J.B. Birks, *Photophysics of Aromatic Molecules*, Wiley-Interscience, New York, **1970**.

(19) J. A. Riddik, W.B. Bunger, T.K. Sakano, *Organic Solvents*, John Wiley & Sons **1986**.

(20) T. Yoshinaga, H. Hiratsuka, Y. Tanizaki, *Bull. Chem. Soc. Jap.* **1977**, 50, 2548.

(21) T. Azumi and S.P. McGlynn, *J. Chem. Phys.* **1962**, 37, 2413.

(22) In ref. 1, a dipole moment of 21 D has been derived for 9DP using eq. 3-9 only with $F_2(\varepsilon,n)$ values larger than 0.25 and omitting the data for butylacetate and tetrahydrofuran. This plot gives a better correlation coefficient of -0.988 and takes into account the change of the emitting state in 9DP at $F_2>0.25$.

(23) L. Onsager, *J. Am. Chem. Soc.* **1936**, 58, 1486.

(24) Z.R Grabowski, K. Rotkiewicz, A. Siemiarczuk, D.J. Cowley, W. Baumann, *Nouv. J. Chim.* **1979**, 3 443.

(25) S. Delmond, J.-F. Létard, R. Lapouyade, W. Rettig, *J. Photochem. Photobiol. A: Chem.* **1997**, 105, 135.

(26) G. Herzberg, E. Teller, *Z. Phys. Chem.* **1933**, B21, 410.

Chapter 10

Possible Applications of D-A Biphenyls as Fluorescent Probes

Abstract:
The ability and structural requirements of donor-acceptor (D-A) biphenyls showing photoinduced intramolecular charge transfer to serve as fluorescent probes for different microenvironmental properties is investigated. An adiabatic photoreaction promotes the twisted compound to probe sudden changes of the microviscosity as demonstrated by changes of the fluorescence lifetimes at the glass transition of polar solvents. In nonpolar solvent mixtures, the appearance of a long lifetime below a certain temperature for the three D-A biphenyls investigated is interpreted by the formation of self-complexes indicating a change of the matrix structure or order. Hydrogen bonding effects of the twisted D-A biphenyl may be exploited to sense traces of protic solvents. The D-A biphenyls being planar in the CT excited state are well suitable as pH sensing fluorescent probes with a dynamic pH dependent, self-calibrating fluorescence ratio signal varying over four orders of magnitude. The reasons for the twist angle dependent usefulness of the D-A biphenyls as fluorescent probes are discussed.

Keywords: fluorescent probes, pH, microenvironment, photonic molecular device, hydrogen bonding

10.1 Introduction

A photon-driven (or photonic) molecular device (PMD) is an assembly of molecular components capable of performing light-induced functions, i.e. a device powered by light or elaborating light.[1,2] In a perspective view, supramolecular PMD's could serve as a combination of highly efficient (solar) light energy collectors (antenna function), transporters (e.g. by electron or energy transfer) and converters into a suitable energy form (e.g. heat, electricity or chemical potential). Alternatively, PMD's can fulfill the task to explore a certain property of a macroscopic sample by performing a photoinduced process which depends on the microenvironment. A fluorescent probe is such a "micro-machine". The mode of its operation may be descibed in an illustrative way, as follows:

At first, the macroscopic sample is doped or labelled with the PMD, the fluorescent probe. After switching on the light, the PMD starts to work, i.e. performs its characteristic photoinduced process while interacting with the microenvironment and storing the obtained information into a property of the probe molecule itself. Finally, the PMD transduces the derived information to a detector by a fluorescence signal. Of course, this needs to be a closed photocycle with return of the PMD to the starting ground state.

It depends on the molecular and electronic structure as well as on the intramolecular process activated by the illumination which properties of the microenvironment can be probed. Here, the probe molecules investigated are the donor-acceptor biphenyls **I-III** and the photoinduced process sensing the microenvironment is intramolecular charge transfer which is

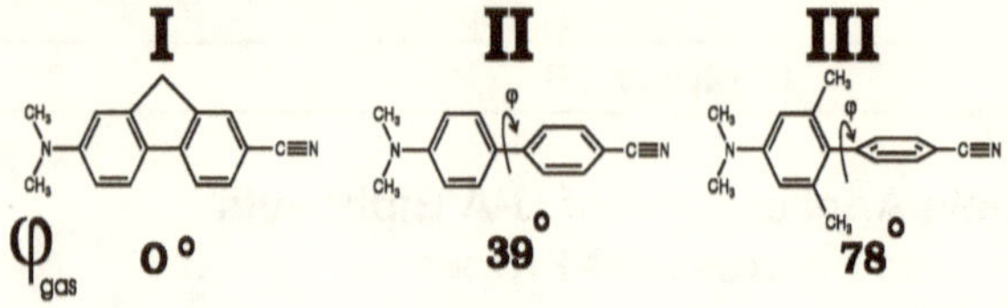

Scheme 10.1 *Molecular structures of the biaromatic model compounds (I-III) which are used to investigate the conformation dependent ability to serve as fluorescent probes.*

especially for **III** accompanied by a slow conformational photoreaction (ch.3,4,6,7).[3-5] The aim of this study is to investigate the usefulness, molecular requirements (here the twist angle φ) and possibilities to apply such biaromatic molecules in combination with intramolecular charge transfer as fluorescent probes.

Fluorescent probes have been reviewed several times[6-10] and their leading advantage versus other methods is a high sensitivity combined with the possibilty of specific molecular design. From the large field of microscopic properties which can be sensed by fluorescent probes (polarity, phase transitions, fluidity, order, hydrogen bonds, pH, ions, viscosity, electric potential, temperature, heavy atoms, quenchers), four parameters are selected in the present study which are briefly introduced in the following.

Solvent polarity. The meaning of polarity is not trivial, since it reflects the complex interplay between different types of dielectric solute-solvent interactions which can be a bulk property of the dielectric field (e.g. Onsager theory)[11] or which can be of specific molecular nature such as hydrogen bonding. Moreover, the interactions are time-dependent (dynamic polarity) due to viscosity controlled reorientation of the solvent dipoles to the photoinduced change of the solute dipole moment which itself depends on the rate of the intramolecular photorelaxations. It is therefore not surprising that a number of different empirical scales of polarity, e.g. ET(30),[12] SPP,[13] Z(100),[14] $\Delta\chi$,[15] or π^*[16] have been suggested in literature as alternatives to the macroscopic and theoretically derived Onsager[11] polarity functions $F_2(\varepsilon_r, n)$ or $F(\varepsilon_r)$ where ε_r is the static dielectric constant and n is the refractive index (see 3.3.2). The fact that a unified polarity scale has not yet really been established, demonstrates the necessity of electronically well-defined molecules the optical properties of which can definetely be traced back to polarity effects. Needless to say, that solvent polarity can be probed with optical methods using dyes with large dipole moments in the ground or excited state due to polarity dependent electronic transition parameters such as the transition energy or intensity. Here, only the static bulk polarity will be considered which requires the absence of strong specific interactions in the ^{1}CT state and a fast equilibration of the solvent-solute interactions within the excited state lifetime of **I-III**. Both requirements are fulfilled, if only fast relaxing aprotic solvents (τ_s< 5ps) are considered. A combination of the measurement of stimulated emission and excited state absorption will be shown to be capable of probing polarity.

Fluidity and glass transition temperature. The change of fluidity between a liquid and a rigid surrounding can be well identified by fluorescent probes performing a photoinduced geometrical relaxation to a conformation with different emission properties. In practice, it has been shown that the fluorescence intensity of such a PMD probe can be used to monitor the progress of a polymerization reaction.[7,17,18] A sharp increase of the fluorescence intensity and more slightly of the emission energy can be observed where the molecular weight or the degree of polymerization increases drastically. Obviously, the immense increase of the microviscosity closes the channel to the adiabatic photoreaction of the probe, either through a suppression of the assistant solvent relaxation[18] or directly through a reduction of the free volume needed for the intramolecular relaxation.[7,19] Such a strong increase of the microviscosity also occurs in solvents at the transition from the liquid to the rigid glass and therefore, following the fluorescence (here, its lifetime) as a function of temperature allows to check the applicability of **I-III** as a sensitive fluorescent probe of large viscosity changes.[20]

Hydrogen bonds Traces of hydrogen bonding impurities or solvents (protic solvents) can have a strong effect on photophysical properties of a dye[21-28] and can therefore lead to a possible misinterpretation in a photophysical study.[24] The importance of intramolecular electron transfer for the sensibility to hydrogen bonding has been demonstrated several times.[21-27] Usually, the occurrence of hydrogen bonding interaction conceivable by a partial proton shift (PSh) to the dye is testified by a shift of transition energy and a decrease of the fluorescence quantum yield through an enhanced nonradiative decay as compared to the behaviour in a similarly polar but aprotic solvent. To elucidate to what extent such a specific mechanism is active in **I-III** fluorescence experiments in ethanol at different temperatures and by addition of protons are analyzed.

pH The advantages of pH sensing with fluorescent probes as compared to conventional pH measurements are mainly the high sensitivity down to the single molecule and the consequent subnanometer spatial resolution.[6,9] So called fluorescent photoinduced electron transfer (PET) pH sensors[29] are based on the fluorescence enhancement (FE) that is „switched on" upon protonation of the probe molecule due to closing the channel to the less emissive electron transfer state.[9,30] However, this approach needs controllable conditions such as homogeneous and constant probe concentration, a constant optical path, and the absence of quenching agents, which are hard to meet in small samples (e.g. intracellular situation or microspots) where fluorescence pH sensing is superior to other methods. To avoid these complications, the PET concept was expanded to a second generation of sensors which show two emission bands with opposite proton-sensitive response and thereby allow to read out the pH directly by using the „self-calibrating"[31] emission intensity ratio.[6,9,30] A different strategy to probe the pH by fluorescence will be proposed in the present work combining direct optical population of a highly emissive charge transfer state as well as optical excitation of an emissive locally excited state to yield a high pH dependent intensity ratio.

10.2 Experimental

Details of the fluorescence steady-state and lifetime experiments are given in ch. 3, 4 and 8.

10.2.1 Measurements of pH

The pH solutions were prepared in a 1:1 (v/v) ethanol-water mixture with a solute concentration of 1.1×10^{-5}M for **I** and 2.0×10^{-5}M for **II**. For the absorption and fluorescence titration, traces of 10M (37% by weight) HCl were added and the pH was monitored using a digital pH meter equipped with a glass electrode. The pH calibration was performed with standard aqueous solutions of pH = 1.68 and pH = 9.18. To account for the differences in the liquid junction potential (ΔU_j) and the proton activity coefficients ($\Delta \log \gamma_{H+}$) between the EtOH/H$_2$O (1:1) sample solution and the aqueous solution the measured value of pH^{mes} is corrected to the real pH by[32]

$$pH = pH^{mes} - Ü_j + \log \gamma_{H^+} = pH^{mes} - 0.21.$$

From repeated titrations, the error of the measured pH values between 1 and 6 is estimated to be less than 0.2 units. The pH value for the separately prepared highly acidified solution is directly calculated from the analytical concentration of HCl in 5ml EtOH mixed with 5ml of a mixture of 55% 10M HCl and 45% H$_2$O to be -0.44 without correction for the activity coefficient ($\gamma_\pm$). Since $\gamma_\pm$ amounts to 0.7 for the relevant EtOH/H$_2$O (1:1) mixture for the range of HCl concentration 0.05M to 0.1M,[33] the used value can be regarded as a lower limit. A more realistic value close to 0 would even strengthen the interpretation in the text that proton-induced CT quenching occurs for pH<1 (cf. Fig. 10.9).

10.3 Results and Discussion

10.3.1 Micropolarity

The aim of this brief section is to qualitatively discuss that, in principle, the photoinduced intramolecular charge transfer in donor-acceptor biphenyls is suitable to probe changes of the polarity through associated variations of transition energies and intensities. It has recently been reviewed that two principal ways exist to use a dye as a sensor of local polarity by monitoring a polarity-induced (i) shift of the fluorescence energy or (ii) a change of the fluorescence intensity.[6] As regards the intensity method (ii), advantageous fluorescence molecular systems are self-calibrating, which means that by application they do not require a reference measurement in an environment of known polarity.[31] Self-calibration is given in compounds showing polarity dependent dual fluorescence or vibration structure (Ham-effect).[34]

The strong electron transfer character of S_1 in **I-III** connected with a large difference of the dipole moments in S_0 and S_1 indeed allows us to apply the first approach (i) as testified by the linear solvatochromic plots in Fig. 3.3. For **III**, the polarity dependence of the fluorescence intensity may additionally be applied, but in case of possible specific solvent interactions the radiative rate constant is needed by measuring both quantum yield and lifetime. Latter possibility is therefore too complicated in practice and lacks the advantage of self-calibration. Since probing of polarity with charge transfer fluorescence is a well known phenomenon[7,8,13] and the polarity dependent emission of **I-III** has already been intensively treated in the previous chapters, a more detailed discussion is abandonded and instead of that an alternative method to probe polarity by interrogating polarity dependent transiton energies and intensities is suggested. Thereby, the spectroscopy of fluorescence gain is combined with excited state absorption (ESA) of the electronically well-defined charge transfer biphenyls (cf. ch. 4).

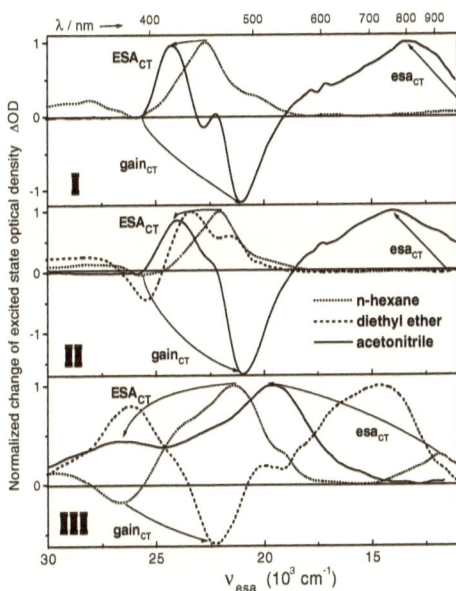

Fig. 10.1 *Relaxed (time delay 400 ps) excited state absorption and fluorescence gain spectra of **I-III** in three solvents of strongly different polarity. The arrows follow the absorption bands ESA_{CT} and esa_{CT} as well as the stimulated fluorescence band $gain_{CT}$ with increasing polarity. With the aid of computer processing the surrounding polarity can be gauged by a multidimensional analysis of the three bands which exhibit polarity dependent transition energies and intensities. The possible ratioing approach needs neither a reference experiment nor a known probe concentration (self-calibration).*

In Fig. 10.1, the gain and ESA spectra of **I-III** in solvents of different polarity, i.e. n-hexane, diethylether and acetonitrile, are plotted. The delay time of 400 ps between pump and probe pulse ensures that solvent and intramolecular relaxations are completed (see ch. 4: Tab.4.3). Fig. 3.3 shows that the shapes of the ESA spectra are considerably more sensitive to changes of polarity than those of the emission spectra because three bands are observed (ch.4).[4] These are the high and low energy absorption bands ESA_{CT} ($^1CT \rightarrow {}^1B_a$) and esa_{CT} ($^1CT \rightarrow {}^1L_a$) as well as the stimulated fluorescence band gain$_{CT}$ ($S_0 \leftarrow {}^1CT$). In particular, the esa$_{CT}$ band varies strongly on the solvent polarity, since the electronic interaction between the ^1CT and ^1L_a, which are both of equal symmetry, sizeably depends on the polarity controlled twist angle φ and their energy difference (4.4.1). Thus, with increasing polarity the esa$_{CT}$ band shows not only a hypsochromic shift, but also a relative increase of intensity, which renders it visible for **I** and **II** only in highly polar ACN. Note, that the polarity dependent intensities and energies of the three ESA spectra have the advantage of being self-calibrating. Hence, by combination of fluorescence, observed here as negative signalling gain, and ESA **I-III** are capable to probe polarity. It is noteworthy to stress that the ESA method is based on a photon-driven process, since a polarity dependent photoreaction takes place in **III** and a polarity dependent excited state stabilization for **I-III** would additionally allow to probe dynamic (time-dependent) polarity. Both features are not involved in the polarity sensing by static $S_0 \rightarrow S_n$ absorption.[12]

10.3.2 Transition Temperature from Liquid to Solid (Tg)

At the temperature of the transition (Tg) where the liquid transforms either into a rigidly transparent or a partially crystalline and opaque glass, a sudden and large increase of the microviscosity occurs.[19,35] This section deals with the influence of Tg on the fluorescence lifetimes of **I-III** in order to derive the ability of the fluorescence of **I-III** to probe sudden microviscosity changes.

In polar solvents, we could expect a stronger impact of Tg on the fluorescence lifetimes of the highly twisted compound **III** as compared to those of **I** and **II** because of the viscosity dependent conformational photoreaction from **CT** to **CTR** occurring only for **III**. In nonpolar solvents, on the other hand, the **CTR** species in **III** is not active, which lets us anticipate similar behaviour of **I-III** at Tg.

Nonpolar Solvents

The emission decay behaviour of **I-III**, which is illustrated in Fig. 10.2 by a plot of the longest exponential fluorescence lifetime, is investigated in two similar alkane solvent mixtures consisting of methylcyclohexane (MCH) and isopentane (Ip) in a volume ratio of 3:1 (MIp) and 1:4 (IpM). Both low-temperature mixtures are expected to form transparent glasses and to be fluid down to their glass transition temperatures (Tg(MIp)=98 K and Tg(IpM)=77 K)[36] lower than the freezing temperatures (T_f) of the composites MCH (T_f=146.5 K) and Ip (T_f=113.2).[37]

As expected, the fluorescence decays of **I-III** are monoexponential and remain practically constant ($\tau_f(\mathbf{I})$ = (1.45±0.1) ns; $\tau_f(\mathbf{II})$ = (1.3±0.1) ns and $\tau_f(\mathbf{III}$ in IpM) = 1.4 ns at 298

K up to 2.3 ns at 121 K) in the temperature range of the liquid phase. The slight temperature dependence of τ_f for **III** (E_a=2.2±0.1 kJ/mol) can be attributed to a thermally activated ISC process to the close lying 1^3B state (see ch. 5). With lowering the temperature, the unchanging and monoexponential radiative decay behaviour suddenly becomes biexponential with the addition of a long lifetime component (τ_2(**I**)=20-25 ns, τ_2(**II**)=10-12 ns; τ_2(**III**)=4ns). This reproducible change occurs for all D-A biphenyls exactly at the same temperature, namely at 110 K in IpM and at 98 K in Mlp. The observed effect must therefore be due to a change of the matrix structure which must be connected with the transition from the liquid to the solid form of the solvent. The comparison with the literature values of Tg shows a perfect agreement for the observed transition temperature in Mlp but not with that in IpM. The experimental transition temperature of solidification in IpM is only 3 K below the cited freezing point T_f of Ip[37] which directly reveals the crystallization of the solution before the tabulated Tg is reached at 36 K less than T_f.

It is also reasonable to assume that a partial crystallinic order of the alkane matrix is formed below the observed temperature Tg in Mlp. Such behaviour is well known for frozen n-alkanes (Sh'polskii effect).[38] Further systematic investigations with different nonpolar solvents could clarify that point. Without doubt, the sudden increase of the solvent density or the matrix order induces a molecular and electronic change in **I-III**.

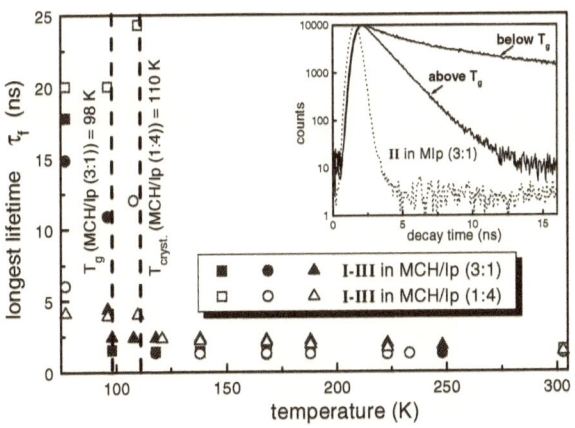

Fig. 20.2 *Temperature dependence of the longest lifetime component obtained for* **I-III** *in two different methylcyclohexane/isopentane mixtures. The inset displays example decays of* **II** *in Mlp(3:1) for a temperature slightly above and below the glass transition temperature.*

Ch. 10 Application as Fluorescent Probes

Which mechanism of the probe molecules **I-III** is responsible for the additional long lifetime and what is the role of the photoinduced intramolecular charge transfer process? For unsubstituted biphenyl or fluorene such a behaviour is not known[40] indicating the importance of the donor and acceptor substituents. The formation of self-complexes (SC) is consistent with a forbidden emission character and the insolubility of polar solutes in a nonpolar matrix which promotes self-aggregation. To elucidate whether the SC are formed in the ground or excited state and to investigate the influence of the molecular structure (φ), the fluorescence decays of **I-III** in Mlp at 77 K are analyzed (Fig. 10.3, Tab. 10.1). The observation that no risetimes are present in the long wavelength emission region but a dependence on the excitation wavelength (see different decay of **I** excited at 320 nm and 350 nm in Fig. 10.2) is found, indicate a prearrangement for the formation of SC already in S_0. The relative emission yield (ϕ_2 in Tab. 10.1) of the SC increases with the observation wavelength λ_{em} which has likewise been observed for SC emission from dimethylaminobenzonitriles supporting the hypothesis of SC formation.[40]

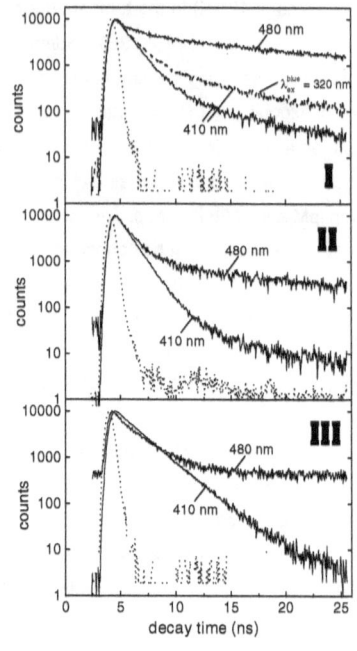

Fig. 10.3 Emission times traces of **I-III** in Mlp observed at 410 nm and 480 nm by excitation into the absorption maxima. The response functions are plotted in dotted line and for **I**, a decay curve obtained by excitation at 320 nm is also shown.

Tab. 10.1 Short (τ_1) and Long (τ_2) Fluorescence Lifetimes, Relative Yields of Counts for τ_2 ($\phi_2=\alpha_2\tau_2/(\alpha_1\tau_1+\alpha_2\tau_2)$), Counts of Background (= Phosphorescence) Relative to the Decay Maximum (BG), Fit Quality Parameters (χ^2) and Steady-State Emission Anisotropies (r) for **I-III** in 3:1 Methylcyclohexane/Isopentane (Mlp) at 77 K Excited at λ_{ex} and Observed at λ_{em}, respectively.

	λ_{ex} (nm)	λ_{em} (nm)	τ_1 (ns)	τ_2 (ns)	ϕ_2	BG	χ^2	r
I	320	410	1.35	8.58	37.6 %		1.29	
I	350	410	1.33	8.14	12.2 %	0.0 %	1.18	0.10[a]
I	350	480	1.21	17.79	90.0 %	0.35 %	1.29	0.06[a]
II	335	410	1.18	7.79	2.5 %	0.0 %	1.18	0.13
II	335	480	1.18	14.64	46.5 %	0.35 %	1.24	0.03
III	315	410	2.06	4.12	3.1 %	0.0 %	1.08	0.46
III	315	480	2.03	4.17	10.0 %	4.5 %	1.06	0.30

[a] slightly opaque glass

The decrease of ϕ_2 in Tab. 10.1 from **I** to **III** (ϕ_2 at 480 nm: 90%, 46.5 %, 0-3 %) as well as the excellent correlation with the melting point of the compounds (m.p.:226°C, 217°C, 132°C) proves that the efficiency of SC formation is related to the intermolecular interaction forces and is therefore determined by the flatness of the molecular shape of the probe molecules. A further consequence of the improved molecule-to-molecule spatial arrangement from **III** to **I** should be a stronger intermolecular π-orbital interaction which can explain the increase of the long lifetime τ_2 (see Tab 10.1) in the same order by a pertubation of the allowed ^1L_a-type $S_1 \rightarrow S_0$ transition. In addition, the aggregation in Mlp leads to a loss of emission anisotropy as compared to that in ethanol (ch. 5) which can be due to electronic interaction but also due to invisible microcrystals acting as „depolarizing" centers. The overall yield of triplet formation and radiation is very roughly quantified in Tab. 10.1 (BG) by the background intensity observed prior to the excitation pulse relative to the maximum decay intensity (Fig. 10.3). However, it seems to be rather unaffected by the SC, since a similar correlation of the phosphorescence intensity between **III** and **I** or **II** is found as in ethanol (ch. 5).

The observed effects are attributed to nonzero intramolecular ET interaction in S_0 which are instantaneously increased by absorption of a photon. Hence, in a simplified model, the intermolecular attraction in the ground state prearranges the relative orientation of the molecules to each other, which after photoinduced intramolecular charge separation leads to considerably stronger Coulombic intermolecular interactions. This process is not oberved either in a nonpolar fluid medium because of solvent and solute diffusion nor in a polar glass because the solutes can still be dielectrically saturated by the surrounding solvent dipoles.

In summary, the efficient SC formation in planar or weakly twisted donor-acceptor biphenyls may be exploited to probe nonpolar environmental microstructures, which has here been demonstrated for the transition from the viscous liquid to the glassy state of a nonpolar solvent mixture. A promising field of application is fluorescence sensing of liquid crystals where SC formation of cyanobiphenyl side groups has been found to be dependent on the order of the polymers investigated.[41]

Polar Solvents

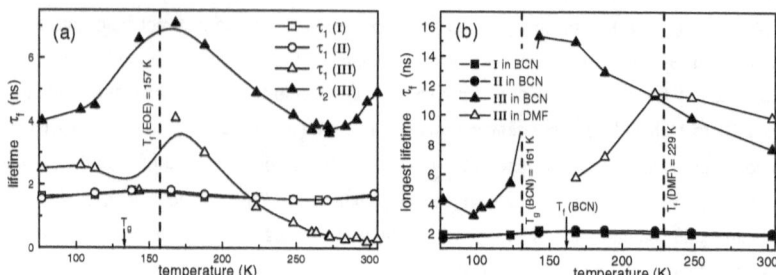

Fig. 10.4 Temperature dependence of fluorescence lifetimes of **I-III** in (a) the medium dipolar solvent diethylether (EOE) and (b) in highly dipolar butyronitril (BCN) and dimethylformamide (DMF) to determine the transition from the liquid to the rigid glass. The glass or freezing transition temperatures (T_f or T_g) taken from ref. 36 or 37, respectively, which agree with the experimentally determined transition temperatures, are indicated by dashed lines.

Fig. 10.4 shows that the fluorescence lifetimes obtained from exponential fitting[42] of the emission decays do not change for **I** and **II** within 0.1 ns in EOE and 0.3 ns in BCN. By observing the transparancy of the cuvette upon cooling it is found that the liquid transforms into a solid form which is a rigidly transparent glass (Tg) in BCN but a crystalline and opaque glass in EOE and DMF. According to the invariant decay behaviour of **I** and **II**, the photophysical and kinetic behaviour in the liquid and solid is essentially the same and can sufficiently be described by a singly emitting **CT** species in S_1 populated by the Franck-Condon state with a time constant beyond the scope of the experimental time resolution (100-200 ps). On the contrary, the two lifetimes observed for **III** reach a maximum with cooling (see also ch. 6) at a temperature which is close to the cited value of the freezing temperature T_f for EOE and DMF. As soon as the glassy or solid state is formed, the values of both τ_1 (shown only for EOE)[43] and τ_2 suddenly drop to considerably lower values, especially in the highly polar solvents BCN (from 15.3 ns at 143 K to 5.4 ns at 123 K) and DMF (from 11.5 ns at 223 K to 7.2 ns at 188 K), where the less emissive **CTR** species of **III** dominates the long lifetime component. This indicates a drastic and matrix induced change of the intramolecular relaxations. Note, that an additional lifetime below the determined transition temperature of solidification is not observed like in nonpolar solvents which excludes the formation of a new emissive species.

The smooth increase of the lifetimes with lowering the temperature above Tg has been demonstrated in ch. 6 to arise from the kinetically controlled photoreaction from **CT** to **CTR** in EOE, which can likewise explain the behaviour of **III** in the highly polar solvents. In the undercooled solution or solid (below T_f), however, the temperature dependence of the fluorecence decay behaviour is more complicated connected with a less continuous change of

Ch. 10 Application as Fluorescent Probes

the decay associated spectra (DAS(ν_f) see eq. 6-8) than above T_f. Since the lifetimes of **I** and **II** are only weakly affected by the glass transition, the FC→CT reaction seems to be rather unhindered by the strong increase of viscosity at Tg or T_f. Consequently, the freezing of the equilibrium between **CT** and **CTR** as well as the barrierless relaxations within the potential wells of the **CT** and **CTR** species are responsible for the reduction of the fluorescence lifetimes observed for **III** below Tg/T_f. However, a unique kinetic model for the whole temperature region below T_f can not be derived.[44]

In conclusion, the photoinduced intramolecular relaxations of **III** are such strongly disturbed and restricted by the environmental change at the glass transition that the resulting shortening of the emission lifetimes can sensitively probe the experimental temperature of solidification which may significantly differ from the literature values for Tg, because of partial crystallization which reduces Tg and broadens the temperature range for the glass transition.[35] On the other hand, the relatively large lifetime for **III** in BCN at 143 K reveals that the solution is still liquid ("undercooled liquid") below the value of T_f (161 K).[37] The agreement with the literature value of Tg for BCN (133K)[36] indicates the formation of a transparent glass as verified by inspection "by eye".

10.3.3 Protic Solvents

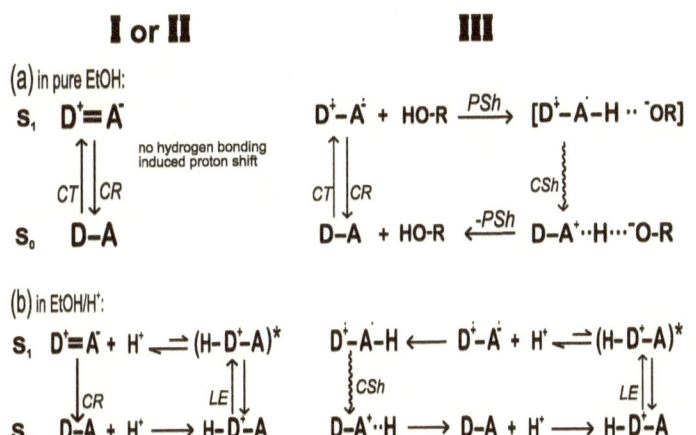

Scheme 10.2 *Simplest model for the photoinduced protonation cycles of* **I-III** *(a) in pure EtOH and (b) in EtOH/H⁺. Abbreviations: CT Charge Transfer excitation; CR Charge Recombination emission; CSh Nonradiative Intramolecular Charge Shift; PSh Intermolecular Proton Shift; LE Local Excitation and emission of cyanobiphenyl chromophore (see text)*

The fluorescence quenching mechanism in EtOH

Comparing the photophysical behaviour of **III** in ethanol (EtOH) at 298 K with that of **III** in acetonitrile (ACN) and that of **I**, **II** and other donor-acceptor biphenyls known from literature, the high nonradiative rate constant only in the protic solvent could be attributed to a dynamic quenching process arising from a proton shift (PSh) from EtOH to the nitrile group of **III** in the ^{1}CT state. Omitting nonradiative IC deactivation to the ground state S_0, the mechanism involved is depicted in scheme 10.2a. For **I** and **II**, the charge transfer excitation (CT) is followed by charge recombination (CR) fluorescence,[45] while the proton shift and the subsequent nonradiative charge shift (CSh) is found only for **III**. This outstanding mechanism takes place only for **III**, because the primary populated **CT** species rapidly transforms to the **CTR** species characterized by a reactive biradicaloid electronic structure and a strongly localized negative charge density on the benzonitrile unit (A) in the twisted S_1 geometry.

Effect of temperature

The temperature dependent photophysical properties of **I-III**, as shown in Fig. 10.5 for the fluorescence quantum yields $\Phi_f(T)$, the longest observed lifetimes $\tau_f(T)$ and the non radiative rate constants $k_{nr}(T)=(1-\Phi_f)/\tau_f$ of **III**, support the proposed scheme 10.2.[46] In accord with the simple photophysical model for **I** and **II** in scheme 10.2a, the lifetimes of **I** and **II** remain constant around (2.1 ± 0.1) ns in the temperature range considered in Fig. 10.5 for **III**.[47] The almost parallel plots of Φ_f and τ_f for **III** in EtOH indicate only a moderately temperature dependent fluorescence rate constant from 3×10^7 s^{-1} at 298 K to 11 $\times 10^7$ s^{-1} at 130 K which is probably due to the excited state equilibrium between the **CT** and the dominating **CTR** species in polar solvents. This equilibrium is slightly shifted towards the precursor **CT** species with lower temperatures/higher viscosities (see Ch. 6). The increase of Φ_f and τ_f results from a freezing of the additional photochemical nonradiative channel (PSh in scheme 10.2a)

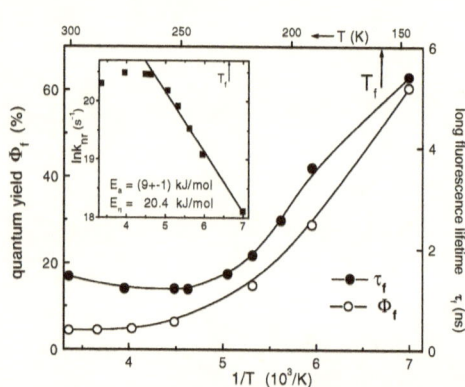

Fig. 10.5 *Temperature dependence of the fluorescence lifetimes (τ_f) and quantum yields (Φ_f) of **III** in ethanol. The corresponding Arrhenius plot of the nonradiative rates ($k_{nr}=(1-\Phi_f)/\tau_f$) is shown in the inset. The derived Arrhenius barrier (E_a) and the activation energy for the viscous flow of the solvent are also given.*

connected with a remarkable activation barrier of $E_a = 9$ kJ/mol as shown in the inset of Fig. 10.5. The Arrhenius barrier E_a is two times smaller than that for the viscous flow of EtOH ($E\eta = 20.4$ kJ) which is conceivable with a photoreaction inside the prearranged solvent cage. The temperature dependence of the photoreaction derives partly from a less efficient formation of **CTR** from **CT** at lower temperatures and, on the other hand, from an energy barrier for the proton shift from EtOH to **CTR**. The influence of the CT/CTR equilibrium is possibly also reflected by the smaller nonradiative rates at high temperatures than expected from the Arrhenius plot. In this high temperature region, the back reaction from **CTR** to **CT** may be accelerated, such that less fluorescence can be quenched. The distinction between a low (reversible CT→CTR reaction) and high temperature region (irrreversible CT→CTR reaction) has been worked out in ch. 6. If the proposed photoinduced proton shift is active, an enhanced fluorescence quenching can be anticipated for **III** in acidified solution. This effect is investigated in the following to further manifest the proton affinity of **III** in the excited state.

Effect of added protons

In Fig. 10.6, the drastic change of the fluorescence and excitation spectra of **I-III** in EtOH after acidification of the solutions (0.3 M HCl in EtOH for solute concentration of about 10^{-5} M yielding 10000:1 excess of HCl) obviously reveals the strong influence of free protons on the photophysics of all three donor-acceptor biphenyls. Dual fluorescence is observed for **I** and **II** in EtOH/H$^+$ provoked by the excited state prototropic coexistence of the CT (base) and a protonated (monocationic acid) form. On the other hand, the proton induced dissapearance of the CT band in the excitation as well as in the absorption spectra indicates that almost only the protonated form exists in S_0 (strictly speaking a residual concentration of the basic forms yields a reduction of the excitation intensities in the CT absorption band not to exactly 0% of the original intensity in the absence of protons but to 0.75% for **I**, 1.5% for **II** and 0% for **III**). Moreover, the blue-shifted first absorption band demonstrates that the proton is accomodated at the donor side (D), in particular occupying the free electron pair of the amino nitrogen. The molecular and electronic structures involved are discussed in the next section.

Since the protonation of **I-III** takes places in the ground state, the lower fluorescence yield of the **CT** species in acidified EtOH is due to static quenching (Fig. 10.6). Further, one can see that in the case of proton addition, the intensity of the **CT** fluorescence band is higher as it would be expected, if solely the free base absorbs at the excitation wavelength used (cf. Fig. 10.6: intensity of the CT excitation spectrum at maximum of LE/H$^+$ excitation spectrum is less than the CT fluorescence intensity obtained from excitation at maximum of LE/H$^+$). Consequently, the **CT** fluorescence band arises after deprotonation in the excited state. This can be substantiated by the pK_a^* values for the hyphothetical case of an established equilibrium which are given by

$$pK_a^* - pK_a = \frac{\Delta G^* - \Delta G}{RT\ln 10} \quad (10\text{-}1)$$

and can be derived from the Förster cycle[48] using the absorption and fluorescence energies to evaluate the acid-base (= LE/H$^+$-CT) free energy differences (assuming a cancelling of entropy changes) ΔG and ΔG^* in S_0 and S_1 according to

$$pK_a^* - pK_a = \frac{hc}{kT\ln 10}\left(\frac{\nu_a(CT) - \nu_f(CT)}{2} - \frac{\nu_a(LE/H^+) - \nu_f(LE/H^+)}{2}\right) \quad (10\text{-}2)$$

Taking into account the small pK_a values in S_0 obtained in the next section, pK_a^* values of the three D-A biphenyls are extremely negative pointing to a driving force ($\Delta G^* < 0$) of deprotonation in S_1. From these two experimental facts, it can analogously be concluded that **III** is likewise deprotonated in S_1 leading to the same ^{1}CT state that is active in pure EtOH. Similarly to the situation in pure EtOH, strong fluorescence quenching occurs only for **III**. This implies that after deprotonation the fluorescence quenching mechanism in EtOH described

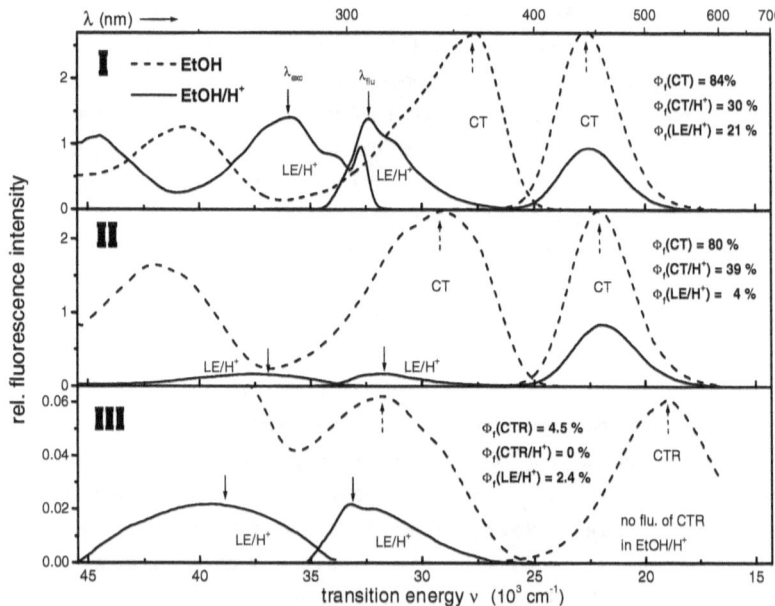

Fig. 10.6 *Fluorescence and excitation spectra of I-III in EtOH prior to (- - -) and after (——) acidification at constant concentration of the solute probes. The optical densities of I-III in EtOH at the CT absorption maximum (dashed arrow) were equal, such that all fluorescence intensities are relatively related to each other. The arrows indicate the excitation and emission energies chosen.*

above also takes place in EtOH/H⁺. The protonation cycles in the proton-saturated alcohol solution (0<pH<pKa, see 10.3.4) are schematically summarized in scheme 10.2b.

Tab. 10.2 pK_a^*-pK_a Values for the Prototropic Equilibria of the LE (acid) and CT (base) Species in **I-III** Determined by the Förster Cycle (eq. 10-2) Using the Absorption (v_a) and Fluorescence Energies (v_f) of LE/H⁺ and CT.

		v_a (10³ cm⁻¹)			v_f (10³ cm⁻¹)			pK_a^*-pK_a		
		I	II	III	I	II	III	I	II	III
EtOH	LE/H⁺	32.8ᵃ	37.8	39.4ᵇ	32.4ᵃ	31.6	33.2	-16	-19	-23
	CT	27.8	28.9	32.0	22.5	21.9	18.7ᶜ			
EtOH/H₂O	LE/H⁺	32.8ᵃ	37.7	n.d.	32.4ᵃ	31.5	n.d.	-17	-20	n.d.
(1:1)	CT	27.7	28.8	n.d.	21.6	20.9	n.d.			

ᵃ 0-0 energy ᵇ from fluorescence excitation ᶜ emission of CTR species

Since the dynamic fluorescence quenching of **III** in EtOH/H⁺ (Φ_f < 0.2%) is much stronger than in EtOH (Φ_f = 4.5%), assuming a similar rate of the ^{1}LE→^{1}CT process as for **I** and **II** (reduction of $\Phi_f(^1CT)$ by a factor of ½), it can be deduced that the efficiency of the proton-induced dynamic quenching process of **III** is larger in the acidified EtOH solution. This reflects the higher proton activity in EtOH/H⁺ and therefore proves the sensibility of **III** on proton and hydrogen bonding interaction. Furthermore, the mechanism of fluorescence quenching instead of a mechanism induced by bulk properties of EtOH was verified by addition of small traces of EtOH to acetonitrile, which indeed showed the expected reduction of fluorescence intensity in spite of unchanged optical density.

Accordingly, this behaviour promotes **III** as a sensitive fluorescent probe to detect the presence of protic solvents. One can imagine to use the fluorescence intensity of **III** as a monitor of small traces, e.g., of alcohols or water formed in a chemical condensation process. In practice, a calibration by Stern-Vollmer experiments measuring the fluorescence intensity of **III** as a function of the concentration of the protic agent in the solution to be probed has to be performed.

10.3.4 Sensitive and Self-Calibrating Sensing of pH in Aqueous Solution

This section aims to work out the possibilty to gauge pH utilizing the fluorescence of the donor-acceptor biphenyls. Due to the strong fluorescence quenching of **III** by protons, we are restricted to the planar (in S_1) compounds **I** and **II**. To enable the measurement of pH the pure ethanol solvent used above is exchanged by an aqueous solution of 50:50 EtOH/H₂O from which relatively high mean activity coefficients of HCl ($\gamma_\pm \geq 0.68$ for pH > 1.4)[33] are known and which is a suitable solvent mixture for practical purposes. A straightforward sensing in the very acid pH range of pK_a^* is, of course, impossible because of too low proton activity coeffients at pH<<0 and possible polyprotonation. On the other hand, probing in the pH range around the ground state pK_a, which is usually between 2 and 5 for aromatic amines[10,33,49] seemed to be promising.

Therefore, the pka of **I** and **II** is determined in the following by the absorption method combined with the equivalent fluorescence experiment to investigate the applicability of the more sensitive emission measurements. Before suggesting a practicable fluorescence sensing of pH, the molecular and electronic structures in acidic solutions are analyzed which leads to a verification and extension of the protonation cycle proposed in scheme 10.2b.

Determination of pK_a by absorption and fluorescence titration

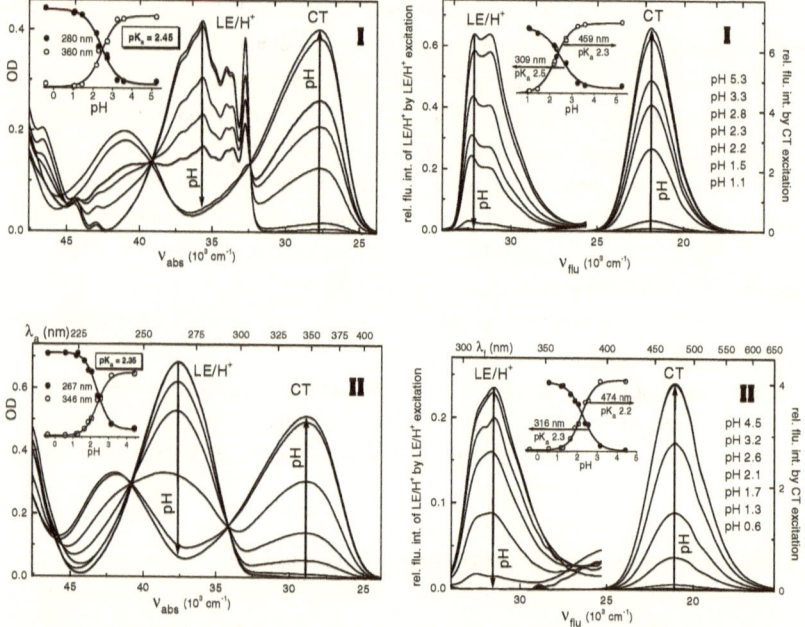

Fig. 10.7 pH dependent absorption spectra and titration curves of **I** and **II** in 1:1 EtOH/H$_2$O (v/v).

Fig. 10.8 pH dependent fluorescence spectra and titration curves of **I** and **II** in 1:1 EtOH/H$_2$O (v/v). The pH values given are also valid for Fig. 10.7.

Fig. 10.7 and 10.8 present the absorption and the relevant fluorescence spectra of **I** and **II** as a function of pH together with the corresponding titration curves shown as insets. Obviously, with increasing pH the CT absorption and fluorescence intensity is favoured, whereas at high proton concentrations the LE bands are enhanced. The isosbestic aborption points at 308 nm and 255 nm for **I** and at 292 nm and 245 nm for **II** reflect the acid-base equilibrium between the protonated form H-$^+$D-A responsible for the LE/H$^+$ bands and the neutral form D-A responsible for the CT bands. From the inflection points of the LE/H$^+$ and CT absorption titration curves, the pK_a value in S_0 is determined to be 2.65 for **I** and 2.55 for **II** independent of the band monitored. This value is somewhat lower than that for various aromatic amines (cf. pK_a of aniline is 4.5)[10] because of the electron withdrawing benzonitrile unit A. The obtained values are in agreement with the literature value for ADMA (pK_a=2.5)[49] where the acceptor unit of **II** is replaced by the anthracene moiety. Interestingly, the LE/H$^+$ and CT fluorescence titration curves obtained by excitation of the corresponding LE and CT absorption maxima, yield practically the same results within the pK_a error limit of ±0.2 (discussion see below) and also reproduce the tendency that the pK_a of **II** is slightly smaller than that of **I**. The latter observation may indicate a slightly better π-delocalization of the dimethylamino nitrogen electron lone pair in **II**.

As an empirical conclusion, the proton titration and consequently the sensing of pH by absorption can be substituted by the respective fluorescence method. But let us now turn again to mechanistic matters in order to try a refinement of scheme 10.2b for the prototropic behaviour of **I** and **II**.

Molecular and electronic structures in acidic solution

To come straight to the results the proposed prototropic mechanisms are illustrated in scheme 10.3.

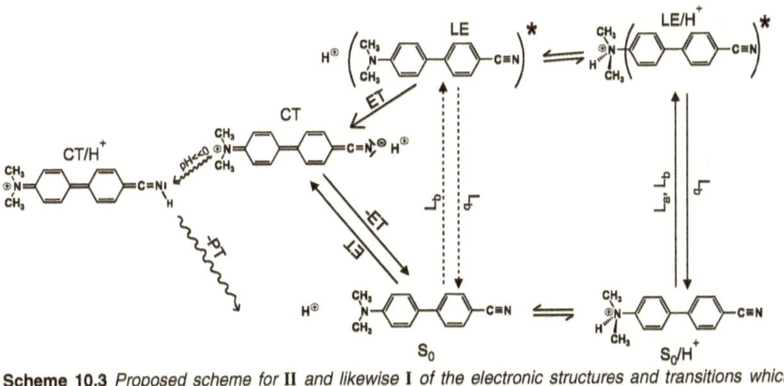

Scheme 10.3 *Proposed scheme for **II** and likewise **I** of the electronic structures and transitions which accounts for the observed photoinduced prototropic behaviour. Abbreviations: L_b, L_a electronic absorption and emission transitions according to Platt nomenclature;[50] ET, -ET electronic transition with strong electron transfer character; PT, -PT proton transfer between probe and solvent*

Absorption and emission

The low energy absorption and fluorescence emanates from the S_0-^1CT electronic transition which is reflected by the sizeable red shift of emission by increasing the solvent polarity from EtOH to EtOH/H_2O (Tab. 10.2). Due to the fact that a blue shift is not observed in the absorption ensures that 1:1 hydrogen bonded complexes with H_2O at the dimethylamino nitrogen are of minor importance or even completely absent. On the other hand, the absorption and fluorescence behaviour of the cationic LE species is more difficile, but contains a wealth of important information. The comparison of the LE/H^+ absorption maximum of **II** (v_a=37800 cm^{-1}) with the $S_0 \rightarrow {}^1L_a$ maxima of unsubstituted biphenyl (v_a=39700 cm^{-1}),[51] p-dimethylaminobiphenyl (v_a=33900 cm^{-1}),[52] p-cyanobiphenyl (v_a=37500 cm^{-1})[53] identifies the latter molecular entity in **II** as the part where the main photon energy is deposited. This is not surprising since the electron lone pair of the amino nitrogen is occupied by the proton and thus is no more available for the π-electron system. The shape of the structured absorption band for the donor-acceptor fluorene **I** differs from the parent fluorene[51] in such a way that (i) the global maximum is not the sharp 0-0 band at 32800 cm^{-1} (33100 cm^{-1} in pure fluorene) but the vibronic band at 35700 cm^{-1} and (ii) the ^1L_a band of fluorene at 40000 cm^{-1} is not visible in **I**. In analogy to **II**, it can be deduced that the parent ^1L_a band of fluorene is bathochromically shifted in **I** to almost the same spectral position as the ^1L_a band in **II** leading to the overlap with the structured ^1L_b band in **I**. Similar to the behaviour of the parent pair of diphenyls,[51] only the ^1L_b absorption band of the fluorene D-A derivative is strongly intense (details see ch. 2, note also that the perfect agreement of the 0-0 band with the minimum of the second derivative $\epsilon''(v_a)$ spectrum in acetonitrile proves that the previous assignment of the hidden structure to the ^1L_b transition was correct.).[54] This implies that LE/H^+ emission of **I** and likewise **II** occurs from the ^1L_b type state. Indeed, assuming similar rates for the LE/H^+ to CT transformation in **III** (see below), the distinctly smaller fluorescence quantum yields of **II** ($\Phi_f(LE/H^+)$=4%) and **III** ($\Phi_f(LE/H^+)$=2.4%) as compared to **I** ($\Phi_f(LE/H^+)$=23%) are consistent with emission from ^1L_b, which is strong only for **I** because of the symmerty reasons discussed in ch. 2(Fig. 10.6.).[54] Note, that for the sake of simplicity and clearness, scheme 10.2 does not account for the vertical internal conversion (no molecular change) from ^1L_a to the lower lying ^1L_b. Therefore, the general term LE is used in the text instead of the more precise notation of a local ^1L_a or ^1L_b absorption and a local ^1L_b emission species.

Slow prototropic equilibrium in S_1

The occurrence of dual fluorescence of the acid and base species over a large pH range (Fig. 10.6,10.8, 10.9) as well as the similarity of the absorption and fluorescence titration curves with the same inflection point at pK_a point to a slow prototropic equilibrium in S_1 where the rate of fluorescence competes with the proton transfer processes. However, how to explain the apparent discrepancy to the strongly negative pK_a^* values connected with a strong thermodynamic driving force for the acid-to-base or LE/H$^+$ to CT deprotonation? While the lack of LE/H$^+$ emission after pure CT excitation is conceivable with a too endergonic base to acid (CT to LE/H$^+$) reaction, the slow acid to base (LE/H$^+$ to CT) reaction is not expected on the basis of the considerably negative free enthalpy change obtained by the Förster cycle (Tab. 10.2). The only explanation for this phenomenon seems to be that the prototropic equilibrium does not occur between the LE/H$^+$ and CT species directly but suggests the involvement of an intermediate deprotonated species denoted in scheme 10.3 by LE. The release of the proton then extends the π-electron delocalization from the p-cyanobiphenyl chromophore to the original π-system of the whole molecule the electronic structure of which has been characterized by a mixing of $^1L_b(A)$ and $^1L_b(D)$ type configurations in ch.2.[54] In spite of the better π-electron delocalization, the remaining ^1L_b electronic nature allows only a weak energetic stabilization accompanied with the LE/H$^+$→LE process (note again the agreement between the 0-0 energy for the S_0→L_b absorption of the protonated form (Fig. 10.7) and the free base (h$_2$ in Fig. 2.3)), such that the observed slow prototropic equilibrium becomes well understandable. Similarly slow LE/H$^+$→LE deprotonation reactions have been reported for the charge transfer processing biaryls 1-(p-aminophenyl)pyrene and ADMA.[33,49]

In comparison to the deprotonation reaction, the subsequent electron transfer to the CT species is quite rapid due to strong exergonicity. The latter process is reminiscent to the excited state electron transfer ^{1}FC→^{1}CT interconversion observed by time-resolved absorption in ch. 4.[4] According to scheme 10.3 and scheme 4.3, the LE species populated by the protonated excited state species is equivalent to the FC species populated after vibrational FC relaxation. Both experiments, pH dependent fluorescence and time-resolved transient absorption in ch.4, consequently agree with the view that the primary populated species has to be regarded as an independent and stable electronic state. It would therefore be an interesting project for future investigations to study the electron transfer step in S_1 which should occurr with a delay in acidified solutions, e.g. by means of picosecond-resolved transient absorption.

Dynamic quenching of CT fluorescence in I and II

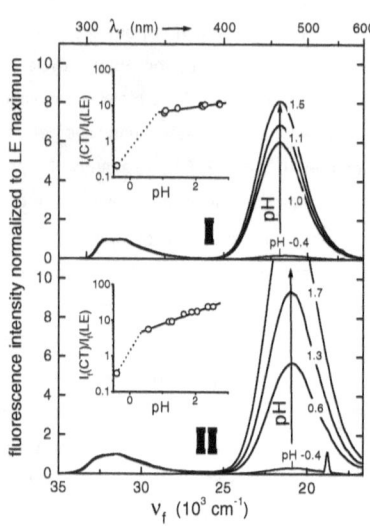

Fig. 10.9 *Normalized Fluorescence spectra of I and II in EtOH/H$_2$O/H$^+$ at pH<pK$_a$ obtained by excitation of the LE/H$^+$ absorption maximum. The plot of the intensity ratios in the inset deviates from linearity at pH<1.*

Fig. 10.9 shows that the logarithmic intensity ratio of CT to LE/H$^+$ fluorescence linearly decreases from high pH values down to pH≈1 but that the last measured point for very acid conditions at pH≈-0.4 deviates from the linear plot being much less than expected from an extrapolation of the data at pH>0.8. This indicates a quenching process of the unprotonated CT species which most probably reflects the same quench mechanismsm that is active for compound III in acidified EtOH (Fig 10.6) and may therefore be attributed to a proton deposition at the partially negatively charged nitrile group leading to the protonated counterpart of the CT species (scheme 10.3). The quenching process of CT as testified by a CT to LE/H$^+$ fluorescence intensity ratio I$_f$(CT)/I$_f$(LE/H$^+$) of less than 0.3 is not that drastically observed for I (0.7) and II (5.1) in the acidified EtOH without addition of 50% H$_2$O shown in Fig. 10.6. Though the exact pH is not known in EtOH/H$^+$, the 14-times smaller HCl concentration and the expectable smaller proton activity coefficients are consistent with less effective proton-induced quenching pointing to a pH in EtOH/H$^+$ (Fig. 10.6) larger than zero (cf. calibration line in Fig. 10.10).

Fluorescence sensing of pH using I and II

Pursuing a self-calibrating, easy-to-measure and highly sensitive fluorescence method to probe pH using I and II leads to the proposal to measure the ratio of fluorescence intensity I$_{ex}$I$_{flu}$(CT)/ I$_{ex}$I$_{flu}$(LE/H$^+$) observed at the LE/H$^+$ emission maximum by excitation of the LE/H$^+$ absorption maximum versus the intensity observed at the CT emission maximum by excitation of the CT absorption maximum (cf. Fig. 10.7 and 10.8). This ratioing method is alternative to the usual measure of fluorescence enhancement (FE) or of emission ratio measurements with excitation at a single wavenumber, i.e., the isosbestic point.[6,9,30] The superiority of the present method for the case of well separated absorption and fluorescence bands which due to unfavourable fluorescent probe behaviour could yet not be applied succesfully,[55] shall be demonstrated in the following.

The first requirement of self-calibration is achieved since, in contrast to the single band FE measurement, the ratioing approach does not depend on the detector sensitivity or probe concentration (photo- or thermal bleaching of the probe) and hence needs no reference pH calibration. Taking into account the excitation and emission correction, which is generally implemented in modern fluorescence spectrometers, this measurement yields the same ratio values independent of the fluorimeter used. Secondly, the method is easy to handle because only two intensity values have to be measured and a linear calibration line can be applied (Fig. 10.10). The third requirement of a high sensitivity and a large variation of the intensity ratio is usually hardest to meet. However, Fig. 10.10 demonstrates an extraordinary favourable change of the ratio over four

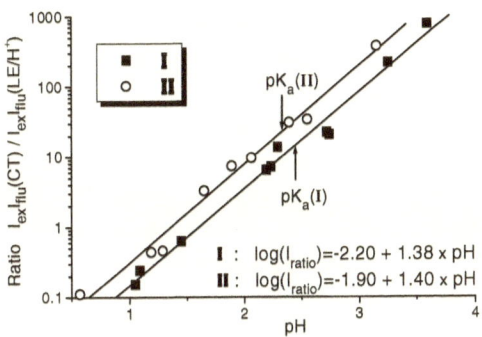

Fig. 10.10 pH calibration curves using the fluorescence intensity ratio between CT fluorescence maximum excited at CT absorption maximum and LE/H$^+$ fluorescence maximum excited at LE/H$^+$ absorption maximum. The correlation coefficients of both linear fits are R=0.993.

orders of magnitude from 1/10 to 1000 in a comparably wide dynamic pH range over three pH units from 1 to 4. Such large variations are due to the combination of excitation and fluorescence intensity of separated bands and are thus scarcely to achieve by the usual procedure considering the ratio of fluorescence intensity only. Note also, the linear relationship between the logarithmic ratio and pH yielding linear regression lines with correlation coefficients better than 0.99. Needless to say, that the line for **II** lies somewhat higher than for **I** because the intrinsic fluorescence yield from LE/H$^+$ in the case of absent prototropic reaction is lower for the biphenyl derivative **II** (short-axis polarized ^1L_b fluorescence with small k_f) than for the fluorene derivative **I** (long-axis polarized ^1L_b fluorescence with large k_f). The advantages of the specific molecular pH probes **I** and **II** are summarized below.

Summary of the advantages of pH sensing with D-A biphenyls and their reasons

The donor-acceptor biphenyls **I** and **II** are extraordinary suitable for fluorescence sensing of pH because of various advantageous reasons:

(i) The fluorescence intensity ratio $I_{ex}|_{flu}(CT)/I_{ex}|_{flu}(LE/H^+)$ obtained at $pH>pK_a$ is considerably larger than that at $pH<pK_a$ such that the prerequisite of high sensitivity is by far fulfilled (Fig. 10.10).

(ii) A linear relationship between the logarithm of the measured ratio and the pH is extremely valuable for practice use because it minimizes errors. (Fig. 10.10)

(iii) the frequently encountered complex problem of charge transfer fluorescence quenching by protons is not meet for **I** and **II**, at least in the pH range above unity (Fig. 10.9).

(iv) Large extinction coefficients for the basic as well as for the acid form ($\varepsilon_{max}(\mathbf{I}) \approx 4 \cdot 10^4$ M^{-1}cm^{-1}, $\varepsilon_{max}(\mathbf{II}) \approx 3 \cdot 10^4$ M^{-1}cm^{-1}) and sizeable fluorescence quantum yields and especially for **I**, ε_{max} as well as Φ_f of the pure basic (CT) and acid (LE/H$^+$) form are comparably large (Fig. 10.6-10.8).

(v) The complete separation of fluorescence AND (!) absorption bands of the basic and acid form allows to our knowledge the first time to apply succesfully the method to measure the intensity ratio between LE/H$^+$ excited LE/H$^+$ fluorescence and CT excited CT fluorescence. Such conditions are not fulfilled in various biaryls based on anthracence[49] and pyrene[33] exhibiting charge transfer emission.

(vi) The pH range over which the fluorescence is strongly pH dependent due to the excited state prototropic equilibrium is far removed from that determined by the ground state equilibrium ($pK_a^* << pK_a$) providing the linear intensity relationship with pH on a logarithmic scale.

All these advantages can be traced back to the specific electronic and conformational properties of **I** and **II**:

The zeroth-order electron transfer state of **I** and **II** in polar solvents like H$_2$O is considerably lower lying than the next locally excited state. Such a situation is difficult to meet in biaryls with larger polycyclic aromatic subunits (see ch. 11, low lying LE states!) like anthracence or pyrene (cf. ch.11) showing completely overlapping CT and LE absorption bands.[33,49] As a consequence, separated CT and LE/H$^+$ levels are achieved for **I** and **II** giving rise to the advantages ii, v, vi.

The planar conformation of **I** and **II** in the ^{1}CT state is responsible for a appreciable HOMO-LUMO overlap associated with unusally large S$_0$-CT electronic transition moments (ch.8). Accordingly, high absorption and fluorescence intensity and the delocalized electronic structure of CT lead to the advantages i, iii and iv.

10.4 Conclusion

Photoinduced intramolecular charge transfer in donor-acceptor (D-A) biphenlys could principally be shown to be useful in sensing different microenvironmental properties by fluorescence. However, the molecular design of the probe compounds, which is varied in this study through the twist angle φ between the donor and acceptor aryl moiety, determines which specific microenvironmental property can be probed. The D-A biaryls **I** and **II** being planar in the excited state (S_1) show complementary abilities of fluorescence probing with respect to the strongly twisted D-A biaryl **III**, such that two types of biaryl probes can be distinguished by conformational, electronical and relaxation properties in the CT excited state:

D-A biaryls of the first type (**I** and **II**) characterized in S_1 by a planar conformation, delocalized charge and a structural relaxation with weak influence on the fluorescence properties are favourably sensible (i) to polarity due to their large dipole moments and viscosity/temperature independent excited state electronic structure, (ii) to a change of the matrix order due to their stronger aggregation tendency and (iii) to pH due to various reasons such as well separated and intense LE and CT absorption and fluorescence bands as well as absent quenching of CT emission by protons.

Complementary to that, D-A biaryls of the second class (**III**) characterized in S_1 by a strongly twisted conformation, sizeably localized charges on the subunits and an environment dependent structural relaxation observable through fluorescence are uniquely sensible (i) to sudden changes of microviscosity due to the accompanied on-off switching of the adiabatic photoreaction towards a more stabilized and less emissive charge transfer species and (ii) to specific interactions and potential quenchers such as protic solvents and protons due to a possible attack of charged atomic centers.

Molecular engineering of D-A biaryls can therefore principally control specific fluorescent probing ability by tuning the twist angle. However, specific substituents effects and donor/acceptor energetics determining the energy gap between LE and CT states have to be considered as well.

10.5 References and Notes

(1) Supramolecular Photochemistry; Balzani, V., Ed.; Reidel: Dordrecht, **1987**.

(2) Scandola, F.; Bignozzi, C.A.; Chiorboli, C.; Indelli, M.T.; Rampi, M.A. *Coord. Chem. Rev.* **1990**, 97, 299.

(3) Maus, M., Rettig, W., Bonafoux, D.; Lapouyade, R. *submitted*.

(4) Maus, M.; Rettig, W.; Jonusauskas, G.; Lapouyade, R. and Rullière, C. *J. Phys. Chem.* **1998**, 102, 7393.

(5) (a) Maus, M.; Rettig, W.; Lapouyade, R. *J. Inf. Rec.* **1996**, 22, 451; (b) Rettig, W.; Maus, M. *Ber. Bunsenges. Phys. Chem.* **1996**, 100, 2091. (c) Maus, M.; Rettig, W.; *J. Inf. Rec.* **1998**, 24(5/6), 461.

Ch. 10 Application as Fluorescent Probes

(6) Rettig, W.; Lapouyade, R. in: *Topics in Fluorescence Spectroscopy*, Vol. 4: Probe Design and Chemical Sensing, ed. J.R. Lakowicz; Plenum Press, New York, **1994**.

(7) Rettig, W.; Baumann, W. in *Progress in Photochemistry and Photophysics*, Vol. VI, ed. J.F. Rabek, CRC Press, Boca Raton, **1992**.

(8) Valeur, B. in: *Molecular Luminescence Spectroscopy*, Part 3, ed. S.G. Schulman, John Wiley & Sons, **1993**.

(9) De Silva, A.P.; Gunaratne, H.Q. N., Gunnlaugsson, T.; Huxley, A.J.M.; McCoy, C.P.; Rademacher, J.T.; Rice, T.E. Chem. Rev. **1997**, 97, 1515.

(10) Wehry, E.L. in: *„Practical Fluorescence"*, pp.127, ed. G.G.Guilbault, Marcel Dekker Inc., New York, **1990**.

(11) Onsager, L. *J. Am. Chem. Soc.* **1936**, 58, 1486.

(12) Dimroth, K; Reichardt, C.; Siepmann, T.; Bohlmann, F. Liebigs Ann. Chem. **1963**, 661, 1.

(13) Catalán, J.; López, V.; Pérez, P.; Martín-Villamil, R.; Rodriguez, *J.G. Liebigs. Ann.* **1995**, 241.

(14) Kosower, E.M. *J. Am. Chem. Soc.* **1958**, 80, 3253.

(15) Drago, R.S.; Hirsch, M.S.; Ferris, D.C.; Chronister, C.W. *J. Chem. Soc. Perkin Trans.* **1994**, 2, 219.

(16) Kamlet, M.J.; Abboud, J.L.M.; Taft, R.W. *J. Am. Chem. Soc.* 1977, 99, 6027; Prog. Phys. Org. Chem. **1981**, 13, 485.

(17) Loutfy, R.O. *Pure Appl. Chem.* **1986**, 58, 1239.

(18) Yan, Z.-L.; Wu, S.-K. *J. Photochem. Photobiol. A: Chem.* **1996**, 96, 199.

(19) Fritz, R. Ph.D thesis *„Temperatur- und druckabhängige Untersuchungen zur Bestimmung des freien Volumens in niedermolekularen organischen Gläsern und Polymeren mittels adiabatischer Photoreaktionen fluoreszierender Sondenmoleküle"* TU Berlin; Köster Verlag, Berlin, **1994**, ISBN 3-929937-57-2.

(20) In principle, the photoreaction of III is also able to probe moderate changes of microviscosity as has been shown by the pressure dependent fluorescence in ch. 7. However, the influence of temperature due to an intrinsic thermal barrier and micropolarity due to the high dipole moments is considerably stronger and since one of these effects accompany a viscosity change, a straightforward determination of the viscosity is not possible.

(21) Martin, E.; Weigand, E.; Pardo, A. *J. Luminesc.* **1996**, 68, 157.

(22) López Arbeloa, T.; López Arbeloa, F.; Hernández Bartolomé, P.; López Arbeloa, I. *Chem. Phys.* **1992**, 160, 123.

(23) López Arbeloa, T.; López Arbeloa, F.; López Arbeloa, I. *J. Luminesc.* **1996**, 68, 149.

(24) Cazeau-Dubroca, C.; Peirigua, A.; Brahim, M.B.; Nouchi, G.; Cazeau, P. *Chem. Phys. Lett.* **1989**, 157, 393.

(25) Kadiri, A.; Kabouchi, B.; Benali, B.; Cazeau-Dubroca, C.; Nouchi, G. *Spectrochimica Acta* **1994**, 50A, 1.

(26) Herbich, J.; Waluk, J. *Chem. Phys.* **1994**, 188, 247.

(27) Kim, Y.H.; Cho, D.W.; Song, N.W.; Kim, D.; Yoon, M. *J. Photochem. Photobiol. A: Chem.* **1997**, 106, 161.

(28) Ikeda, N.; Miyasaka, H.; Okada, T.; Mataga, N. *J. Am. Chem. Soc.* **1983**, 105, 5206.

(29) The term PET is somewhat misleading since in the underlying mechanism the electron is not transferred directly by light, but follows the excitation into a locally excited state.

(30) De Silva, A.P.; Gunaratne, H.Q. N., Gunnlaugsson, T.; Lynch, P.L.M. *New. J. Chem.* **1996**, 20, 871.

(31) The term „self-calibrating" means that only once a calibration curve has to be measured which can be used for the analyses. The signal of fluorescent probes without self-calibration depends on several additional factors, such as geometry of the sample or setup sensitivity, which requires a calibration with standards for each analytic measurement.

(32) Galster, H. *pH-Messung*, VCH Verlagsgesellschaft, Weinheim **1990**.

(33) Hagopian, S.; Singer, A. *J. Am. Chem. Soc.* **1985**, 107, 1874.

(34) Kalyanasundaran, K.; Thomas, J.K. *J. Am. Chem. Soc.* **1977**, 99, 2039.

(35) Ivin, K.J. in "Encyclopedia of Polymer Science and Technology 2", pp. 745, John Wiley & Sons, Inc., New York, **1977**.

(36) Pestemer, M. *Anleitung zum Messen von Absorptionsspektren im ultravioletten und sichtbaren Spektralbereich*, Thieme Verlag, Stuttgart 1964.

(37) Riddick, J.A.; Bunger, W.B.; Sakano, T.K. *Organic Solvents*, John Wiley & Sons **1986**.

(38) Lamotte, M.; Merle, A.M.; Jousset-Dubien, J.; Dupuy, F. *Chem. Phys. Lett.* **1975**, 35, 410.

(39) Swiatkowski, G. *Ph.D thesis "Intramolekulare Torsionsrelaxationsprozesse sterisch gehinderter unpolarer Aromaten"* **1982**, Berlin.

(40) Rotkiewicz, K.; Leismann, H.; Rettig, W. *J. Photochem. Photobiol. A:Chem* **1989**, 49, 347.

(41) private communication with Dr. J. Stumpe; Poster 60, MAFS Conference, Berlin, Germany, 21.- 24.9.1997.

(42) The addition of a short lifetime (τ_s) less than 200 ps for all compounds below Tg is needed to achieve wavenumber dependent fit curves with a global $\chi^2<1.3$. This component with a decay time within the limit of the time-resolution of the instrument is due to solvation processes following the FC→CT interconversion (see also ref. 47).

(43) To avoid confusion a plot of τ_1 of III in BCN and DMF is not added to Fig.10.3. In BCN, τ_1 drops from 4.6 ns at 143 K to 3.0 ns at 123 K and in DMF, it changes from 2.3 ns at 223 K to 1.8 ns at 188 K. The accuracy of this short time component is limited at higher temperatures due to small amplitudes α_1 which results from a fast photoreaction from CT to CTR and partial direct population by the Franck Condon state at these highly polar conditions.

(44) Particularly, the analysis of the amplitude spectra (DAS) at 143 K in the temperature region of the undercooled solution revealed rather complicated behaviour because only in this range an additional lifetime is needed to describe the kinetics by global analysis adequately. The same behaviour has been found in ref 19 for 9,9'-bianthryl in ethanol.

(45) This is, of course, a simplified picture, since ch. 4 and 5 have revealed additional electronic pathways, to a $^1L_b/^1CT$ type ^1FC and the triplet manifold, which are, however, of no importance for the present study and occur similarly for I-III.

(46) Since in the polar solvent EtOH the adiabatic photoreaction CT→CTR is definitely in the irreversible regime, the long lifetime directly represents the intrinsic lifetime of the CTR species $\tau_0(CTR)$. However, $\Phi_f=[CT]\Phi_f^{CT}+[CTR]\Phi_f^{CTR}$ becomes slightly higher by an enhanced contribution of Φ_f^{CT} at low temperatures. Because an appropiate correction of Φ_f with respect to the contribution of Φ_f^{CT} would decrease $k_{nr}(T)$, such that the activation barrier Ea becomes even somewhat larger. Accordingly, the interpretation in the text is not affected by the neglection of Φ_f^{CT} or would even be supported by a correction.

(47) Similarly to the decay behaviour in butyronitrile (ref. 42), a short-time component τ_s, which however becomes rather slow at low temperatures (e.g. 1 ns at 168 K), is needed to fit the time traces adequately. The well-known solvation probe coumarin 153 was simultaneously measured at 253 K and 168 K and showed exactly the same short time components. This and the fact that the „normal" lifetimes τ_1 of I and II in EtOH are similar to those in diethylether ensures that τ_s can be attributed to solvent relaxation processes.

(48) (a) Förster, Th. Z. Elektrochem. *Angew. Phys. Chem.* **1950**, 54, 42, 531. (b) Lippert, E. in: *Organic Molecular Photophysics*, ed. Birks, J.B.; Wiley Sons, London **1975**.

(49) Shizuka, H.; Ogiwara, T.; Kimura, E. *J. Phys. Chem.* **1985**, 89, 4302.

(50) Platt, J.R. *J. Chem. Phys.* **1949**, 17, 484.

(51) J. Saciv, A. Yogev and Y. Mazur *J. Am. Chem. Soc.* **1977**, 99, 6861.

(52) Zhu, Y; Schuster, G.B. *J. Am. Chem. Soc.* **1990**, 112, 8583.

(53) David, C.; Baeyens-Volant, D. *Mol. Cryst. Liq. Cryst.* **1980**, 59, 181.

(54) Maus, M.; Rettig, W. *Chem. Phys.* **1997**, 218, 151.

(55) In ref. 30 it is noted that the use of two separate excitation wavelength has been suggested elsewhere to sense pH (Bright, G.B.; Fisher, G.W.; Rogowska, J.; Taylor, D.L.; *Meth. Cell. Biol.* **1989**, 30, 157.), but that the fluorescent probes employed were practically unuseful because they show mainly ionsensitive excitation band shifts rather than on-off fluorescence switching. The reason for the successful application of dual excitation of the present compounds are the combination of (i) separated absorption and emission bands, (ii) the absence of wavelength shifts and (iii) an excited state equilibrium which is either slow and therefore not established during the excited state lifetime or it is quite similar to that in the ground state. Note also, that a pH analysis using the integrated band areas of **I** and **II** is possible which renders both molecular systems unaffected for the case of spectral band shifts in other solutions than investigated.

Chapter 11

Implications to Solar Energy Conversion

Abstract:
The unusual property of planar or moderately twisted electron donor-acceptor biaryls to combine high fluorescence quantum yields with large dipole moments resulting in large Stokes-shifts is demonstrated by experimental results and quantum chemical calculations and proposed for use in fluorescent solar concentrators. On the other hand, almost full charge separation in the excited state of strongly twisted donor-acceptor biaryls provides the possibility of electron transfer initiated photocatalysis. Corresponding perspectives for the use of solar energy are highlighted.

Keywords: fluorescent solar collector or concentrator, solar energy, chemistry and conversion

11.1 Introduction

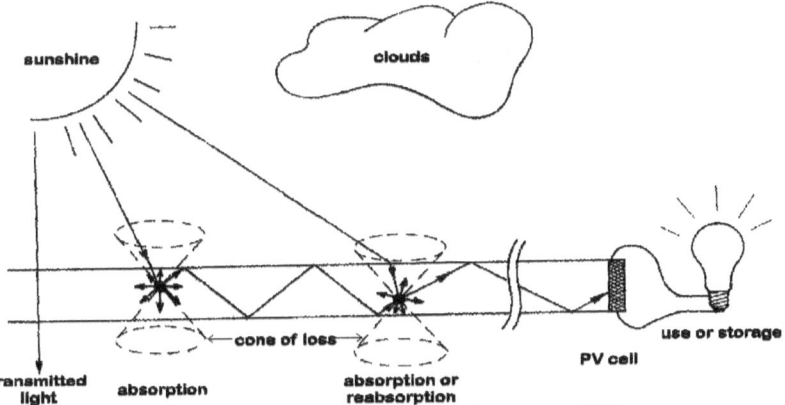

Scheme 11.1 *Schematic representation of a Fluorescent Solar Concentrator (FSC).*

Fluorescent solar concentrators (FSC) are based on the absorption of sunlight by fluorescent dyes imbedded in a transparent plate with a high refractive index n (Fig. 1).[1-6] The absorbed photons are transformed into fluorescence photons about of three quarters of which (more exactly $(1-n^{-2})^{1/2}$ according to Snells law) are trapped inside a plate due to total internal reflection and guided to the edge of the plate, where an efficient photovoltaic cell can be mounted to convert the light into electrical energy. Some of the advantages of optimized FSC versus direct sunlight-collecting by the photovoltaic material (PVM) itself are:

- collection of the whole UV/VIS solar spectrum using FSC multilayers,
- use of less PVM (cost effective),
- less heating of PVM,
- and in contrast to light concentration by lenses and mirrors, FSC also collect diffuse radiation (cloudy sky).

Ch. 11 Solar Energy Conversion

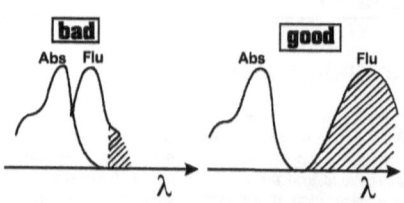

Scheme 11.2 *Overlapping (small Stokes-Shift) and decoupled (large Stokes-shift) absorption and fluorescence spectra. In the case of small Stokes Shift, only a fraction of the fluorescence photons can reach the edges of a FSC plate.*

For the development of useful FSC, photostable dyes are necessary with a high overall efficiency η_{dye} given by

$$\eta_{dye} = \eta_{abs}\eta_{flu}\eta_{Stokes}\eta_{reabs} \quad (11\text{-}1)$$

where η_{abs} is the fraction of solar energy absorbed, η_{flu} is the fluorescence quantum yield Φ_f in the FSC, η_{stokes} is related to the loss of energy by the Stokes shift and η_{reabs} accounts for the loss of energy by the reabsorption of emitted photons.

Experimental and theoretical studies in former[6-12] and recent [4,13-16] years have shown that especially the effiency loss η_{reabs} due to reabsorption of the fluorescence photons inside the plate is very crucial. A possible way to avoid this important loss is the decoupling of absorption and fluorescence spectra (scheme 11.2). For this purpose, dyes can be used which undergo intramolecular adiabatic photoreactions such as proton transfer or isomerization.[14,16] An alternative way is the application of dyes with high excited state dipole moments which exhibit red shifted fluorescence due to a change of interaction between the dye and the surrounding.[12]

Donor-Acceptor Biaryls

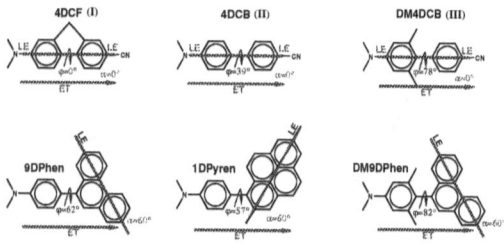

Polymers

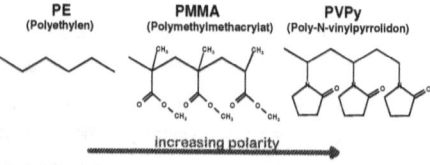

Scheme. 11.3 *Structures of donor-acceptor biaryls (upper) and polymers (below) investigated. The orientation of the transition moment directions α and the calculated equilibrium twist angles φ in S_0 are also indicated.*

Donor-acceptor (D-A) biaryl compounds are known to undergo photoinduced intramolecular charge transfer.[17-22] Here, D-A biaryls are divided into three classes of different electronic nature. The experimental study of the absorption and fluorescence properties of D-A biphenyls in solvents and polymers aims to demonstrate that D-A biaryl dyes with a planar or partially twisted structure can cope with the challenge to achieve a sufficient Stokes-shift without loosing the high fluorescence efficiency. Finally, perspectives of D-A biaryls in solar applications are discussed taking into account experimental investigations in different fields of solar energy.

11.2 Experimental

Details of the fluorescence steady-state and lifetime experiments are given in ch. 3, 4 and 8.

11.2.1 Preparation of Dye Doped Polymer Films and Silica gel

As a rule, no more than 1 mg of the dye per 3.5 g polymer was used for the dye doped polymer films yielding a dye concentration of less than 10^{-3} M. Both polymer and dye were dissolved and stirred in a proper solvent (methanol, chloroform or butylacetate) for one day and then cast on a flat glass plate which was covered by a small dish to allow a slow evaporation of the solvent in its own atmosphere. After air drying for several weeks, drying of the films under reduced pressure had to be done carefully since the films easily broke which renders them difficult for optical measurements. The thickness of the finally dried films were measured to be between 40 to 150 µm. The polymers polymethylmethacrylate (PMMA), poly-n-butylmethacrylat, (PBMA) and Poly-N-vinylpyrrolidone (PVPy) were used as purchased from Aldrich. Commerial polyethylene (PE) was also used, but the dye was incorporated by diffusion from a polar solution in which the dye was dissolved.

To incorporate the biphenyl dyes into the pores (240 Å) of a silica gel powder (FLUKA) which was a gift from Prof. Findenegg/Technical University Berlin, a concentrated acetonitrile solution (10^{-3} to 10^{-2} M) was prepared and after 2h mixing, the solvent was removed with a pipette as far as possible. Then the powder was washed with ACN four times. After further 2h drying by air a possible solvent residue was removed by a stream of hot air. Fluorescence measurements were performed by irradiation of the powder contained in a 1 cm^2 quartz cuvette.

11.3 Results and Discussion

11.3.1 Requirements for Low Lying and Highly Fluorescent Charge Transfer States in D-A Biaryls

At first glance, it seems contradictory that the emitting state combines a high dipole moment, necessary to achieve a large Stokes-shift, together with a large electronic transition moment, needed to obtain high fluorescence quantum yields through a large radiative to nonradiative rate ratio. An appropiate mixing of locally excited (^{1}LE) and strongly solvent stabilized electron transfer (^{1}ET) states as illustrated in scheme 11.2 is necessary to reach such a situation for a low lying ^{1}CT state. Note also, that the coupling of ^{1}ET with the ground state (S_0, not shown in scheme 11.4), which increases with more planar twist angles (φ), additionally increases the transition moment of ^{1}CT. In order to demonstrate the crucial mixing of states the calculated gas phase and the experimental solvent phase situation are compared in Tab. 11.1 and 11.2 for three different D-A biaryl types.

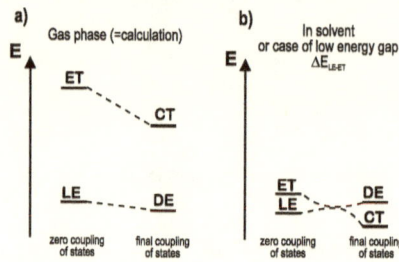

Scheme 11.4 *Excited states electronic interaction model a) in the case of high energy gap between locally excited and electron transfer states yielding higher lying CT states (in the gas phase) and b) in the case of low energy gap yielding a lowest lying CT state with large oscillator strength.*

Tab. 11.1 Gas Phase Transition Moments M and Energies ν^{gas} for the mixed States Obtained with ZINDO/S-CI.

D-A Biaryl	φ (deg)	μ_{ET} (D)	$\langle\mu_{flu}\rangle$ (D)	^1L_b ν^{gas} (10^3 cm^{-1})	M (D)	^1L_a ν^{gas} (10^3 cm^{-1})	M (D)	^1CT ν^{gas} (10^3 cm^{-1})	M (D)
9DPhen	62	29.2	17	29.6 (S$_1$)	0.6	33.0 (S$_2$)	3.8	37.4 (S$_5$)	3.7
1DPyren	57	32.3	17[b]	27.4 (S$_1$)	1.1	29.2 (S$_2$)	5.6	35.9 (S$_5$)	3.7
I (4DCF)	0		21	32.9 (S$_2$)	2.0	41.0 (S$_4$)	1.2	31.7 (S$_1$)	5.5
II (4DCB)	39	32.7	22	33.5 (S$_1$)	1.0	41.4 (S$_4$)	0.4	33.9 (S$_2$)	5.7
III (DM4DCB)	80		27	33.2 (S$_1$)	0.9	39.6 (S$_4$)	4.1	36.3 (S$_3$)	2.8

[a] μ_{ET} (for decoupled geometry with twist angle φ=90°) and φ are calculated using AM1. ν^{gas}, μ and f are obtained with ZINDO/S.
[b] calculated with eq. 3-7 from data of ref. 19 using μ_g from optimized AM1 geometry and the same Onsager radius a=0.57 nm as for 9DPhen in Tab. 9.2.

Tab. 11.2 Experimental Fluorescence Transition Moments M_f (in D)

		class I		class II		class III	
solvent[a]	$F_2(\varepsilon_r,n)$[b]	I	9DPhen[c]	II	1DPyren[d]	III	DM9DPhen[e]
HEX,IO	0.1	5.1	1.6	6.2	5.2	4.3	0.6
EOE	0.25	6.3	4.4	6.4		3.4	4.4
ACN	0.38	6.6	4.0	6.6	5.3	2.3	1.4

[a] the solvents are n-hexane (HEX), iso-octane (IO), diethylether (EOE) and acetonitrile (ACN)
[b] polarity function $f(\varepsilon_r,n)=(\varepsilon_r-1)/(2\varepsilon_r+1)-(n^2-1)/(2n^2+1)$ with static dielectric constant ε_r and refractive index n.
[c] calculated from ref. 17 [d] calculated from ref. 19 [e] calculated from ref. 18

The D-A biaryls of the first class (I), **I** (4DCF) or 9DPhen, are characterized by an initial increase of the fluorescence transition moment (M_f) from nonpolar to polar solvent (Tab. 11.2). This behaviour is comprehensible with a strong ^1L_b character of the emitting state in a surrounding where the ^{1}ET and the "oscillator strong" local ^1L_a states are not sufficiently low lying. In that respect the calculation results in Tab. 11.1 show that the ^1L_b-type S_1 state of 9DPhen has the largest energy difference to the upper ^1L_a and ^{1}CT states for the compounds considered. Note, that **I** (4DCF) is somewhat outstanding due to strong symmetry induced mixing of all states as discussed in the previous chapters (cf. ch.2).

The D-A biaryls of the second class (II), **II** (4DCB) and 1DPyren, exhibit high fluorescence transition moments irrespective of the solvent polarity. Since the calculations predict a weakly allowed ^1L_b transition to be active in the gas phase, the energetical stabilization of the ^{1}ET state introduces ^1L_a and ^{1}CT electronic properties to S_1 already in a nonpolar solvent. Both ^1L_a and ^{1}CT have comparable transition moments (Tab. 11.1) such that a solvent-induced change of S_1 between ^{1}CT and ^1L_a character does not alter the emission probability significantly. Only the experimental permanent dipole moments $\langle \mu_{flu} \rangle$ averaged for the solvent polarity range between HEX and ACN (Tab. 11.1) reveal that S_1 of 1DPyren ($\langle \mu_{flu} \rangle$ =17 D) contains more local ^1L_a character compared to 4DCB, which has a higher value $\langle \mu_{flu} \rangle$ (22 D). This is also consistent with the calculation results in Tab. 11.1, where the the ^1L_a state of the pyrene derivative is found very low lying, while for the biphenyl derivatives, the interacting ^1L_a and ^{1}CT states have already exchanged their character in the gas phase (cf. Fig. 9.1: for 9DPhen and DM9DPhen in solvents, ^{1}CT is below ^1L_a). The reason for the relatively large experimental values of M_f for the D-A biphenyls in Tab. 11.2 are due to the facts that (i) the parallel transition moment directions (α in scheme 11.3) of the ^1L_a and ^{1}ET allows a very effective excited state coupling in spite of a large energy gap $\Delta E^1_{La-}{}^1_{ET}$ in polar solvents (scheme 11.4) and (ii) the atomic π orbital coefficients at the linkage position of the acceptor unit are larger which enables an improved coupling of ^{1}ET with S_0 (cf. eq. 8-11).

Contrary to class I and II, D-A biaryls of class III (Tab. 11.2) reveal an ongoing decrease of the fluorescence transition moments with increasing solvent polarity (^1L_b character is, however, also observed for DM9DPhen in IO). The ^{1}ET contribution to S_1 of these compounds is considerably larger than in the previous ones as testified by the high dipole moments of 4DDMCB (27 D) and DM9DPhen (23 D).

Tab. 11.3 Fluorescence Quantum Yields Φ_f

		class I		class II		class III	
solvent	$F_2(\varepsilon_r,n)$	4DCF(I)	9DPhen[a]	4DCB(II)	1DPyren[b]	DM4DCB(III)	DM9DPhen[c]
HEX,IO	0.1	0.57	0.36	0.71	0.82	0.35	0.16
EOE	0.25	0.83	0.49	0.84	0.85	0.55	0.21
ACN	0.38	0.81	0.85	0.79	0.95	0.21	0.22

[a] from ref. 17 [b] from ref. 19 [c] from ref. 18

From these considerations, we can expect that D-A biaryls of class II, which combine large permanent and transition dipole moments for the emitting ^{1}CT state, are best suitable for application in FSC. Indeed, the experimental fluorescence quantum yields determining the efficiency of FSC (eq. 11-1) are particularly high for the D-A biaryls of class I. Note also, that the compounds of class I have similarly large values of Φ_f in strongly polar solvents because they can be regarded as to be transformed into biaryls of class II, i.e. emitting from a state with ^1L_a-type CT character. Hence the calculations are a usefool tool to predict biaryls with an emissive and polar excited state. It should be stressed, that the above considerations are similarly valid, if heteroatoms are included in the framework of the biaryls, which, on the one hand, can bathochromically shift the spectra and, on the other hand, introduce a weakly emissive $n\pi^*$ state instead or additionally to the ^1L_b state considered above.

In order to predict biaryls with improved overall properties including absorption efficiency (η_{abs}) which synthetic aspects have to taken into account?

The molecular parameters that control the mixing behaviour and therewith the electronic property of the S_1 state are (i) the linkage position, (ii) the angle α between the transition moment directions of the ^1L_a and ^{1}ET state, (iii) the energy gap between the ^1L_a and ^{1}ET state $\Delta E^1_{L_a-{}^1ET}$ and (iv) the interannular twist angle φ:

(i) A linking position has to be selected where large atomic orbital coefficients are present. Otherwise, the consequent weak electronic coupling of ^{1}ET with the ground state yields an undesirably low transition moment for the ^{1}CT state (see 8.3.3). (ii) To achieve constructive coupling between ^1L_a and ^{1}ET an angle α as close to zero as possible is favourable, of course, without breaking rule i. (iii) The experimental results reveal that a small $\Delta E^1_{L_a-{}^1ET}$ is more important than a small α (cf. strong ^{1}ET-^1L_a coupling in spite of $\alpha=60°$ for 9DPhen and 1DPyren) such that a low lying ^{1}ET state controllable through the redox potentials of the aryl subunits can enhance the coupling rather efficiently. (iv) Finally, a larger twist angle φ can further increase the ^{1}ET contributions to S_1, but because of the accompanying reduction of the coupling between ^{1}ET and S_0 yielding smaller transition moments, a compromise with a moderately twisted structure should be chosen. Fortunately, the fluorescence quantum yields are still large, in spite of the sizeably twisted structure in various D-A biaryls (Tab. 11.3).

The next section focuses on the question, which host surrounding or matrix should be used to achieve optimum results in FSC using biaryls based on charge transfer emission.

11.3.2 Which Host Environment is favourable for a FSC?

FSC are usually made of polymeric material because of a high refractive index connected with a somewhat better trapping efficiency of the fluorescence photons, but the rigidity of such a matrix usually prevents sufficient molecular relaxations necessary for the decoupling of the absorption and fluorescence spectra of the imbedded dye.[1-4] In Fig. 1, the absorption and fluorescence spectra of the differently twisted donor-acceptor biphenyls **I-III** (4DCF, 4DCB, 4DMDCB), which can be regarded as model compounds for larger biaryl systems, in the weakly polar polymer polyethylene (PE) and the polar polymer poly-N-vinylpyrrolidone (PVPy) are compared with the relevant spectra in the low viscosity solvents n-hexane (HEX) and acetonitrile (ACN), which have been discussed in detail in ch.3. Because of the higher polarity of PVPy with respect to PE the absorption and even more strongly the fluorescence spectra are shifted to lower energy. Thus the desired increase of the Stokes shift is observed also in the rigid matrices both with increasing polymer polarity and with increasing dipole moment of the compound incorporated.

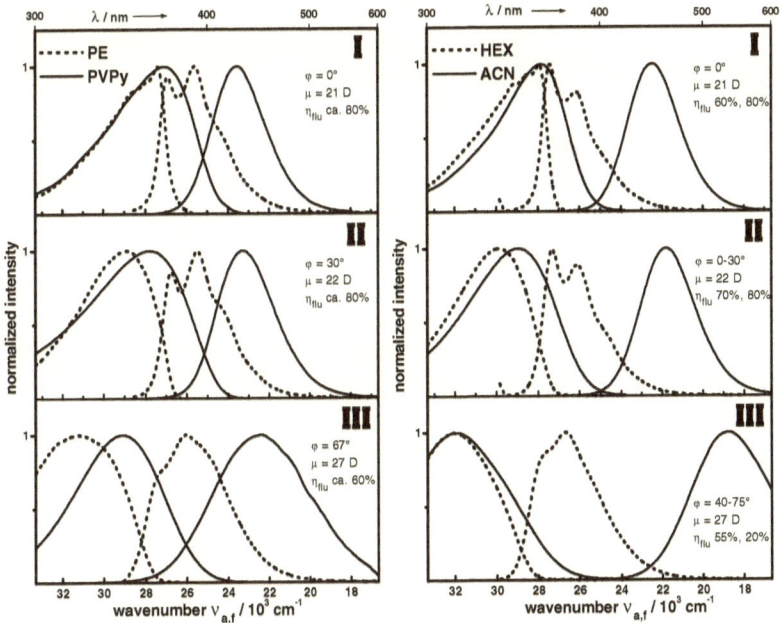

Fig. 31.1 *Comparison of absorption and fluorescence spectra of* **I-III** *in the rigid matrices PE (- - -) and PVPY (———) on the left with those in the liquid solvents HEX (- - -) and ACN (———) on the right. The dye parameters twist angle φ, dipole moment μ and fluorescence efficiency η_{flu} are measured for the solvents, but only estimated for the polymers from the photophysical behaviour of the dyes in low-temperature solvent glasses.*

Relaxations in polymers ? Regarding the question which kind of relaxations take place in the stiff polymer surrounding, the distinctly stronger polarity-induced red shift of the absorption spectra in the polymers compared to those in the solvents, is undoubtedly due to a higher refractive index responsible for a stronger effect of the instantaneous electronic polarization (eq. 3-2). However, the fluorescence spectra in the polar solvent ACN are considerably more red-shifted than those in PVPy. This indicates that in the rigid and polar polymer matrix much less dielectric relaxation occurs which means that the effective dielectric constant ε_r or the Onsager term $f(\varepsilon_r)$ in eq. 3-2, respectively, is much smaller than in ACN. This is understandable, if full polymer dielectric relaxation is not achieved within the lifetime of fluorescence. The discrepancy between the red shift in the polymers and solvents is even larger for the twisted compound **III** where a large amount of the red-shift derives from a conformational relaxation (ch.6: CT→CTR). To investigate whether this geometry relaxation is supressed in polymers the fluorescence maxima in different polymers are compared in Tab.11.4. The red shift of the fluorescence maxima of **I** and **II** from PE to PBMA (polybutylmethacrylate) to PMMA (polymethylmethacrylate) to PVPy reveals the expected increase of the polymer polarity. For **III**, the same order is observed with a red shift somewhat stronger because of its higher dipole moment in S_1. The increase of the „free volume" or the microviscosity, respectively, from more polar PMMA to less polar PBMA was reported to lead to an enhancement of the conformational relaxation from the locally excited to the charge transfer species in 4-N,N-diethylaminobenzonitrile, in spite of the fact that this photoreaction is usually strongly accelerated by the polarity of the environment.[23] Such a microviscosity effect is not found for **III** supporting the view that the CT→CTR relaxation of **III** is not important in the polymers investigated.

Tab. 11.4 Fluorescence Maxima (in 10^3 cm^{-1}) of D-A Biphenyls in Polymers.

polymer matrix	I	II	III
PE	26.8	26.9	27.3
PBMA	26.7	26.5	26.7
PMMA	26.2	26.0	25.3
PVPy	23.6	23.3	22.3

Evaluation of FSC performance in polymers and solvents

An inspection of the spectra in Fig. 11.1 by eye implies a good separation of the absorption and fluorescence spectra in the polymers, but distinctly better in the polar solvent ACN. To obtain a more quantitative basis for the discussion the efficiencies η_{reabs} and η_{Stokes} accounting for the photon loss due to the reabsorption process and the energy loss due to the Stokes-shift are estimated as follows. Similarly to ref. 18, η_{Stokes} is simply estimated by

$$\eta_{Stokes} = v_f/v_a \qquad (11\text{-}2)$$

The calculation of η_{reabs} is based on the assumption that all those emitted photons were reabsorbed AND lost which possess an energy v_f larger than a certain threshold given by the absorption spectrum. Here, the absorption energy $v_a(I_a=3\%)$ was chosen as threshold where the normalized absorption (red edge) intensity I_a amounts to 3% of the band maximum.

$$\eta_{reabs} = \frac{\int_{v_f=0}^{v_f=v_a(I_a=3\%)} I_f(v_f)dv_f}{\int_{v_f=0}^{\infty} I_f(v_f)dv_f} \qquad (11-3)$$

Comparing various published fluorescence spectra obtained from front surface emission with those obtained at the edge of a large scale dye-doped plate,[4,6,10,11,17,18] the calculated η_{reabs} seems to be a reasonable approximation for the part of the emission spectrum that would be transmitted through the plate of a FSC. To convey an impression of the 3% borderline this intensity value corresponds to the intersection of the absorption and emission band for II in ACN. Furthermore, since we are mainly interested here in efficiency losses due to the spectral overlap, the same fluorescence quantum yields for the ACN and HEX solutions are used for PVPy and PE, respectively. The efficiencies η_{Stokes} and η_{reabs} derived theoretically from eq. 11-2 and 11-3 for a dye perfomance in FSC are collected in Tab. 11.5.

Tab 11.5 Efficiencies of Dye (I-III) Performance in a FSC According to Eqs. 11-1 to 11-3.

host	η_{reabs}			η_{stokes}			Φ_f			η_{dye}/η_{abs}		
	I	II	III	I	II	III	I	II	III	I	II	III
HEX	0.66	0.75	0.89	0.96	0.89	0.84	0.57	0.71	0.35	0.36	0.48	0.26
PE	0.64	0.79	0.88	0.95	0.90	0.83	"	"	"	0.35	0.50	0.26
ACN	0.98	0.99	1.00	0.81	0.75	0.59	0.81	0.79	0.21	0.64	0.59	0.12
PVPy	0.65	0.78	0.84	0.87	0.84	0.76	"	"	"	0.46	0.52	0.14

As an encouraging results, the almost complete decoupling of the absorption and fluorescence bands in ACN of the three biphenyls yields optimum values of practically unity for η_{reabs}. Taking into account the energy loss due to the Stokes-shift, best results are achieved for I and II in ACN. In the case of III in ACN, the Stokes-shift is even too large, since the better η_{reabs} of negligible 1-2% compared to η_{reabs} of I and II is confronted with a more crucial energy loss in η_{Stokes}. In addition, the smaller fluorescence quantum yield Φ_f of III yielding an internal dye efficiency $\eta_{dye}/\eta_{abs} = \Phi_f \eta_{Stokes} \eta_{reabs}$ (cf. eq. 11-1) of less than 30% in nonpolar and less than 15% in polar environments renders the twisted compound as less useful for application in FSC. On the other hand, the internal efficiencies of I and II are rather good (ca 60% in ACN) and can well compete with the relevant experimental values reported for the best dye so far for FSC based on a perylenebis(dicarboximide) structure (BASF 241),[24-27] or for the frequently studied dye DCM as well as for rhodamine 640. The literature values of η_{dye}/η_{abs} for these "landmarking" FSC dyes are 32%,[4] 46%[18] and 32%[18] only.

One might be dissapointed by the lower values of η_{reabs} in the polar polymer compared to ACN, but these values are understandable because of the accompanying bathochromic shift of the absorption band. Note also, that due to the better values of η_{Stokes} the total efficiency in the polymer is only 7% less than in the solvent. Another surprising point that speaks in favour of D-A dye molecules for FSC is the similarity of η_{reabs} in the polar and nonpolar polymer, such that the higher η_{Stokes} in the nonpolar PE leads to a higher combined efficiency $\eta_{reabs} \cdot \eta_{stokes}$. The comparably good decoupling of absorption and emission bands can be traced back to the strong electron transfer interactions in the emitting state already in the nonpolar matrix which allows a sizeable stabilization by the polarizable medium.

In any case, we have to conclude that polar solvents, here especially in combination with the planar D-A biphenyl **I** (η_{dye}/η_{abs}=64% and $\eta_{reabs} \cdot \eta_{stokes}$=79%), are best suitable for FSC mainly because of the extremely high values for η_{reabs} (compare η_{reabs}=32% for BASF 241 from ref.4). However, the photon trapping efficiency must be optimized, e.g. by using a solvent/vessel combination of similar refractive index.

Seeking for a solution to nevertheless combine the advantages of a solid matrix (better trapping efficiency and photostabilizing effect on the dyes due to the prevention of photochemical dye reactions, e.g., with oxygen or a second dye molecule) together with a more favourable Stokes-shift, preliminary fluorescence experiments were performed using a silicagel 240 with pore sizes of 240 Å^3 where the solute dissolved in the polar solvent ACN have been incorporated together. Indeed, the comparison of the fluorescence spectra of **I** and **II** in pure ACN and in silicagel 240 doped with ACN exhibits the wanted effect to match up the spectra in the solid matrix with those obtained in the pure solvent (Fig. 11.2). A similar concept of dye inclusion into zeolite microcrystals (but without solvent inclusion) has been proposed by Calzaferri[28] who suggested interesting improvements of such an artificial antenna device by making use of the highly anisotropic dye assemblies inside the linear channels of the zeolite.

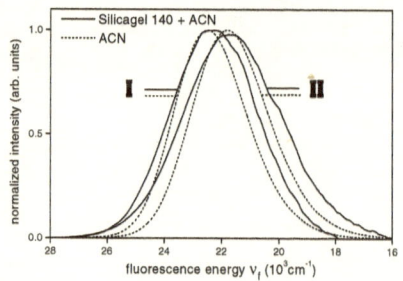

Fig. 11.4 *Fluorescence spectra of* **I** *and* **II** *in a silica gel with pore sizes of 240 Å^3 doped with ACN (——) compared to those in pure ACN (- - -).*

11.4 Solar Perspectives

11.4.1 Fluorescent Solar Concentrators Using Planar or Moderately Twisted D-A Biaryls

Donor-Acceptor Biaryls are interesting candidates for application in Fluorescence Solar Concentrators (FSC) because they exhibit unusually large Stokes-Shifts in combination with high fluorescence efficiencies. The analysis of the electronic states shows that the effective mixing of ^{1}ET with allowed ^{1}LE transitions is responsible for these properties. The efficiency of absorption and fluorescence as well as the Stokes-Shift can be controlled by structural parameters such as the twist angle φ, the connectivity position, the orientation of allowed transition moment directions α and the size of the aromatic moieties. Hence this theoretical and experimental study indicates that donor-acceptor biaryls with large fluorescence quantum yields (Φ_f) and dipole moments (μ_f) can foreseeably be designed. In particular, the model D-A biphenyls **I-III** are evaluated to achieve internal dye efficiencies (η_{dye}/η_{abs}) of about 60% in FSC consisting of a polar host environment taking into account Φ_f, the energy loss by reabsorption (η_{reabs}) and the Stokes-shifted emission (η_{stokes}). Such efficiencies could cope with the relevant values of dyes which currently exhibit the best performance in FSC. However, for the application of donor-acceptor biaryls in FSC, longer wavelength chromophores have to be introduced in order to shift the absorption spectrum into a more profitable part of the solar spectrum. Promising aryl units are based on a perylene or better perylene diimide molecular framework because of their quantum yields close to unity.[24-27] The problem of too low lying locally excited states preventing efficient electron transfer interactions may be overcome by pretwisting the units to each other. An alternative to increase the dipole moment of S_1 without loosing the radiative strength of the charge transfer state (necessary coupling of ^{1}ET and S_0) could be reached, if the distance between both aryl chromophores is enlarged by an electron conducting chain of phenyl groups.[29]

Regarding the host material for D-A biaryls in FSC, polar solvents are advantageous to polymers because less reabsorption losses can be expected. Liquid FSC[30,31] are also ecologically desirable because of the simple recycling possibility. Even the use of environment-friendly water is possible since the model D-A biaryls **I** and **II** are found (i) to be soluble in H$_2$O/EtOH mixtures and (ii) to keep high fluorescence quantum yields due to the absence of strong specific hydrogen bonding quenching (ch.10). In this case, however, the photon trapping efficiency of the vessel/solvent has to be optimized. For special purposes solid matrices with voluminous pores or channels hosting a polar solvent together with the florescence dye can be favourable. This variant would also not contradict the environment aspect because after photodegradation the dyes can be rinsed out and replaced by new dyes.

The potential use of such particularly designed fluorescent biaryl dyes is, of course, not only restricted to FSC but can be extended, e.g., to the application of solar lasers.[32-34] They are of immense interest for chemical industry, since up to now by using conventional lasers „one photon is much more expensive than the molecule produced".[34]

11.4.2 Towards Electron Transfer Initiated Photocatalysis (ETIP) Using Twisted D-A Biaryls

Although no direct experimental results are given in this work, it may be allowed at this point to discuss some potential application aspects of photoinduced charge transfer in twisted donor-acceptor D-A biaryls.

The molecular photodiode behaviour and photosynthesis model character of the adiabatic photoreaction towards a biradicaloid decoupled conformation in the ^{1}CT state of the twisted D-A biphenyl III has been demonstrated in ref. 20

The radical ion-pair electronic character of the CT excited states of twisted D-A biaryls can be expected to allow <u>electron transfer initiated photocatalysis or photosensitization</u> (ETIP). In fact, this has experimentally been proven the first time by Habib Jiwan and Sourmillion[35] and afterwards by Schopf, Rettig and Bendig.[36] The landmarking and effective work in ref. 35 not only demonstrates the reactivity of the donor as well as of the acceptor aryl unit in the ^{1}CT state but also the variability of ETIP as it may be employed for photochemical upgrading (energy storage) and photodetoxification of pollutants.

Very recent investigations in these two application fields have revealed the unexpected importance of ETIP in photochemical reactions where usually photoinduced triplet sensitization was accepted to be the most effective mechanism.[37] For example, the valence photoisomerization of norbornadienes, which is potentially useful for solar chemical energy storage,[38] is now discussed to occur more efficiently via ETIP.[39,40] Solar detoxification based on dye sensitization usually proceeds after singlet oxygen production.[40-42] Surprisingly, no singlet oxygen was detected in novel and relatively efficient solar photooxidations using the above mentioned perylimide dyes.[41,44,45] A very recent study evidences that these reactions also occur after ETIP of the pollutant.[44] Thus, it seems logical that an enhancement of the intramolecular electron transfer interactions in these dye sensitizers could lead to an even more efficient ETIP mechanism. This might be an interesting challenge for future application of donor-acceptor biarylic compounds based on well-known chromophores.

11.5 References

(1) A. Goetzberger and W. Greubel, *Appl. Phys.* **1977**, 14 123.

(2) Weber, W.H.; Lambe, J. *Appl. Opt.* **1976**, 15, 3544.

(3) Stahl, W.; Zastrow, A. *Physik in unserer Zeit* **1985**, 16, 167.

(4) Vollmer, F. Ph.D. thesis „*Spektroskopische und quantenchemische Untersuchungen von Farbstoffen mit großer Stokes-Verschiebung der Fluoreszenz auf der Basis von adiabatischen Photoreaktionen*" TU Berlin1995, Köster Verlag, ISBN 3-89574-048-9.

(5) Reisfeld, R.; Jørgensen, C.K. in: „*Structure and Bonding 49*", Springer Verlag, Berlin, 1982.

(6) Batchfelder, J.S.; Zewail, A.H.; Cole, T. *Appl. Opt.* **1979**, 18, 3090.

(7) Olson, R.W.; Loring, R.F.; Fayer, M.D. *Appl. Opt.* **1981**, 20, 2934.

(8) Drake, J.M.; Lesiecki, M.L.; Sansegret, J.; Thomas, W.R.L. *Appl. Opt.* **1982**, 21, 2945.

(9) Roncali, J.; Garnier, F. *Appl. Opt.* **1984**, 23, 2809.

(10) Sansregret, J.; Drake, J.M.; Thomas, W.R.L.; Lesiecki, M.L. *Appl. Opt.* **1983**, 22, 573.

(11) Levitt, J.A.; Weber, W.H. *Appl. Opt.* **1977**, 16, 2684.

(12) R.E. Rah, *J. Lum.* **1981**, 24/25, 869.

(13) Sóti, R.; Farkas, É.; Hilbert, M.; Farkas, Zs.; Ketskeméty, I. *J. Lum.* **1996**, 68, 105.

(14) Vollmer, F.; Rettig, W. *J. Photochem. Photobiol. A: Chem.* **1995**, .

(15) El-Daly, S.A.; Hirayama, S. *J. Photochem. Photobiol. A: Chem.***1997**, 110, 59.

(16) W. Rettig, *Nachr. Chem. Tech. Lab.* **1991**, 39, 398.

(17) Onkelinx, A.; De Schryver, F.C.; Viane, L.; Van der Auweraer, M.; Iwai, K.; Yamamoto, M.; Ichikawa, M.; Masuhara, H.; Maus, M.; Rettig, W. *J. Am. Chem. Soc.* **1996**, 118, 2892.

(18) Onkelinx, A.; Schweitzer, G.; De Schryver, F.C.; Miyasaka, H.; Van der Auweraer, M.; Asahi, T.; Masuhara, H.; Fukumura, H.; Yashima, A.; Iwai, K. *J. Phys. Chem. A* **1997**, 101, 5054.

(19) Wiessner, A.; Hüttmann, G.; Kühnle, W.; Staerk, H. *J. Phys. Chem.* **1995**, 99, 14923

(20) Rettig, W.; Maus, M. *Ber. Bunsenges. Phys. Chem.* **1996**, 100, 2091.

(21) (a) Maus, M.; Rettig, W.; Lapouyade, R. *J. Inf. Rec.* **1996**, 22, 451

(22) (a) Herbich, J.; Kapturkiewicz, A. *J. Am. Chem. Soc.* **1998**, 120, 1014.

(23) Al-Hassan, K. A.; Azumi, T.; Rettig, W. *Chem. Phys. Lett.* **1993**, 206, 25.

(24) Seybold, G.; Wagenblast,G. *Dyes and Pigments* **1989**, 11, 303.

(25) Demmig, S.; Langhals, H. *Chem. Ber.* **1988**, 121, 225.

(26) Icli, S.; Icil, H. *Spectroscopy Lett.* **1996**, 29, 1253.

(27) Burgdorff, C.; Löhmannsröben, H.-G.; Reisfeld, R. *Chem. Phys. Lett.* **1992**, 197, 358.

(28) Calzaferri, G. Int. Symp. on Solar Chemistry, Villigen/Switzerland 1997.

(29) Fiebig, T. Ph.D. thesis „*Theoretische und experimentelle Untersuchungen zur Elektronenübertragung in kovalent verbrückten Bichromophoren*", Universität zu Göttingen **1996**.

(30) Filloux, A.; Mugnier, J.; Bourson, J.; Valeur, B. *Rev. Phys. Appl.* **1983**, 18, 273.

(31) Kondepudi, R.; Srinivasan, S. *Solar Energy Materials* **1990**, 20, 257.

(32) Reisfeld, R. *J. Non-Crystalline Solids* **1990**, 121, 254.

(33) Arashi, H.; Cooke, D.; Naito, H. *Jpn. J. Appl. Phys.* **1995**, 34, 4795.

(34) Yogev, A. Int. Symp. on Solar Chemistry, Villigen/Switzerland 1997.

(35) Habib Jiwan, J.L.; Sourmillion, J.Ph. *J. Photochem. Photobiol. A: Chem.* **1992**, 64, 145.

(36) Schopf, G.; Rettig, W.; Bendig, J. *J. Photochem. Photobiol. A: Chem.* **1994**, 84, 33.

(37) Hammond, S.; Wyatt, P.; Deboer, C.D.; Turro, N.J. *J. Am. Chem. Soc.* **1964**, 86, 2532.

(38) Scharf, H.-D.; Fleischhauer, J.; Leismann, H.; Ressler, I.; Schleker, W.; Weitz, R. *Angew. Chem.* **1979**, 91, 696.

(39) Wang, X.S.; Zhang, B.W.; Cao, Y. *J. Photochem. Photobiol. A: Chem.* **1996**, 96, 193.

(40) Maafi, M.; Lion, C.; Aaron, J.-J. *New. J. Chem.* **1996**, 20, 559.

(41) Gerdes, R.; Wöhrle, D.; Spiller, W.; Schneider, G.; Schnurpfeil, G.; Schulz, G.; Poster at Int. Symp. on Solar Chemistry, Villigen/Switzerland 1997., submitted to J. Photochem. Photobiol. A: Chem.

(42) Parent, Y. *Solar Energy* **1996**, 56, 429.

(43) Funken, K.-H.; Hoffschmidt, B.; Demuth, M.; Schmücker, M.; Sturzenegger, M.; Bahnemann, D.; Ortner, J. *Sommerschule „Solare Chemie und Solare Materialforschung"*, Köln, 13./14.10.1997.

(44) Chen, L.; Lucia, L.A.; Gaillard, E.R.; Icil, H.; Icli, S.; Whitten, D.G. *submitted to J. Luminesc.*

(45) Private Communication with Siddik Icli, Poster 216, Vth Int. Conference on Methods and Applications of Fluorescence Spectroscopy, Berlin, 21.-24.9.1997

Chapter 12

Final Conclusions

The differently twisted 4-N,N-dimethylamino-4'-cyanobiphenyls **I**, **II** and **III** (interannular twist angle φ increases from **I** to **III**) which are regarded as model compounds for larger donor-acceptor biaryls have been investigated by optical spectroscopy methods, such as absorption, luminescence, transient absorption, polarization. Temperature, pressure, delay time and solvent polarity were the main parameters varied in the experiments. CNDO/S, ZINDO/S and AM1 quantum chemical calculations were also employed to support the conclusions. Photoinduced intramolecular charge transfer in donor-acceptor biaryls was similarly studied in a pair of differently twisted dimethylanilino-phenanthrenes.

The electronic structure and transitions of the donor-acceptor biphenyls **I-III** can be well interpreted in terms of a composite molecule model on the basis of the dimethylanilino donor (D) moiety and the benzonitrile acceptor (A) moiety. According to the configuration analysis the lowest lying singlet states are two ^1L_b-type, two ^1L_a-type states and an intramolecular charge transfer state (^1CT) which is strongly stabilized with decreasing twist angle φ by the electronic interaction with the higher lying ^1L_a states. In analogy to the parent unsubstituted biphenyl (BP), linear dichroic absorption and polarized fluorescence excitation spectra allowed to assign the absorption bands from 33000 cm^{-1} up to 55000 cm^{-1} to benzenoid electronic transitions. These assignments were rather helpful in the interpretation of the transient absorption spectra. The additional low energy absorption band is due to the long-axis polarized $S_0 \rightarrow {}^1CT$ transition connected with a partial electron transfer from the highest occupied molecular orbital of the donor unit to the lowest unoccupied molecular orbital of the acceptor unit. The steady state fluorescence corresponds to the same transition (optical back charge transfer) as evidenced by the agreement of the surprisingly large transition moments M^1_{CT} as well as of the polarization degrees in case of practically absent solvent and structural relaxations. The high excited state dipole moments derived from the solvatochromic fluorescence shifts unambiguously characterize S_1 as an ^1CT state which has increasing electron transfer character with increasing twist angle:

The state analysis with ground state coupled reference states shows that the large values of M^1_{CT} arise from the extremely strong and twist angle dependent coupling of the electron transfer state (^1ET) with the ^1L_a states latter of which borrow almost completely their

$S_0 \rightarrow {}^1L_a$ transition intensity to the $S_0 \rightarrow {}^1CT$ transition for $\varphi > 60°$. This is testified by the fact that the ${}^1L_a$ absorption band is obviously visible only in the highly twisted D-A biphenyl **III** for which a ground state twist angle of about 70° is derived. In accordance with recent studies of Herbich and Kapturkiewicz on the transition moments of various donor-acceptor biaryls a substantial contribution to M^1_{CT} derives directly from the coupling of the zeroth-order ${}^1ET$ state with S_0. The unusual strong coupling of the zeroth-order ${}^1ET$ with S_0 especially at more planar twist angles is responsible for the fact that the observed fluorescence transition moments M_f of the D-A biaryls **I** and **II** do not decrease with solvent polarity in spite of an increased energy gap to the ${}^1L_a$ states. On the other hand, the twisted structure of **III** prevents a strong contribution to M_f by direct coupling with S_0 such that the solvent-induced decoupling from the ${}^1L_a$ state is reflected by a decrease of M_f with increasing polarity. The same effects are found for the parallely investigated other pair of differently twisted biaryls which stresses the generality of these phenomena and the above interpretation. However, the comparative analysis of quantum chemically and experimentally determined fluorescence transition moments reveals that a change of the coupling between ${}^1ET$ and S_0 must also be involved indicating polarity dependent excited state conformations (see below).

New insights were obtained regarding *electronic relaxations in donor-acceptor biaryls*. For the first time for biaryls, dual spectrally separated stimulated fluorescence bands with clear precursor-successor relationship on a sub-ps timescale could be observed for the D-A biphenyls **I-III**. This could be interpreted with the excited state electron transfer interconversion from a primary populated and vibrationally relaxed state (${}^1FC$) with mixed ${}^1L_b/{}^1CT$ character to the more ${}^1L_a$-type ${}^1CT$ state seen in the steady state experiments. The involvement of a ${}^1L_b$-type state has also been concluded for the phenanthrene derivative due to the low transition moments in weakly polar solvents.

A solvation dependent transformation of the electronic structure is proposed to be a more appropriate description than a simple Jablonski diagramm to account generally for the photoinduced charge transfer processes in donor-acceptor biaryls. This model predicts that three electronically and conformationally different species can accompany successively the photoinduced charge transfer process: (i) a species with essentially localized (nonpolar) electronic character of ${}^1L_a$ and/or ${}^1L_b$-type, (ii) a species of ${}^1L_a/{}^1ET$ character strongly coupled with the S_0 due to a photoinduced more planar conformation and (iii) a species with dominantly ${}^1ET$ character assisted by a highly twisted conformation.

Whether or not dual or even triple fluorescence from such species is observed associated with rather discrete conversion rate constants and energy minima depends on the strength of electronic interaction between the electronically and conformationally different species. The low symmetry of 9DPhen and the consequent strong state mixing may lead to a single conformational minimum in a given solvent, while the higher symmetry of the D-A biphenyls can account for the observation of discrete species (**FC** and **CT**) with different zeroth-order electronic nature. A similar comparative conclusion may be drawn for the symmetric biaromatic amine ADMA and the unsymmetric 1DPyrene since only for ADMA, a short time component of 850 ps in Hexanol slower than the characteristic reorientation time of the solvent could recently (Wiessner, A.; Hüttmann, G.; Kühnle, W.; Staerk, H. J. Phys. Chem. 1995, 99, 14923) be observed and attributed to an electronic change during the intramolecular rotation. This electronic relaxation may have been unobserved in the earlier experiments on ADMA because the electronic change from the primary ^1L_a-type species to the final ^{1}CT (both polarized along the interannular axis) is less pronounced than the ET reaction from **FC** (partially ^1L_b-type) to **CT** in the D-A biphenyls.

The intersystem crossing process in **I-III** plays only a minor role in polar solvents but in nonpolar or frozen solvents a level reversal between the ^{1}CT and a ^3L_b-type triplet state is suggested which accounts for enhanced nonradiative rate constants in nonpolar solvents. Analogously to the parent biphenyl, the twisted D-A biphenyl shows a distinctly higher triplet yield as revealed by smaller emission anisotropy values and a larger contribution of the phosphorescence background to the fluorescence decay curves at 77K.

The initial *photoinduced conformational relaxation* in the donor-acceptor biphenyls occurs towards a more planar structure with improved π-electron delocalization. The transient absorption measurements and the quantum chemical calculations of solvent dependent twist potentials and transition moments support this view. In addition, **II** in solvents of different polarity and **III** in nonpolar solvents exhibit monoexponential fluorescence decays (liquid solvents) and transition moments of fluorescence larger than those of absorption which reveals that for these conditions the molecules in the ^{1}CT excited state remain in a more planar conformation than in the ground state. On the other hand, the fluorescence of the strongly pretwisted D-A biphenyl **III** in dipolar solvents decays with an additional time component which must be due to a further intramolecular relaxation. Various reasons speak in favour for a relaxation towards a conformation more twisted than in the ground state: (i) the fluorescence transition moment is smaller than expected if only a decoupling from the ^1L_a state by the increased energy gap is active (ii) the final transient absorption spectrum in acetonitrile resembles the sum of the dimethylaniline cation and benzonitrile anion spectra (iii) the calculated ^{1}CT twist potential in polar solvents suggests an energy minimum for perpendicular rotamers (iv) dynamic fluorescence quenching by hydrogen bonding is explainable with a highly polar and biradicaloid electronic structure, (v) from the analysis of the excited state equilibrium in diethylether a smaller radiative rate constant ($k^f_{CTR}/k^f_{CT} \approx 0.7$) and a larger dipole moment was derived for the more relaxed charge transfer species **CTR** (μ_{CTR}=30D) than for the precursor species **CT** (μ_{CTR}=26D) consistent with a stronger ^{1}ET character of **CTR** induced by an enhanced twist angle.

The temperature dependent fluorescence band-shape analysis according to Marcus' semiclassical electron transfer theory and the temperature independent radiative rate constants indicate that the initial relaxation towards planarity is not stopped at low temperature in the liquid phase. This was used as the presumption in the kinetic analysis of the excited state conformational equilibrium between the **CT** and **CTR** species of **III**. The species associated spectra and the temperature dependent reaction rate constants could be derived by a global analysis of fluorescence decays. The complete thermodynamic characterization shows that the fluorescence bands of both species are not separated as it is the case for DMABN because the reaction energy is small (ΔH=-2.5 kJ/mol) and the higher ground state reorganization energy ($\Delta\Delta E_{FC}$=7.5 kJ/mol) of the **CTR** species counteracts in such a way that similar fluorescence energies of **CT** and **CTR** are the consequence ($\Delta v_f \approx$850 cm^{-1}). A large energy barrier for the adiabatic photoreaction from the **CT** to **CTR** species (E_a=14 kJ/mol) is responsible for the observed small reaction rate constants and the fact that the activation barrier for the solvent mobility (E_η(EOE)=7kJ/mol) is considerably less substantiates that the photoreaction is thermally activated by an intrinsic energy barrier $E_i^{\#}$. This is also predicted by the calculated twist potential of **III** in EOE due to the counterbalance of the purely electronic and the solvent stabilization energy. The existence of an intrinsic barrier has directly been verified by the comparative analysis of pressure and temperature dependent reaction rate constants in triacetine (E_a = 32 kJ/mol, $E_i^{\#} \approx$ 9kJ/mol) which is only slightly more polar than EOE. The results of the steady-state and time-resolved fluorescence both as a function of temperature and pressure further reveals the viscosity control of the CT→CTR photoisomerization with an exponential viscosity power law parameter α of about 0.2-0.3 which is consistent with a large amplitude motion such as the proposed torsional relaxation.

In the last two chapters 10 and 11, the knowledge of the above conclusions was used to discuss **applicational aspects**. Highly twisted (**III**) and moderately twisted (**II**) or planar (**I**) donor-acceptor biaryls are shown to have complementary application possibilities which can mainly be traced back to the difference in the extent of the photoinduced charge separation. For example, the back charge transfer fluorescence of **I** and **II** is relatively inert to hydrogen bonding or sudden viscosity changes whereas the more polar **CTR** species of **III** can be quenched by a proton shift from alcohol solvent or by a complete proton transfer from a strong acid. Moreover, the viscosity controlled adiabatic photoreaction renders **III** as a fluorescent probe of drastic viscosity changes occuring, e.g., in polymerizations or in melting processes. On the other hand, the absence of proton-induced excited state quenching, the large fluorescence quantum yields and the separated LE/H$^+$ and CT absorption and fluorescence bands makes **I** and **II** well suitable as fluorescent pH sensors. The advantageous signal-ratioing method yields a large ratio range over four orders of magnitude for the pH range from 1 to 4.

Donor-acceptor biaryls of the type of **I** or **II** are interesting candidates for application in Fluorescent Solar Concentrators because they can foreseeably be designed to combine large fluorescence quantum yields with high dipole moments introduced through electron transfer interactions between the aryl moieties. The latter property can produce a large Stokes-shift which helps to solve the reabsorption problem frequently met in fluorescent dyes. Twisted D-A biaryls are suggested to act in Electron Transfer Initiated Photocatalysis or Photosensitization which may serve as a primary process in solar chemical upgrading or detoxification. First experimental results from literature are rather encouraging in this new field of solar energy.

Appendix A

The obtained (CNDO/S) wavefunctions of the LCAO-MOs of the composite donor-acceptor (DA) molecules are grouped into the atomic orbital contributions of the donor (χ_d) and the acceptor (χ_a) part according to

$$\phi_n(DA) = \sum_d c_{nd}\chi_d + \sum_a c_{na}\chi_a \quad \text{(A-1)}$$

where c_{nd} and c_{na} denote the LCAO coefficients on D and A, respectively, for the n th. MO.

The character of a singly excited configuration $^1\Phi_{n \to n'}$ promoting one electron derived from $\phi_n \to \phi_{n'}$ can then be separated into two types of locally excited (LE) and electron transfer (ET and reverse ET) contributions, respectively, as given by eqs. A-2:

$$\%LE(D) = \sum_d c_{nd}^2 \cdot \sum_d c_{n'd}^2 \quad \text{(A-2a)}$$

$$\%LE(A) = \sum_a c_{na}^2 \cdot \sum_a c_{n'a}^2 \quad \text{(A-2b)}$$

$$\%ET(D^+A^-) = \sum_d c_{nd}^2 \cdot \sum_a c_{n'a}^2 \quad \text{(A-2c)}$$

$$\%RET(D^-A^+) = \sum_a c_{na}^2 \cdot \sum_d c_{n'd}^2 \quad \text{(A-2d)}$$

The singlet state wavefunctions of the spectroscopic states S_x are linear combinations of the excited state configurations given by

$$^1\Psi_{S_x}(DA) = \sum_n \sum_{n'} C_{n \to n'} {}^1\Phi_{n \to n'} \quad \text{(A-3)}$$

For the lowest excited singlet states mainly those delocalized (mixed !) configurations contribute which converge to pure ^1L_b, ^1L_a and ^{1}ET at $\varphi=90°$.
The weights of these typical but mixed configuration types to S_x can be summarized by

$$\%^1L_b - type = C_{1\to-2}^2 + C_{3\to-1}^2 + C_{4\to-1}^2 + C_{2\to-4}^2 \quad \text{(A-4a)}$$

$$\%^1L_a - type = C_{1\to-1}^2 + C_{3\to-2}^2 + C_{2\to-1}^2 + C_{4\to-4}^2 \quad \text{(A-4b)}$$

$$\%^1CT(D^{\delta+}A^{\delta-}) = C_{2\to-1}^2 \quad \text{(A-4c)}$$

for **I-III** and by

$$\%^1L_b - type = C_{1\to-3}^2 + C_{3\to-4}^2 + C_{2\to-2}^2 + C_{4\to-1}^2 \quad \text{(A-5a)}$$

$$\%^1L_a - type = C_{1\to-4}^2 + C_{3\to-3}^2 + C_{2\to-1}^2 + C_{4\to-2}^2 \quad \text{(A-5b)}$$

$$\%^1CT(D^{\delta+}A^{\delta-}) = C_{1\to-1}^2 \quad \text{(A-5c)}$$

for **9DP** and **DM9DP** with the shape, energy and symmetry of the orbitals presented in Fig. 2.1 and 9.4, respectively.

Appendix A

Furthermore, in ch. 8 and 9 it is distinguished between the spectroscopic states S_x (e.g. dominating ^1CT, ^1L_b-type, ^1L_a-type) and the reference states ($^1ET, ^1L_b, ^1L_a$) which are described by pure electron configurations in terms of the composite-molecule model, e.g. HOMO(D)→LUMO(A) for ^1ET. The contributions of the pure reference states $^1L_b(A)$, $^1L_b(D)$, $^1L_a(A)$, $^1L_a(D)$, $^1ET(D^+A^-)$ and $^1RET(D^-A^+)$ to a given spectroscopic state S_x are then calculated by

$$\%^1L_a(D)+\%^1L_a(A)=C_a^2=\sum_\alpha C_\alpha^2 \sum_d c^2_{n(\alpha)d}\cdot\sum_d c^2_{n'(\alpha)d}+\sum_{ai}C_\alpha^2\sum_a c^2_{n(\alpha)a}\cdot\sum_a c^2_{n'(\alpha)a} \quad (A-6a)$$

$$\%^1L_b(D)+\%^1L_b(A)=C_b^2=\sum_\beta C_\beta^2\sum_d c^2_{n(\beta)d}\cdot\sum_d c^2_{n'(\beta)d}+\sum_\beta C_\beta^2\sum_a c^2_{n(\beta)a}\cdot\sum_a c^2_{n'(\beta)a} \quad (A-6b)$$

$$\%^1ET(D^+A^-)=C_e^2=\sum_\gamma C_\gamma^2\sum_d c^2_{n(\gamma)d}\cdot\sum_a c^2_{n'(\gamma)a} \quad (A-6c)$$

$$\%^1RET(D^-A^+)=C_r^2=\sum_\delta C_\delta^2\sum_d c^2_{n(\delta)a}\cdot\sum_a c^2_{n'(\delta)d} \quad (A-6d)$$

where the configurations

$\sum_\alpha n(\alpha)\to n'(\alpha)$ contain ^1L_a character,

$\sum_\beta n(\beta)\to n'(\beta)$ contain ^1L_b character,

$\sum_\gamma n(\gamma)\to n'(\gamma)$ contain HOMO(D)→LUMO(A) ^1ET character and

$\sum_\delta n(\delta)\to n'(\delta)$ contain HOMO(A)→2UMO(D) reverse ^1ET character (latter for **I-III**).

Note, that the above notation differs from the usual exciplex notation, since ground state interactions are not treated separately by this method (CNDO/S basis), but are inherently included in the reference states. The present notation of the electronic wavefunctions may be translated to the original exciplex language as follows:

$$^1\Psi_{ET}(\varphi_{D-A}=90°)=\,^1\Psi_{ET^0} \quad (A-7a)$$

$$^1\Psi_{ET}(\varphi_{D-A})=e_1\,^1\Psi_{ET^0}+e_0\,^1\Psi_{S_0} \quad (A-7b)$$

$$^1\Psi_{CT}(\varphi_{D-A})=c_1\,^1\Psi_{ET^0}+c_0\,^1\Psi_{S_0}+\sum_i c_i\,^1\Psi_{S_i^0} \quad (A-7c)$$

where $^1ET^0$ and S_i^0 denote the zeroth-order states in a pure exciplex definition. If it is explicitly referred to the *zeroth-order ^1ET state* in the text (e.g. scheme 9.2), this state is then directly related to the exciplex zeroth-order state $^1ET^0$.

Glossary of Abbreviations

CT	observed kinetic species in the ^{1}CT state attributed to a conformer more planar than in S_0.
CTR	observed kinetic species in the ^{1}CT state attributed to a conformer more twisted than in S_0.
FC	observed kinetic species in the primary populated state ^{1}FC attributed to a conformer more planar than in S_0.
^{1}CT	charge transfer state (here intramolecular)
^{1}ET	electron transfer state (here intramolecular)
^{1}FC	primary populated Franck Condon state after IC
^{1}LE	locally excited state without (or with small) charge separation
^{1}A	state symmetry polarized along the symmetry axis within molecular point group C_2
^{1}B	state symmetry polarized perpendicular to the symmetry axis within molecular point group C_2
^1L_b	state assignment according to Platt nomenclature
^1L_b-type	spectroscopic state with properties of ^1L_b state, but mixed with non-^1L_b electron configurations
^1L_a	state assignment according to Platt nomenclature
^1L_a-type	spectroscopic state with properties of ^1L_a state, but mixed with non-^1L_a electron configurations
CT	transition connected with fractional transfer of charge (here intramolecular)
ET	transition connected with full transfer of an electron (here intramolecular)
LE	local transition without (or with small) net charge transfer
CSh	Charge shift
CS	Charge separation
CR	Charge recombination
S_0	singlet ground state
S_1, S_2	first and second electronically excited singlet state
T_1, T_2	first and second triplet state
IC	internal conversion
ISC	intersystem crossing
D	molecular moiety acting as electron donor
A	molecular moiety acting as electron acceptor
D-A	composite molecule consisting of a donor (D) and acceptor (A) moiety linked by a single bond.
MO	Molecular Orbital
HOMO	Highest Occupied Molecular Orbital; HOMO(D) is the HOMO of the local D unit
LUMO	Lowest Unoccupied Molecular Orbital, LUMO(A) is the LUMO of the local A unit
CI	Configuration Interaction
SCF	Self Consistent Field
SCRF	Self Consistent Reaction Field
→	forward or initial process (e.g. photon absorption)
←	backward process (e.g. photon emission)
α_n	amplitude of time component n
τ	(apparent) lifetime; (mean) time needed for a first-order (multiple-order) process to decrease the concentrations of the initial species to 1/e of its original value.
τ_n	exponential lifetime of process n or component n
τ_0	intrinsic lifetime for absent photoreaction (in ns)
τ_r	radiative lifetime (in ns)

Abbreviations

ν	wavenumber (in cm^{-1})
λ	wavelength (in nm)
ε	decadic molar extinction (or absorption) coefficient (in 10^4 l mol^{-1} cm^{-1})
ε_r	static dielectric constant
$\eta(\varphi)$	rotamer distribution function
η	viscosity
n	refractive index
a,abs	index for absorption a
f,flu	index for fluorescence
a	Onsager cavity radius
h	Planck constant
k_B	Boltzmann constant
k_n	rate constant of process n
k_f	fluorescence rate constant
k_{SB}	Strickler Berg rate constant
f	oscillator strength
log	decadic logarithm
ln	natural logarithm
pK_a	negative decadic logarithm of acidity constant
pK_a^*	pK_a in the excited state
μ_e	dipole moment in S_1
μ_g	dipole moment in S_0
Φ_f	fluorescence quantum yield
φ, φ_{D-A}	interannular twist angle
$\langle\varphi_{kf}\rangle$	spectroscopic φ derived from k_f
$\langle\varphi_M\rangle$	spectroscopic φ derived from M
$\langle\varphi_\eta\rangle$	mean φ convoluted with $\eta(\varphi)$
$E_i^\#$	intrinsic energy barrier
E_a	Arrhenius activation barrier
$\Delta G^\#$	(thermal) free energy barrier
ΔG	change of Gibbs free energy
ΔG_{CT}	ΔG of charge transfer process (charge separation)
ΔG_{-CT}	ΔG of back charge transfer process (charge recombination)
λ_s	outer shell solvent reorganization energy
$\lambda_{i'}$	inner shell reorganization energy of low frequency modes
λ_0	reorganization energy for all low frequency modes (= $\lambda_s + \lambda_{i'}$)
λ_i	reorganization energy for all high frequency vibrational modes
$\langle\nu_i\rangle$	average wavenumber of high frequency vibrational modes active in an electronic transition
λ_{sum}	sum of reorganization energy roughly estimated from Stokes shift
f_e	force constant for the solute-inner solvation shell interaction in S_1
f_g	force constant for the solute-inner solvation shell interaction in S_0
HW	full width at half of maximum intensity (fwhm)
HW'	HW determined from reduced spectrum $I_f(\nu_f)/\nu_f^3$
ADMA	4-(9-anthryl)-N,N-dimethylaniline
BA	9,9'-bianthryl
BP	biphenyl
DMABN	4-N,N-Dimethylaminobenzonitrile
9DP	9-(4-N,N-Dimethylanilino)phenanthrene
DM9DP	9-(4-3,5,N,N-tetramethylanilino)phenanthrene
1DPyren	1-(4-N,N-Dimethylanilino)pyrene

List of Publications, Status Reports, Oral and Poster Presentations

Publications

M.S. Gudipati, J. Daverkausen, M. Maus, G. Hohlneicher; Chem. Phys. 186 (1994), 289-301.
Higher Electronically Excited States of Phenantrene, Carbazole and Fluorene

M.S. Gudipati, M. Maus, J. Daverkausen, G. Hohlneicher; Chem. Phys. 192 (1995), 37-47.
Higher Excited States of Aromatic Hydrocarbons. III. Assigning the in-plane Polarized Transitions of Low-Symmetry Molecules: Chrysene and E-stilbene

M. Maus, W. Rettig, R. Lapouyade; J. Inf. Rec. 22 (1996), 451-456.
Photodynamics of a Pretwisted Donor-Acceptor Biphenyl

A. Onkelinx, F.C. De Schryver, L. Viane, M. Van der Auweraer, K. Iwai, M.; Yamamoto, M. Ichikawa, H. Masuhara, M. Maus, W. Rettig; J. Am. Chem. Soc. 118 (1996), 2892-2902.
Radiative Depopulation of the Excited Intramolecular Charge-Transfer State of 9-(4-(N,N-Dimethylamino)phenyl) phenantrene

W. Rettig, M. Maus, R. Lapouyade; Ber. Bunsenges. Phys. Chem. 100 (1996), 2091-2096.
Conformational Control of Electron Transfer States: Induction of Molecular Photodiode Behaviour

M. Maus, W. Rettig; Eurosun 96, Proceedings, 520-524.
Donor-Acceptor Biaryls Dyes as New Candidates for Application in Fluorescence Solar Concentrators

M. Maus, W. Rettig; Chem. Phys. 218 (1997), 151-162.
The Electronic Structure of 4-(N,N-Dimethylamino)-4'-cyano-biphenyl and Its Planar and Twisted Model Compounds

M. Maus, W. Rettig, S. Depaemelaere, A. Onkelinx, F. C. De Schryver, K. Iwai; Chem. Phys. Lett. 292 (1998), 115-124.
Solvent and Conformation Dependent Electron Transfer Interactions in Flexible Biaromatic Compounds: The Case of 9-(dimethylanilino) phenanthrene

Maus, W. Rettig; J. Inf. Rec. 24-5/6 (1998) 461-468.
Donor-Acceptor Biphenyls: Electronic and Conformational Pathways after Photoexcitation

M. Maus, W. Rettig, G. Jonusauskas, R. Lapouyade, C. Rulliére; J. Phys. Chem. **1998**, 102, 7393-7405.
Sub-picosecond Transient Absorption of Donor-Acceptor Biphenyls. Intramolecular Control of the Excited State Charge Transfer Processes by a Weak Electronic Coupling.

M. Maus, W. Rettig, D. Bonafoux, R. Lapouyade; submitted.
Photoinduced Intramolecular Charge Transfer in a Series of Differently Twisted Donor-Acceptor Biphenyls as Revealed by Fluorescence

Publications

Status Reports

M.S. Gudipati, M. Maus, J. Daverkausen, G. Hohlneicher; BESSY Annual Report 1993.
Photoselection Spectroscopy on Aromatic Hydrocarbons in Argon Matrices at 15K Using Synchrotron Radiation

M. Maus, W. Rettig; BESSY Annual Report 1994.
Photophysical and Conformational Effects of D-A Biphenyls due to a Low-Lying Excited CT State

W. Rettig, J. Springer, R. Fritz, M.Maus; BMFT Final Report 1995 (05 5KT FAB9)
Fluoreszenzsonden basierend auf der Dynamik adiabatischer Photoreaktionen

M. Maus, W. Rettig; BESSY Annual Report 1995.
Intramolecular Photodynamics of a Pretwisted Donor-Acceptor Biphenyl

W.Rettig, M.Maus; DFG Final Report 1997 and Application for 1 Year Extension
(Re 387/9-1 and 2)
Dynamische Prozesse von Fluoreszenzsonden in organischen Gläsern und in Polymeren

M. Maus, W. Rettig; BESSY Annual Report 1997.
Temperature-dependent Study of the Conformational Equilibrium in the Excited State of a Pretwisted Donor-Acceptor Biphenyl Using Time-Resolved Emission Spectroscopy

Oral Presentations at Conferences

30.07-04.08.95 London
"XVIIth International Conference on Photochemistry"
Picosecond Spectroscopy of Donor-Acceptor Biphenyls

22.-24.11.95 Dresden
"GDCh Tagung der FG Photochemie"
Photodynamik eines vorverdrillten Donor-Acceptor Biphenyls

03-08.08.97 Warsaw
"XVIIIth International Conference on Photochemistry"
Viscosity Control of the Equilibrium Between Two Excited State Charge Transfer Species in a Pretwisted Donor-Acceptor Biphenyl

19.-21.11.97 Cologne
"15. GDCh Tagung der FG Photochemie"
Donor-Acceptor Biphenyls: Electronic and Conformational Pathways after Photoexcitation

Publications

Poster Presentations

17.-22.7.94 Prague
"XVth IUPAC Symposium on PHOTOCHEMISTRY"
Conformational Effects on Photophysical Parameters in Donor-Acceptor Biphenyls

11.94 Berlin
"93. Bunsentagung -Photochemie-"
Photophysik von Donor-Akzeptor Biphenylen

14.-18.7.96 Riccione
"Intra-and Inter-Molecular Photoprocesses of Conjugated Molecules"
The Separation of Electron Transfer and Conformational Changes as Evidenced by Dual and Triple Fluorescence in Donor-Acceptor Biphenyls

16.-19.9.96 Freiburg
"EUROSUN 96"
Donor-Acceptor Biaryls Dyes as New Candidates for Application in Fluorescence Solar Concentrators

01.-06.01.97 Cairo
"4 th International Conference on Solar Energy Storage and Applied Photochemistry"
Donor-Acceptor Biaryls: Charge Transfer Interactions and their use in Fluorescence Solar Concentrators

21.-24.09.97 Berlin
"V^{th} International Conference on Methods and Applications of Fluorescence Spectroscopy"
Microenvironmental Effects on the Charge Transfer Fluorescence of Donor-Acceptor Biphenyls

19.-24.7.98 Barcelona
"XVIIth IUPAC Symposium on PHOTOCHEMISTRY"
Photoinduced Relaxations in Donor-Acceptor Biphenyls Towards Intramolecular Charge Separation

09.-14.08.98 Berlin
"Twelfth International Conference on Photochemical Conversion and Storage of Solar Energy"
Photoinduced Intramolecular Electron Transfer in Donor-Acceptor Biaryls and its Potential Use for Solar Applications

Coauthor Poster contributions:

16.-18.5.96 Jena
"95. Bunsentagung der Bunsengesellschaft -Primärprozesse der Photosynthese-"
Konformationskontrolle von Elektron Transfer Zuständen

17.-22.7.97 Berlin
"VW-Stiftung -Elektronen Transfer-"
Intramolecular Relaxation Pathways in Biphenyl Derivatives: Towards the Control of Molecular Photodiode Behaviour

Lebenslauf

Name: Michael Maus
Geburtsdatum: 29.09.1966
Geburtsort: Köln
Familienstand: verheiratet seit 30.08.94

Schulbildung:

1973 - 1977	Grundschule Köln-Vogelsang
1977 - 1986	Albertus-Magnus-Gymnasium Köln-Ehrenfeld
13.06.86	Abitur

Universitätsausbildung:

05.09.86	Immatrikulation an der Universität zu Köln im Studienfach Chemie/Diplom
14.02.89	Diplomvorprüfung
20.01.92	Beginn der Diplomarbeit am Institut für Physikalische und Theoretische Chemie der Universität zu Köln unter der Leitung von Prof. Dr. G. Hohlneicher
	"Höher angeregte Zustände von Triphenylen und anderen polycyclischen, aromatischen Kohlenwasserstoffen"
01.06.92-31.10.92	Studentische Hilfskraft der Universität zu Köln
19.10.92	Hauptdiplom Chemie
08.03.93-31.03.95	Wissenschaftlicher Mitarbeiter am Ivan-N. Stranski Institut für Physikalische Chemie der Technischen Universität zu Berlin
	unter finanzieller Unterstützung des BMFT
01.04.95-30.04.98	Wissenschaftlicher Mitarbeiter am Walther Nernst Institut für Physikalische Chemie der Humboldt Universität zu Berlin
	unter finanzieller Unterstützung der DFG

www.ingramcontent.com/pod-product-compliance
Lightning Source LLC
Chambersburg PA
CBHW030923180526
45163CB00002B/447